Springer Biographies

More information about this series at http://www.springer.com/series/13617

Mary R. Tahan

The Life of José María Sobral

Scientist, Diarist, and Pioneer in Antarctica

 Springer

Mary R. Tahan
Vancouver, BC
Canada

ISSN 2365-0613   ISSN 2365-0621   (electronic)
Springer Biographies
ISBN 978-3-319-88403-5   ISBN 978-3-319-67268-7   (eBook)
https://doi.org/10.1007/978-3-319-67268-7

This Springer imprint is published by Springer Nature
The registered company is Springer International Publishing AG
The registered company address is: Gewerbestrasse 11, 6330 Cham, Switzerland

For K.A. and Rosette.
And for Oxford and Othello.

Ladies:
Sirs:
If I have tired you, if I have abused your benevolence, excuse me that there are no flowers at the Pole, no heat but that which springs from the bosom of the soldier, who on more than one occasion had to seek the colors of his country in the immaculate white of the ice and in the blue of its twilights, in order to cheer his spirit that had become disheartened by fatigue and by the frightening isolation of those regions.

José María Sobral

Foreword

A Legacy: Fifty-Seven Years of His Life Working for Science and Humanity

Mary R. Tahan, one of the international artists selected by the Dirección Nacional del Antártico to perform her work on-site in Antarctica, photographed and videographed environmental, historical, and natural images in several areas of the Antarctic continent, and conducted interviews and research with scientists, historians, and explorers for her films and books.

A Canadian-American writer and researcher, with Egyptian-Lebanese roots, Mary R. Tahan has emphasized the employment and care of dogs in Antarctica in many of her writings. She chose José María Sobral as a subject for this great work, taking as an essential basis what she read from Sobral himself about this topic, which was his affection for these animals in the South Pole region, not only as the dogs that pulled the sledges during the exhausting excursions, but also for being his companions during those two years in the Antarctic.

To write a book about Dr. José María Sobral today, 56 years after his death, and 116 years after his arrival in Antarctica on that scientific expedition that was commanded by the ever-so-important Swedish Doctor of Geology, Otto Gustaf Nordenskjöld, who carried out one of the most important expeditions to the South Pole of the last century, and that marked a milestone in the expeditions to the white continent, is quite an event. By one of life's coincidences, the person who was appointed as the representative of the Argentine nation, being at that time 21 years of age and enlisted in the Argentine Navy with the rank of Alférez de Fragata (Ensign), was our grandfather. He was promoted to Alférez de Navío (Under-Lieutenant) after returning from Antarctica and forever remained in everyone's memory with that same rank, given that the Argentine Navy did not grant him the scholarship that he had requested to study his vocation, which was geology, in Sweden, at Uppsala University. Therefore, he requested a discharge, which the Argentine Navy granted him. In August of 1905, the ship *Drottning*

Sophia took him to Sweden to realize his dream: studying geology, as that doctorate did not yet exist in Argentine universities at that time.

Before that, and after giving several talks about Antarctica, he wrote his book *Dos Años Entre los Hielos, 1901–1903 (Two Years Amidst the Ice)*, the first book about Antarctica written by an Argentine, in the year 1904. Prior to his being discharged, and while still on active duty with the Navy, he enrolled in the natural sciences degree course with specialization in geology at the Facultad de Ciencias Exactas, Fisicas y Naturales (School of Exact, Physical, and Natural Sciences at the University of Buenos Aires) in Argentina. He sat for three exams, which he passed, but the Navy did not allow this double activity, and so he had no choice but to request a discharge, which he obtained on December 30, 1904. Once in Sweden, he passed the admissions exam at Uppsala University in the three subjects that he had passed in Buenos Aires, and, in 1913, he received his graduate degree. Then, in June of that same year, he earned a doctorate. While he was studying in Sweden, in 1906, he married a Swedish woman named Elna Wilhelmina Klingström, with whom he had nine children—four in Sweden and five more in Argentina: two girls and seven boys.

Very little is known about his scientific activity in Argentina, where he was barely recognized. Admiral Guillermo Brown once said about these themes: "It's a pity that he belonged to a country that does not know how to value its heroes" (such a paradox). He was awarded the "Naval Cross for Distinguished Services" medal, just as Guillermo Brown had been awarded this medal in 1827.

He spent his life traveling throughout the country, often times with his family, and other times alone, carrying out studies in both geology and petrology. He held important positions in various state departments, but he would always enter those positions in a "third category" level, and would then be promoted as a result of his great capacity for study and work. He always continued to maintain his friendship with Otto Nordenskjöld, Gösta Bodman, Carl Skottsberg, and other companions from the *Antarctic*, the ship that had marked his relationship with geology and natural sciences. He always liked field and campaign work, and many times rejected laboratory and office work. He was super-active, as can be seen.

From our place, as grandchildren of José María Sobral, we thank Mary R. Tahan immensely for her work and unwavering efforts to position our grandfather in the historical place that he deserves.

Guillermo J. Sobral
Jorge A. Sobral
Grandsons of Dr. José María Sobral

Preface

A Stranger in His Own Land

José María Sobral was a person of science, a patriotic Naval officer, and a pioneer of the great frozen white continent. He explored where others did not dare go, fostering an insatiable curiosity that led him to attempt to understand the nature and the humanity that surrounded him. He was a brave young lieutenant and Antarctic pioneer; a scholar and doctor of geology; a philosopher and student of human behavior; a friend and a foreigner. He desperately wanted to contribute to knowledge and to society. And yet his isolation haunted him throughout his life—in the Navy, in Antarctica, in his adopted home of Sweden, and in his homeland of Argentina. He devoted his life to his work and study at home and abroad, but could not convince his own government to support him. He brought the Argentine homeland to the great white continent, but could not remain with his beloved Argentine Navy. Even after the end of his life, for many years, Sobral was not given the recognition he deserved. It is the author's hope that, through these writings, Sobral's story will be better known, and his personality and achievements will be more closely embraced.

This book is the first publishing of Under-Lieutenant José María Sobral's expedition diary and daily journal, kept during his participation in the Swedish-Argentine Antarctic Expedition of 1901–1903, under the leadership of Otto Nordenskjöld. It portrays the role and achievements of this young pioneer within this historic expedition and provides the raw, real-time account he wrote about his experiences, including personal thoughts he did not later include in his official published book about the expedition. It also retells the remarkable accomplishments of this military officer turned explorer turned scientist, both before and after the expedition—prior to his being selected to participate in the Antarctic expedition, and after he returned from Antarctica and began to pursue geological studies in Sweden.

Sobral was the first Argentine to explore Antarctica, and his expedition was the first scientific expedition to endure two consecutive winters on the Antarctic continent. The young Under-Lieutenant Sobral was the lone "Argentine" in the "Swedish-Argentine" expedition. The other members of the expedition were Swedish scientists and Norwegian sailors. Thus, Sobral was doubly isolated—in his new environment and among his foreign colleagues. Yet, he worked closely with them, learning their language and learning the science. Despite extreme danger and almost unendurable hardships, the expedition made geographical, geological, paleontological, and meteorological discoveries of historic importance along the Antarctic Peninsula.

The Swedish-Argentine Antarctic Expedition was the first scientific expedition to intentionally overwinter in Antarctica and unwittingly became the first to spend two winters in the Antarctic, when their expedition ship, the *Antarctic*, captained by the Norwegian whale and seal hunter Carl Anton Larsen, became trapped in the ice and sank near Paulet Island. Three parties from the 29-member expedition spent the second winter separated by ice and distance, and stripped of any hope for rescue. They fought isolation, deprivation, and near starvation, yet they never ceased their scientific work. The parties were stranded in three separate locations—at Snow Hill Island (Cerro Nevado), Hope Bay (Bahía Esperanza), and Paulet Island. Sobral and his team at Snow Hill continued their scientific studies and specimen collection throughout their second winter, traveling across Seymour Island (Marambio) and the nearby islands, with Sobral writing of these excursions in his personal diary. By this time, the men did not know if a ship would ever appear to pick them up, and still, they pursued their scientific passion, adding to the knowledge of humankind. The expedition was one of the most successful in terms of scientific data and fossil collections of land vertebrates and plants discovered on the Antarctic Peninsula, and provided evidence and further credence to the Gondwana theory of continental drift. It also proved that the Antarctic Peninsula was not an archipelago but was part of the Antarctic continent, and that Antarctica had experienced a climatic trans-formation from a warmer, wetter climate to a frozen region of ice. The expedition members were also the first to sight the Larsen Ice Shelf, which since that time has been greatly reduced, changed, and affected by a now progressively warming climate.

After the expedition parties' miraculous reunion at Snow Hill, and heroic rescue by the Argentine corvette *Uruguay*, Lieutenant Sobral took his newfound love of science and, finding no place for it in his role with the Navy, uprooted to Uppsala, Sweden, where he studied geology with his mentor Dr. Nordenskjöld. Later, he returned to his homeland as the first Argentine scientist to doctor in geology and petrography. In Argentina, Dr. Sobral devoted his life to contributing to the sci-entific fields of geology, petrology, and hydrology. But, whether for political rea-sons, commercial profit, or simple short-sightedness, he was basically ignored at home. Only at the very end of his life did his country muster some appreciation for its accomplished son.

The diary featured in this book provides a compelling, first-hand, personal account of the Antarctic expedition activities, findings, and discoveries, including Sobral's private thoughts, intimate observations, and real-time recounting of events. It reflects the courage and dedication of the scientists. It represents the first-ever continuous recording of weather data in Antarctica. And it sheds further light on the importance of this expedition to world knowledge and to Antarctic history, as well as on the significance of Sobral's role in the expedition.

After the Antarctic expedition, in Sweden and in Argentina, Sobral continued his extensive writings, preparing copious tomes that provide insight to the science and the man. He was at home with his research, his explorations, and his family. But he was still a stranger in his own land.

Sobral was not always an easy man to live with, often quite demanding and exacting in what he felt needed to be accomplished. The under-lieutenant was conscientious almost to a fault. At times in Antarctica, he drove his colleagues crazy, unfairly earning their ridicule and teasingly humorous jokes, and enduring their ostracism of him. At other times, he was their voice of reason, speaking to them in their own language that he had painstakingly learned during the voyage and during the seclusion at Snow Hill. But at all times, he strove to do what he thought was right—following his government's orders, representing his country, minimizing the harm done to the Antarctic environment and its wildlife by the actions of the expedition, and supporting his colleagues in seeking and collecting the science, data, and sampling material to better understand the earth, the ocean, and the meteorological phenomena of our world. He was a meticulous record keeper and a student of human behavior. He was a poet in describing the icy beauty of the frozen continent. And he was a friend to the friendless, attempting to stay the hand of those who would kill more penguins than necessary and advocating for ethical treatment of the sledge dogs. He was isolated, but not an isolationist, seeking companionship with those around him. And he loved his country, torn between serving on the Antarctic expedition and serving in the war that he thought was being waged back home (it was not, but the question of possible war was an agony to him until the very end of the expedition). Later, he was torn between the Navy and science. He chose science and that decision changed his life and took him away from his country. But with his new science degree, he returned to Argentina and made history there. And yet history has not fully acknowledged him, any more than his contemporaries had at the time of his great work. This book is intended to place Sobral in his rightful position in history—as a dedicated Navy man, a devoted scientist, an Antarctic pioneer, and a heroic son of Argentina.

Vancouver, BC, Canada   Mary R. Tahan

Acknowledgements

The author wishes to thank Dr. Jorge Rabassa, Senior Researcher and former Director of CADIC-CONICET (Centro Austral de Investigaciones Científicas—Consejo Nacional de Investigaciones Científicas y Técnicas), and Professor of Geology at the National University of Tierra del Fuego at Ushuaia; Dr. Eugenio Luis Facchin, Capitán de Navío VGM (RE), Argentine Navy Captain, historian, safety advisor, and professor in Buenos Aires; Guillermo José Sobral, grandson of José María Sobral; Jorge A. Sobral, grandson of José María Sobral; Capitán de Fragata Guillermo Sergio Spinelli, director of the Departamento de Estudios Históricos Navales (DEHN) de Casa Amarilla; the staff at the Archivo Histórico of the Departamento de Estudios Históricos Navales (Archivo DEHN); the Dirección Nacional del Antártico (DNA); María Edelia Giró and Edelia "Puchi" Gamarino Giró in Tierra Mayor and Ushuaia, Argentina; SIHN CN (RE) Della Rodolfa Daniel Rubén and the Naval Hydrography Service of Argentina for use of the nautical chart; and the author's translators: Caren Calderòn, Efrain Diaz-Horna, Mario Santiago Diaz, Luz Moreno Campos, Guadalupe "Tuty" Moreno Campos, Emilia Ramunda, and Blanca Paniagua.

The author also thanks international artist K.A. Colorado for the inspiration.

Contents

Overview

The Contribution of José María Sobral to Argentine Antarctic Exploration

Sobral: The Dawn of Antarctic Science in Argentina

The presence of the Navy Under-Lieutenant José María Sobral in the Antarctic expedition was not due to chance, an outburst of authority, or the taking advantage of an opportunity; rather, it was part of a plan and part of special international scientific circumstances that led Argentina to take part in an ambitious international program for the sake of Antarctic knowledge.

In 1899, in Berlin, the VIIth International Geographical Congress was held, whose focus was the last unknown continent, that is, the Antarctic continent.[1] It was recommended that studies be conducted and data collected in the following disciplines: oceanography, geomagnetism, meteorology, geophysics, geology, and biology.[2] As a result of this opportunity, an International Antarctic Commission was created to manage and organize such studies.[3]

With the participation of England, Germany, France, and Sweden, the so-called International Antarctic Expedition was planned, to which the Argentine Republic was invited, a country that was committed to installing a meteorological and terrestrial magnetism station on what is today known as Isla Observatorio

---

[1]Destéfani Laurio, El Alférez Sobral y la soberanía Argentina en la Antártida, Editorial Universitaria de Buenos Aires, Buenos Aires 1979. Page 48.

[2]Quevedo Paiva Adolfo, "hace 90 años en la Antártida", Asociación Polar Pingüinera Argentina, Buenos Aires 1994. Page 18.

[3]Destéfani Laurio, "100 años de un rescate épico en la Antártida. Nordenskjöld – Sobral – Irizar" Instituto de Publicaciones Navales. Buenos Aires, 2003 page 20 and subsequent pages.

(Observatory Island), in the province of Tierra del Fuego, Antarctica, and South Atlantic Islands. Simultaneously, Frigate Lieutenant Horacio Ballvé[4] was sent to Europe to acquire instruments, to coordinate with the organizers the needs for geomagnetic and meteorological observations required for this great worldwide effort, and, above all, to be trained in these sciences, which were of great strategic value for the country. It is worth remembering that, in those times, situational instruments on board were the magnetic compass and the sextant, which, with the resolution of complex mathematical calculations of astronomic navigation, enabled the crew to determine a position. This, combined with the meteorological aspect, formed part of the most important determinants for accurate offshore navigation.

Lieutenant Ballvé[5] was the most enthusiastic proponent of Sobral's presence in the Swedish expedition,[6] as he considered that the participation of an Argentine in an expedition of such magnitude would trigger the awakening of Argentine interest in Antarctica and in Antarctic science in general. Ballvé knew Sobral's personality, and he was convinced that Sobral was the one who was best suited to adapt to the most extreme environmental harshness and to the demands of scientific work with respect to collecting and recording data. In addition, he believed in Sobral's adventurous spirit, his resilience, and his tremendous ability to endure great efforts for the sake of fulfilling the expedition objectives.

The construction of the meteorological and magnetism laboratory, which Lieutenant Ballvé would direct, was carried out promptly and speedily, and in February of 1902, data began to be recorded in a statistical series that lasted over 16 years.

The expeditions that were spurred by the VIIth International Geographical Congress were[7,8] as follows: the British expedition, aboard the *Discovery* (1901–1904), which would study the Ross Sea,[9] under the charge of Robert Falcon Scott; the German expedition, with the *Gauss* (1901–1903), whose members overwintered because they became trapped in the ice on the coast that they named Wilhelm II, and which was led by Erich Dagobert von Drygalski; the Scottish, with the *Scotia* (1903–1904), in the Weddell Sea, commanded by William Speirs Bruce; the Swedish, with the *Antarctic* (1901–1903), to the east of the Antarctic Peninsula,

---

[4]Destéfani Laurio, El Alférez Sobral y la soberanía Argentina en la Antártida, Editorial Universitaria de Buenos Aires, Buenos Aires 1979. Page 49.

[5]Pierrou Enrique Jorge "90 años de labor de la Armada Argentina en la Antártida" Volume 1 Servicio de Hidrografía Naval Publication H 919. Buenos Aires 1975. Page 122 and subsequent pages.

[6]Destéfani Laurio, El Alférez Sobral y la soberanía Argentina en la Antártida, Editorial Universitaria de Buenos Aires, Buenos Aires 1979. Page 62.

[7]Secretaría General Naval – Armada Argentina. "Corbeta Uruguay 1903 Centenario del rescate de la expedición Nordenskjöld 2003" Buenos Aires 2003. Page 24 and subsequent pages.

[8]Quevedo Paiva Adolfo E. "Historia de la Antártida" Ediciones Argentinidad Buenos Aires 2012. Page 50.

[9]Quevedo Paiva Adolfo, "hace 90 años en la Antártida", Asociación Polar Pingüinera Argentina, Buenos Aires 1994. pp. 20 and 21.

under the charge of Nils Otto Gustaf Nordenskjöld; and the French, with the *Français* (1902–1905), to the west of the Antarctic Peninsula, in the Gerlache Strait,[10] led by Jean-Baptiste Auguste Charcot.[11,12]

A Navy Committed to Science and to Concrete Scientific Investigation Goals

The senior management of the Navy, at the end of the nineteenth century, was sensitive to the need for generating knowledge, so that the land of Argentina, and its promise of development and well-being, could achieve sustainable progress. This spirit reigned in the institutions of the national state, and although some organizations did not have great technological development, they were eager for growth and improvement.

The only valid path was that of research—research in geology and other disciplines. Knowledge was the way, and the Navy moved to acquire equipment, train its human resources, and avoid squandering any of the very scarce material resources available.

One of the plans was to pursue geographical knowledge, an area of responsibility for the Naval institution, since that knowledge is directly related to nautical safety, not only for Navy ships, but also for all ships that navigate national waters.

Antarctica was one of the important projects for the obtaining of such knowledge, and for that purpose, Argentina participated in international geographical congresses. Naval officers were sent to attend these congresses, many of which invited the Navy and the country to officially participate.

In the case of the Antarctic expedition, after a very rigorous search, the selection of Sobral, a person with courage, determination, diligence and, above all, with a huge sense of duty, was undoubtedly a wise decision.

Sobral: Conviction, Work, and Determination

At the age of 21, Sobral was appointed to be part of the Swedish expedition, commanded with so much aptitude and determination by Otto Nordenskjöld. He

---

[10]Quevedo Paiva Adolfo, "hace 90 años en la Antártida", Asociación Polar Pingüinera Argentina, Buenos Aires 1994. pp. 21–26.

[11]Destéfani Laurio, El Alférez Sobral y la soberanía Argentina en la Antártida, Editorial Universitaria de Buenos Aires, Buenos Aires 1979. Page 50 and subsequent pages.

[12]Sobral José María "Dos años entre los hielos 1901–1903" Editorial Universitaria de Buenos Aires, Colección Reservada del Museo del Fin del Mundo. Buenos Aires, 2003. Page 71 and subsequent pages.

had only 48 h to prepare and to acquire the equipment required for overwintering. Very few of the items that he found in the storerooms of the Argentine Navy, and that he bought in the shops of Buenos Aires, during the middle of the summer, were of any use.[13] At that time, the Navy completely lacked experience in Polar preparation, and although it was taking part in several scientific commissions in Europe, its structure did not possess a capacity for Polar exploration at the level of leading countries, as it does today.

Sobral's young age and relative inexperience have been discussed in some writings, and it is worth mentioning that his experienced colleagues were not much older than he: The chief of the expedition was 32, Axel Ohlin was 34, Samuel A. Duse was 27, and the rest were much younger, including Carl Skottsberg, who was 21.[14,15] This argument is insubstantial, and the truth, considering the historical and evolutionary context, is that a young naval officer could join the expedition with the same eagerness and uncertainty as any other officer, the difference always being his conviction, his spirit, his natural penchant for work, his determination, and his unbending personality.

In the writings of his companions, these fellow adventurers[16] highlight Sobral's responsibility in measurements and scientific works, which were performed with complete vigor, even in the worst climatic conditions. This was so even during the second year of overwintering, when, in addition to the harsh conditions, there was also great uncertainty as to whether the expedition members would be looked for, whether, despite being sought, they would be found, and other uncertainties in a long series of unanswered questions.[17]

Sobral: A Scientist Among Scientists

Although the first human group to overwinter in Antarctica was that of the ship *Bélgica*, commanded by the Lieutenant of the Belgian Royal Navy, Baron Adrien de Gerlache de Gomery, to the south of Peter I Island, for over a year, it was in an involuntary manner, because their ship had become trapped in the ice.

In the case of the Swedish expedition, this overwintering was planned and prepared for.

---

[13]Destéfani Laurio, El Alférez Sobral y la soberanía Argentina en la Antártida, Editorial Universitaria de Buenos Aires, Buenos Aires 1979. Page 67.

[14]Destéfani Laurio, El Alférez Sobral y la soberanía Argentina en la Antártida, Editorial Universitaria de Buenos Aires, Buenos Aires 1979. Page 51.

[15]Destéfani Laurio, "100 años de un rescate épico en la Antártida. Nordenskjöld – Sobral – Irizar" Instituto de Publicaciones Navales. Buenos Aires, 2003. Page 24.

[16]Destéfani Laurio, El Alférez Sobral y la soberanía Argentina en la Antártida, Editorial Universitaria de Buenos Aires, Buenos Aires 1979. Page 214.

[17]Sobral José María "Dos años entre los hielos 1901–1903" Editorial Universitaria de Buenos Aires, Colección Reservada del Museo del Fin del Mundo. Buenos Aires, 2003. Page 134.

Scientific activities were covered in the following way[18]:

Nordenskjöld covered the area of geology, and the collection and classification of fossils;

Erik Ekelöf, whose main role was to be the expedition physician, was in charge of bacteriology, zoology, and botany; and

Gösta Bodman and Sobral made magnetic, meteorological, and astronomical observations and records.

In the area of basic logistics: Ole Jonassen served as the blacksmith, carpenter, and shoemaker, in addition to caring for and feeding the team of dogs, and Gustaf Åkerlundh served as the cook and the assistant for the expedition.[19,20]

Sobral: The Completion of a Mission

It was the combination of the right person and the right plan that brought success to the mission. Sobral was able to adjust to the circumstances, even the most extreme ones in which he found himself. And the country's plan offered unrestricted support, even though both Argentina and the Naval institution did not have any previous experience with Polar exploration.

Sobral's departure to, and triumphant return from, the Antarctic continent was preceded by many years of work and followed by many more years of analysis and study. Further planning and research campaigns ensured that the efforts and sacrifice of Sobral and his expedition would not fall into oblivion.

For the Navy, Sobral's work and his success were an enormous incentive to continue on the path of knowledge. Sobral had set an important precedent which reaffirmed that research was the only way to acquire that knowledge, and this precedent would not be forgotten.

Sobral: A Scientific Vocation and a Successful Career

On naval ships, there has always been a scientific presence and a permanent surveying of all kinds of data that can feed statistical series, as well as a collecting of oceanographic and hydrographic data for the sake of knowledge, and for the reason that, with the advancement of military technology, these data are necessary for control systems.

---

[18]Destéfani Laurio, "100 años de un rescate épico en la Antártida. Nordenskjöld – Sobral – Irizar" Instituto de Publicaciones Navales. Buenos Aires, 2003. Page 95.

[19]Destéfani Laurio Ibid. Page 95.

[20]Sobral José María "Dos años entre los hielos 1901–1903" Editorial Universitaria de Buenos Aires, Colección Reservada del Museo del Fin del Mundo. Buenos Aires, 2003. Page 133.

The international context in which the country of Argentina existed, after the Swedish expedition, did not make it possible for the Navy to allow one of its scarce human resources to devote himself to the study of geology, as Sobral aspired to do,[21] enthusiastic as he was about what he had learned in Antarctica, and about what his overwintering companions suggested to him, noting his natural inclination toward science and his endeavor to prevail in each challenge that he would undertake. This is why he had to choose between his two great vocations: the sea and science.

Once again, his spirit led him to put aside the security and predictability of the military career and to embark on a new venture. He traveled to Sweden,[22] where his Antarctic colleagues awaited him, and by 1913, he had earned a Ph.D. in geology and petrography.[23,24] Thus, he became the first Argentine to obtain a formal university degree, in addition to also being the first to earn a doctorate in this specific science.

He preferred a humble position in the national public administration[25] rather than others that were better paid in Canada,[26] and it was in this manner that, with professionalism and determination, he became the director of the Dirección Nacional de Minas e Hidrología (National Directorate of Mines and Hydrology).[27] This was the highest position for a professional of that specialty, and it was a position from where he had to struggle against the interests of foreign companies, which, back then, were the only ones that exploited mines in the Argentine country.[28] In addition, his work also made him stand out as a petrologist.[29]

Under Sobral's direction, and in spite of inconveniences, the geological map of the Republic of Argentina was completed in 1923.

In 1926, he was appointed a member of the Royal Scientific Society of Gothenburg, and in 1930, he was honored by the American Geographical Society, who awarded him with "The David Livingstone Centenary Medal."[30]

---

[21]Quevedo Paiva Adolfo, "hace 90 años en la Antártida", Asociación Polar Pingüinera Argentina, Buenos Aires 1994. Page 86.

[22]Capdevila Ricardo, Comerci Santiago "Los tiempos de la Antártida – historia antártica argentina" Editora Cultural Tierra del Fuego. Ushuaia 2013. Page 95 and subsequent pages.

[23]José María Sobral, "Dos años entre los hielos 1901–1903" Eudeba, Colección Reservada del Museo del Fin del Mundo, Buenos Aires, December 2003. Preface by Jorge Rabassa. Page 22.

[24]Destéfani Laurio, El Alférez Sobral y la soberanía Argentina en la Antártida, Editorial Universitaria de Buenos Aires, Buenos Aires 1979. pp. 244 and 245.

[25]Quevedo Paiva Adolfo E. "Historia de la Antártida" Ediciones Argentinidad Buenos Aires 2012. Page 297.

[26]Destéfani Laurio Ibid. Page 245.

[27]Destéfani Laurio, El Alférez Sobral y la soberanía Argentina en la Antártida, Editorial Universitaria de Buenos Aires, Buenos Aires 1979. Page 260.

[28]Destéfani Laurio Ibid. Page 260.

[29]José María Sobral Ibid. Preface by Jorge Rabassa. Page 45.

[30]Destéfani Laurio Ibidem. Page 266.

Political circumstances in Argentina alienated him from his profession, and he was appointed as Argentine Consul in Oslo for a little over a year, until June of 1931. In Norway, he was designated a member of the Geological Society, and upon returning to Argentina, he was employed at Yacimientos Petroliferos Fiscales (Fiscal Oilfields) where, after three years of fruitful work, he retired with a meager retirement pension, and had to take over his family's lands in order to be able to afford his necessities.[31]

Although Sobral had left the Navy, he had never stopped being an important part of Argentine science, and his contributions in matters of mining, metallography, and even petroleum are of huge importance for the country.

His life, then, was always linked to his scientific vocation, but he was far more recognized abroad than in his own country. This caused a small, but poignant disappointment.

There was a fear that time had buried, in his country's eyes, the magnificent feat accomplished by the *Uruguay* in rescuing the expedition in which he had taken part. But this was not so. The director of the Naval Academy, who was responsible for—what at that time was known as—the ammunitions depot pontoon *Uruguay*, gave back its dignity to the noble hull that had accomplished so much for Antarctica, and recovered it for the Argentine historical heritage, restoring it to its original state and returning its honor to the now-and-forever corvette ARA *Uruguay*. Under-Lieutenant Sobral was invited to accompany the transfer of the *Uruguay* from Ensenada to the city of Buenos Aires, a recognition given by the Navy and the entire nation of Argentina to this hero who was full of convictions, wisdom, and moral integrity.

A few months after the event, José María Sobral died, surely with the peace of mind of seeing that his life had not been in vain, and that the Navy, which he had been forced to abandon in order to direct his potential toward other horizons, had forgotten neither him nor his achievement, which had been the genesis of what Antarctica means to Argentina today.

Dr. Eugenio Luis Facchin
Captain, Argentine Navy
Antarctic Historian, Safety Advisor, and Professor
Ph.D. in Political Science; Master's in
Research Methodology, Postgraduate
in Management

---

[31]Destéfani Laurio Ibidem. Page 267 and subsequent pages.

Introduction

The Scientific Accomplishments of José María Sobral and the Swedish Antarctic Expedition, and Their Significance to Global Understanding of Earth Sciences and History

José María Sobral was the greatest Antarctic hero of the Argentine Republic and an example of brilliance, modesty, and courage at the beginning of the twentieth century. With fresh youthfulness and honest scientific investigation, he participated in the Swedish Antarctic Expedition at the age of 21. A dedicated Navy under-lieutenant at the time, he performed beyond expectation. His life of significant historical achievements was unjustly forgotten by the institutions that should have kept his name alive through the passage of time.

In 1980 and 1981, I had the opportunity to work as a geologist commissioned by the Argentine Antarctic Institute, specifically in glacial geology and geomorphology of the James Ross archipelago, in particular the James Ross Island, properly named, and Seymour Island (Marambio Island), and in overflying the rest of the islands of that area, including the Northern portion of the Snow Hill Island (Cerro Nevado Island), where Otto Nordenskjöld—the leader of the Swedish expedition—located his winter station. The descriptions made later by Sobral in his book, *Dos Años Entre los Hielos, 1901–1903* (*Two Years Amidst the Ice*), as well as the ones made by Nordenskjöld in *Viaje al Polo Sur* (*Voyage to the South Pole*) (Madrid, 1904–1905), are so vivid and lucid, and they reminded me step by step of the unforgettable moments in our own expeditions.

Now, with this historic publishing of Sobral's expedition diary, which allows us to read the personal thoughts, ideas, and observations of this remarkable individual as he was experiencing them, we have a true landmark event, and a historic document, as these words have never been published before, in any language—not even in Spanish.

In July of 1901, Otto Nordenskjöld had already contacted the director of the Argentine Observatory at the Isla de los Estados (Staaten Island), east of Tierra del Fuego, where the magnetic observatory on Observatory Island is located in the New

Year Archipelago. The observatory commander, Lieutenant Horacio Ballvé of the Argentine Navy, had agreed to collaborate in the expeditions to the South Pole regions. Nordenskjöld stated that, if possible, he would wish to have an officer from the Argentine Navy incorporated into the expedition representing his government. That officer was José María Sobral, who embarked on the expedition in December 1901.

On their way to Staaten Island to check the instruments of the Argentine observatory, while still sailing in the vicinity of the Malvinas/Falkland Islands, Nordenskjöld decided to make a short visit to that archipelago, where he had planned for Dr. Johan Gunnar Andersson, a Swedish geologist, to study the geology of the islands and compare them to Tierra del Fuego during the following summer, and later join the expedition in the Antarctic Peninsula. Nordenskjöld then made a trip on horseback into the Malvinas/Falkland Islands and was very interested in the curious stony deposits, characteristic of these islands, originally described by Charles Darwin during his visit in 1833. Nordenskjöld correctly interpreted these "rivers of stone," as they are known today, as remnants of an old cold climate period, although not a direct product of a local ice age, because he could not find undeniable evidence of their existence. The absence of such a local ice shell was much later shown by the Scottish researchers Chalmers Clapperton (University of Aberdeen) and David Sugden (University of Edinburgh), who proved that the ice had been restricted to a few cirque glaciers and in other words, restricted to the mountaintops of these islands during the Quaternary Age (Clapperton 1971, 1993).

After leaving Port Stanley (today Puerto Argentino) in the Malvinas/Falkland Islands, the expedition arrived at Staaten Island and New Year Island. There they found a very small island, in the form of an open plateau, sub-circular, covered by peat, with small bays and sub-circular coves, and a steep waterfront, devoid of all significant vegetation. Sobral was asked to talk with his Argentine colleagues from the observatory and explain the goals of the expedition. He visited the observatory and the new lighthouse then under construction. It was Sobral who proudly interpreted between his Naval colleagues and his expeditionary companions, describing the functions of this Naval Observatory, which was an example of the vigorous Argentina of the times that was searching for a breakthrough in concert with advanced nations of the era, investing important resources in scientific investigation and the exploration of its territorial boundaries.

Days later, they sailed to the South Shetland Islands, arriving on January 10, 1902. Nordenskjöld compared the landscape of these islands with past environmental conditions in northern Europe and Scandinavia, as these areas had been covered by a large Pleistocene ice sheet. Now, these islands bear smaller, separated ice caps, one in each island of the archipelago. He used King George Island as a model for understanding the characteristics that Norway would have had when it was covered by ice, and recognized the lowlands eroded by ice, and the sharp peaks, which remained above the surface glaciers. Inside the island, Nordenskjöld found greenish porphyritic rocks, which he compared with similar rocks in Tierra del Fuego, assigning them primarily to the Mesozoic period.

After crossing the Drake Passage and its turbulent waters that demarcate the transition into the Antarctic seas, Sobral astonishingly observed the first iceberg of great magnitude and predicted accurately the risk that those gigantic masses of floating ice represent to navigation. As the expedition eagerly searched for a passage to the east, from the occidental coast of the Antarctic Peninsula, which at the time was mistakenly believed to be a tight archipelago with frozen straights and canals, their return to the northern extreme of the peninsula signified the first great accomplishment of the expedition: to have demonstrated that said passage was nonexistent, and that the distinct previously baptized "lands" were part of the same continental segment. Sobral described the enormous quantity of whales and seals that are most certainly less abundant today. And without previous scientific experience, he accurately discussed the location of the permanent snow line, only 20–30 m above sea level. Today, that line is between 100 and 200 m high, for the same regions—irrefutable proof of the progressive global warming. He also observed correctly the receding that was already manifesting itself in the glaciers from more extreme positions in the recent past, such as is occurring both in the islands and in the peninsula.

The expedition visited Roquemaurel Cape at the eastern entrance of the Orleans Channel. There, Nordenskjöld mentioned that the mountains were formed by layers of gray granite and dark eruptive rock, with many quartz crystals which he could not identify. Then, the expedition continued to Louis Philippe Land.

On January 15, the ship reached Joinville Island, and on the continental land, the explorers found a site that would later become very important for Antarctic exploration: "Hoppet vik," "Hope Bay," or Bahía Esperanza, where a major Argentine Antarctic base currently exists. Nordenskjöld described it as a beautiful bay, covered by glacial deposits, with a magnificent semicircular valley, featuring steep slopes, and very interesting glacial deposits that included two curious erratic blocks, i.e., transported by ice from some remote area, belonging to a type of rock that he could not identify in the surrounding region. South of Hope Bay, the mountains were recognized as a formation composed of a mixture of basaltic lavas, tuffs, and thick volcanic agglomerates very common on the east coast of the peninsula, sometimes in layers, and other times in irregular masses. Nordenskjöld later identified these rocks also on Mount Haddington at James Ross Island and on other islands in the region with flat plateaus and steep sandy cliffs.

The expedition then continued to Paulet Island, which Nordenskjöld correctly interpreted as a "dormant volcano." He described the islands as exclusively composed of basalt, with different contents of olivine, a group of igneous mineral silicates. He correctly identified the small lake inside the island as an ancient crater, with many glassy lava fragments around it. He collected samples of basalt covered by a thin crust, which could be removed by scraping with a knife, which turned out to be composed of calcium phosphate from the guano deposited by the penguins, abundant on the island then, as they are today.

From Paulet Island, they sailed to Seymour Island (Marambio Island), and from there to Snow Hill Island and the Erebus and Terror Bay, so named by Sir James

Ross in the previous century to honor his own ships. It was then that he first observed the massive ice walls of the barrier that lie beyond the island of Seymour.

On January 15, they reached the northern portion of Seymour Island, where petrified molluscs and fossilized wood had been found in 1892, representing the first fossils found in Antarctica. These fossils belonged to the Cretaceous and Tertiary periods, and showed the world that in times past, Antarctica had had much warmer climates, stirring the interest of scientists to investigate. This previous finding was one of the main objectives of the expedition and was one of the reasons that Nordenskjöld had originally decided upon this location for the wintering station. He was very excited to be the first geologist in history to personally collect Antarctic fossils, and made an excursion to the interior of Seymour Island, but, finding meager samples, became a little discouraged. He had no way of knowing that he had landed in that part of the island which had a poorer fossil content, a short walk in any direction, and he would have been in contact with excellent paleontological sites. Given this apparent failure, Nordenskjöld decided not to install the winter base at Seymour Island.

Unable to find a way through the sea ice and iceberg concentrations, the expedition searched for an appropriate site nearby for the winter station. When an exploratory group disembarked on the ice, Nordenskjöld was surprised to find large freshwater lakes on the ice shelf, a fact that had not been mentioned before in the literature on Polar expeditions.

It should be noted that Nordenskjöld correctly observed that in the first years of the decade between 1890 and 1900, there had been abundantly produced icebergs in the Atlantic and Indian oceans, which he correctly interpreted as both the product of the Arctic and Antarctic Polar regions. While Nordenskjöld was unable to present a clear explanation, these conditions were the result of climatic changes related to the final phase of global cooling, an episode known as "The Little Ice Age," around 1850 AD in temperate regions.

In the distance, the expedition members were able to observe a distant mountain range, the Antarctic Andes or Antarctandes ("Antartandes" in Spanish), which they had spotted earlier from the western sector, in the Gerlache Strait, on the opposite coast of the Antarctic Peninsula, while they had been searching for a navigable passageway.

Nordenskjöld wrote many pages in his diary in which he described the icebergs of these Antarctic seas, recognizing the enormous icebergs with flat surfaces, which came from the ice shelves, and irregular icebergs, full of cracks and culminating in tips, which were formed at the end where the continental glaciers reach the sea. He was particularly impressed by the flat icebergs, which he considered hallmarks of the South Pole seas. He described them as ice shelves, 60–70 feet high, sometimes with areas covering thousands of square kilometers, indicating that this could likely have misled previous explorers, because when one looks at many of these flat icebergs together, they could have been interpreted as nonexistent ice barriers. The origin of these icebergs was still difficult to understand for Nordenskjöld and his colleagues, but he rightly concluded that they could have been formed directly from the movement of continental glaciers.

On February 9, the ship was near Seymour Island, and for the first time, they approached Cockburn Island, which was described as a very steep cone of volcanic rock, arranged on older inclined stratified layers. Cockburn Island was of particular interest to Nordenskjöld; it was here that James Ross had first landed in Antarctica. Then, he navigated the inner portion of the Gulf of Erebus and Terror, called Sidney Herbert Bay, which had never been visited before. The mountains around the bay were clearly composed of volcanic rocks and covered by huge glaciers, and the immense mass of Mount Haddington, on James Ross Island, could be seen in the distance.

On February 12, the expedition was able to visually explore Seymour Island, which Nordenskjöld described as quite unique to the region, with its ice-free surface, deeply eroded river valleys, and curious elevations. Nordenskjöld was entirely correct in his description: There is no other large island totally free of ice in the region, which is due to the rain/snow shadow effect exerted by the Antarctandes over the weather that is generated by the westerly winds. Steep towering cliffs made it impossible to land on the southern portion of Seymour Island, so Nordenskjöld decided to explore the northern end of Snow Hill Island, which he considered appropriate for the winter season, with its ice-free coast and protected sites. Upon landing on the island, Nordenskjöld immediately found layers of sand and carriers of basaltic dikes, which due to its reduced thickness, he thought would not interfere with the magnetic observations. Soon these layers provided a large number of marine fossils of various groups and classes, much more abundant than the remains he had found on Seymour Island. Nordenskjöld immediately decided that this was the place for which he had been searching.

The expedition went about setting up their winter quarters. They were well provided for in scientific equipment. They had a "meridiana" to measure longitudes, a prismatic compass, three timers (one portable), a series of instruments for measuring magnetic variations, theodolites (one magnetic), drawing instruments, thermometers, hygrometers, marine barometers, anemometers, barometric altimeter, equipment to measure solar radiation intensity, an aspiration psychrometer by Assmann, six cameras with corresponding accessories, and equipment/materials for bacteriological sample preservation.

Establishing the winter base was difficult. One problem was that Sobral was not fluent in Swedish, and thus, he devoted a lot of time communicating with his expeditionary mates. He tried to learn the language quickly by using a dictionary. By the end of the first year of winter, Sobral spoke Swedish fluently enough to fully interact with his colleagues. Sobral also faced a climate that was vastly different, with permanent snow and ice, which he had not known until then, being from a temperate, almost subtropical region of Argentina. Yet, Sobral helped Nordenskjöld in establishing the base and began working with the scientific instruments.

Here, Nordenskjöld climbed the nearby hills, collecting fossils and walking the high plateau, never before visited by humans. He found a complete desert, without any vegetation and even without sand, which had been deflated by wind. The surface was composed of a cluster of blocks and residual volcanic pebbles.

At their winter station on Snow Hill, Sobral and his expedition accomplished great science. On March 9, Sobral accompanied Nordenskjöld for more than 10 h, to study the surroundings and look for fossils nearby. At that time, they calculated the height of the surrounding elevations at 300 m, only half of what had been estimated in maps available at that time. Nordenskjöld found several ice-free hills, which had not been mentioned in the descriptions of the expedition of James Ross, correctly interpreting that the ice and snow had decreased significantly over the last 60 years, but he found this hard to believe. Today, we know that this was linked to the end of the so-called Little Ice Age, a previously mentioned, global cold climatic event, that developed approximately between the seventeenth century and 1850 AD. Rabassa (1982) found moraines and glacial deposits free of lichens, formed in contact with the ice in front of a debris-covered glacier in Bonita Bay, in the nearby James Ross Island, that were interpreted as landforms left by the ice during that period.

On March 11, Nordenskjöld, Sobral, and Ole Jonassen began a journey by boat to the southern tip of the occupied island where they could visit the Snow Hill glacier. They recognized Cape Hamilton and James Ross, composed of the same sedimentary rocks as Snow Hill, covered by Mesozoic marine sands and late Tertiary volcanic layers. They also visited Lockyer Island, which Nordenskjöld found was composed of coarse volcanic agglomerates, with large fragments of basaltic lavas and pure olivine nodules. This island had been considered by James Ross to be joined to Snow Hill by a mass of ice at low elevations. However, Nordenskjöld found that the ice bridge in 1902 was nonexistent, and correctly interpreted that this was evidence of recent climate change.

Winter approached, and scientific observations continued at the station. At first, Sobral assisted Gösta Bodman with the magnetic and meteorological observations, according to the agreed-upon international scientific program, but soon afterward he was in charge. Nordenskjöld had confidence in Sobral's work, in an activity that he considered very important, since the unpublished results of the magnetic surveys in the southern Polar regions were eagerly anticipated in the Northern Hemisphere. These observations were made for 24 h at a time on the first and fifteenth day of each month.

Sobral also assisted Bodman in the meteorological observations, at first with four daily readings of the instruments. Beginning in April, nocturnal observations were also carried out: One of them stayed up until 2:00 a.m., and the other between 4:00 and 6:00 a.m.

On April 27, on Seymour Island, Nordenskjöld obtained many very well-preserved fossils of molluscs, ammonites, and even gastropods, having believed until then that the latter were not present on the island. This way he correctly recognized that at least part of the island was composed of Mesozoic formations.

On May 8, he launched a major glaciological experiment, truly pioneering for its time. With Sobral's help, Nordenskjöld pierced the Snow Hill glacier, placing a line of bamboo poles, whose location was measured periodically with a theodolite and a pair of thermometers placed on ice, in order to measure the variation of temperature on the ice at various depths. The group made these observations regularly during its long stay on the island. In the summer, they correctly observed that there had been accumulation on the glacier surface, probably due to intense wind deflation.

Spring brought with it the possibility for a long journey by sledge. This spring sledge trip alone, which began on September 30, 1902, is one of the special scientific notes. The distances traveled and data amassed were so amazing that this excursion truly could have been the one and only exploratory activity realized during the entire expedition. On its own merits, this trip deserves a privileged place in the history of Antarctic expeditions for the courage, endurance, and determination of this individual.

During this trip, Sobral, Nordenskjöld, and Jonassen visited James Ross Island (which Nordenskjöld called "Land of Haddington"), discovered the existence of the Prince Gustav Channel, and reached the islands of Robertson and Christensen. Here, Nordenskjöld described a mountain composed exclusively of lava and tuff, an ancient crater, concluding that it had not erupted in a while. Actually, he was alluding to the so-called James Ross volcanics formed in late Tertiary and early Pleistocene times.

On one of these islands, Nordenskjöld found many boulders on the tops of the mountains, which he correctly interpreted as remnants of an ancient glacial period when ice was at least 300 feet thick. He was referring, no doubt, to at least the last ice age, developed over the last 100,000 years, or perhaps even of an older glaciation (Rabassa 1982, 1987). An interesting interpretation on the Antarctic glaciers was proposed by the explorer, who suggested that ice was forming directly on the sea ice and not necessarily as part of a normal glacier generated in continental areas.

At the end of November, a new trip to Seymour Island allowed Nordenskjöld to discover the great valley and cross the layers of lignite coal and other materials found in the southern tip of the island, where he found many petrified plant fossils. On December 3, Nordenskjöld crossed the transverse valley, which is actually a large fault line that separates the Mesozoic strata to the south and the Tertiary rocks to the north. Finally, he reached the high plateau which ends at the sub-horizontal north of the island, where Argentine Vice-Commodore Marambio Base, with its landing strip for large airplanes, is located currently. This plain has glacial origins where one finds a glacigenic baseline (till) consisting of a large glacier piedmont coming from the Antarctic Peninsula itself, judging by the lithology of the boulders transported to the island. The glacier would have spread during times of lower Pleistocene (1.0–1.6 million years ago) or perhaps older, when the deep channels around the islands had not yet formed. At the base of this plain, isolated fossil bones were discovered by Nordenskjöld, perhaps one of the first discoveries of fossil vertebrates in Antarctica, which he correctly interpreted as belonging to a giant fossil penguin, greater even than the current emperor penguin. Seymour Island is well known today for its remnants of Mesozoic reptiles, penguins, and whales from the early Tertiary.

Nordenskjöld discussed the role that the South Polar regions played during certain periods in which land vertebrates and modern plants appeared on Earth. He was referring to the Cretaceous period, which was the final period of the Mesozoic and the beginning of the Tertiary (Cenozoico). He was also referring to the distribution of plants and animals in the continents of the Southern Hemisphere (South America, Africa, and Australia) which led Nordenskjöld to propose, in 1904, the possible existence of physical connections between these land masses, imagining them as a single large continent around the South Pole. Thus, the first concepts about this supercontinent now called "Gondwana" and the "continental drift theory" were proposed in this visionary manner by this prominent Swedish scientist, more than 15 years before the classic work of Alfred Wegener, who popularized these ideas. He also correctly observed that the penguin fossils indicated that the process of specialization of these birds would have occurred in this region during the Lower Tertiary, well before the onset of Antarctic glaciation, suggesting adaptive differentiation and early evolution of this group that long precede the climatic deterioration and are independent of it.

The subsequent search along the transversal valley provided Nordenskjöld with abundant plants, leaves, fruit fossils of conifers, Araucariaceae, southern beeches (Fagaceae, now the dominant species of the Patagonian/Tierra del Fuego forest), and other primitive trees, preserved in volcanic-sedimentary tuff. He correctly interpreted that these regions had been covered with dense forests that were inhabited by large mammals and contained plants similar to their counterparts in South America and Europe. He certainly noticed the abundance of species particularly related to Araucaria and Nothofagus still living, as a test of their relationship with the Patagonian and Magellanic fossils, and even with today's forests. Discussing the origins of these plant remains, Nordenskjöld properly noted that their state of preservation indicated that their origin was directly related to emerging nearby lands and not to materials derived from long distances (e.g., so-called mid-latitude temperate lands, South American or Australian). The study of the complete collection of fossils led Nordenskjöld to recognize that two formations of very different age periods (Cretaceous and Eocene) existed at Seymour Island. He successfully established that younger layers were found in the northern portion of the island, where plants and penguin fossils were discovered, and where the ammonites were absent. In the winter season of Snow Hill Island, the layers were correctly assigned to the Mid-Late Cretaceous period. Layers of the same age were also correctly identified by Nordenskjöld in the lower portions of the James Ross and Cockburn islands.

These discoveries led Nordenskjöld to recognize that the Antarctic land had been under climates quite different from today's climate; at least from the Jurassic to the mid-Tertiary, it was wet, warm, and mild.

During their second winter, Nordenskjöld undertook yet another sledge expedition, in September 1903, to James Ross Archipelago, during which he correctly identified Mount Haddington as an ancient inactive volcano, based on the observations of volcanic agglomerations; other rocks from the same source are now known as James Ross volcanics. He also traveled along the Prince Gustav Channel, confirming the insular nature of James Ross Island. This channel was interpreted by Nordenskjöld as being different in origin from the Gerlache and Orléans channels.

That is when Nordenskjöld decided to baptize the "land of Mount Haddington" as James Ross Island, and the small island located north of this, as Vega Island, honoring a Swedish explorer, who had financially sponsored the Polar expedition of his uncle, A.E. Nordenskjöld, a decade prior. These island names are still in use today.

In October, following the astounding meeting with Johan Gunnar Andersson, Nordenskjöld, Andersson, and Sobral returned to Seymour Island to study sedimentary rocks and fossils, and to make magnetic measurements. Early upon their arrival, Nordenskjöld eagerly showed the newcomer his remarkable findings. They compared the fossil flora and fauna of Seymour Island with the observations that Andersson had made at Hope Bay. Afterward, they did the same in Cockburn Island, where they camped. Andersson and Nordenskjöld carefully reviewed all the paleontological sites that were identified in Snow Hill Island and matched them with those of Cockburn Island, which resulted in them being of the same geological age, but with an even more complete fossil content. In Cockburn Island, above the Cretaceous sedimentary rocks, they found basaltic lava, which gives this island its defining characteristic. Finally, on these castings, they found fossiliferous marine deposits that were much more recent, known today as "Pecten Conglomerate," which was an elevated marine beach from the late Tertiary (Andersson 1906).

The heroic actions of Sobral and the return of the expedition in November 1903 on the *Uruguay* are both fine examples of a remarkable job done by Argentine sailors who, despite the lack of prior Antarctic experience, performed admirably in a very hostile environment, without maps or instruments as we have today.

On December 2, 1903, aboard the Argentine destroyer *Uruguay*, Sobral and the expedition members entered the port of Buenos Aires, where people turned to the docks to view the return of the Swedish-Argentine Antarctic Expedition.

It was the finest hour for José María Sobral, considered a national hero, acclaimed by the people on the streets of Buenos Aires, recognized as a recipient of several awards from the Navy, and praised by Argentina, various public/governmental organizations, scientific institutions, and the press.

Despite the honors, Sobral did not lose his humility and simplicity, showing his greatness and innate wisdom, leaving behind works that describe one of the most remarkable feats of scientific exploration of the early twentieth century.

These remarkable virtues, however, were not truly valued, and he was not appropriately thanked by his superiors. Sobral, shortly after, abandoned the Argentine Navy and was invited by his admired teacher and protector to study geology at Uppsala University in Sweden.

The historical and scientific legacy of José María Sobral, the Swedish expedition, and the journey of the *Uruguay*, are invaluable.

José María Sobral was the first Argentine in Antarctica, the first Argentine with a formal university degree in geology, the first Argentine with a doctorate in this science, the first Argentine with a professional title as Director of the National Directorate of Mines and Hydrology—the predecessor of the current National Geological and Mining Service Institute (SEGEMAR), an eminent petrological investigator, and an honest/efficient administrator. He never had the national recognition for his accomplishments, which was a great injustice, but not surprising in these local groups.

José María Sobral was a meticulous observer, a great storyteller, a brave explorer, an innate scientist, and a fair man who left us a lifetime of wonderful work. This book and the diary contained herein present an honest and intimate account of those heroic times. It is a valuable and historic document for the sake of our knowledge and for the benefit of our future generations.

Dr. Jorge Rabassa
Senior Researcher, CADIC-CONICET, Ushuaia, Tierra del Fuego, Argentina
Professor of Geology, Universidad Nacional de Tierra del Fuego at Ushuaia
Member of the Argentine National Academy of Sciences

References

Andersson JG (1906) On the geology of Graham Land. Bull Uppsala: Geol Inst Univ Uppsala VII:17–71

Clapperton Ch (1971) Evidence of cirque glaciation in the Falkland Island. J Glaciol 10:121–125

Clapperton Ch (1993) Quaternary geology and geomorphology of South America. Elsevier, Amsterdam

Nordenskjöld O (1904–1905) Viaje al Polo Sur, vol 2. Maucci, Barcelona (with contributions from Andersson JG, Larsen K, and Skottsberg C)

Rabassa J (1982) Stratigraphy of the glacigenic deposits in Northern James Ross Island, Antarctic Peninsula. In: Evenson EB, Schlüchter Ch, Rabassa J (eds) Tills and related deposits. A.A. Balkema Publishers, Rotterdam

Rabassa J (1987) Drumlins and drumlinoid forms in James Ross Island, Antarctic Peninsula. In: Menzies J, Rose J (eds) Drumlins, a symposium. A.A. Balkema Publishers, Rotterdam

Editor's Note

The Antarctic Diary of José María Sobral

This book features a complete English-language translation of Under-Lieutenant José María Sobral's original expedition diary, which he kept during his voyage to Antarctica and during his second winter on the continent. The diary was written in Castilian Spanish and peppered with Swedish terms as he came to learn them, as well as English words, which he already knew. This diary currently is housed in the Argentine Naval Archives (Archivo DEHN), to which the author was given special access.

The diary text in this book is an integration of two close translations of the original diary, with interpretation and edits by the author, written for readability and illustrative portrayal. In some instances, it was necessary to paraphrase the original wording in order to convey the intent and clarify the meaning of the diary entries' content.

Given the fact that Sobral was attempting to conserve space, he filled each page completely with his entries, leaving few paragraph indentations to separate his thoughts. That spacing is respected in this book.

The translated diary portion of the book is accompanied by the author's annotated text regarding Sobral's journal entries and how they tied in with historical and concurrent events, explanatory descriptions about Sobral, and information and analysis regarding the expedition's scientific findings and the geopolitical conditions of that time. This is done so as to provide a contextual background and framework for the diary entries and to provide an understanding of the challenges and achievements of the Swedish-Argentine Antarctic Expedition.

Complete names and sentence completions are also provided by the author, in brackets, within the diary text.

Selected samples of Sobral's letters, documents, and speeches, as well as portions of his published book about the expedition, have also been translated and are described and excerpted in this book.

It is the belief of the author that an important perspective can be gained through reading this historic figure's personal account, inner thoughts, scientific work notes, and geopolitical musings, in his own words. It is hoped that these pages will bring Sobral's story to life.

Mary R. Tahan

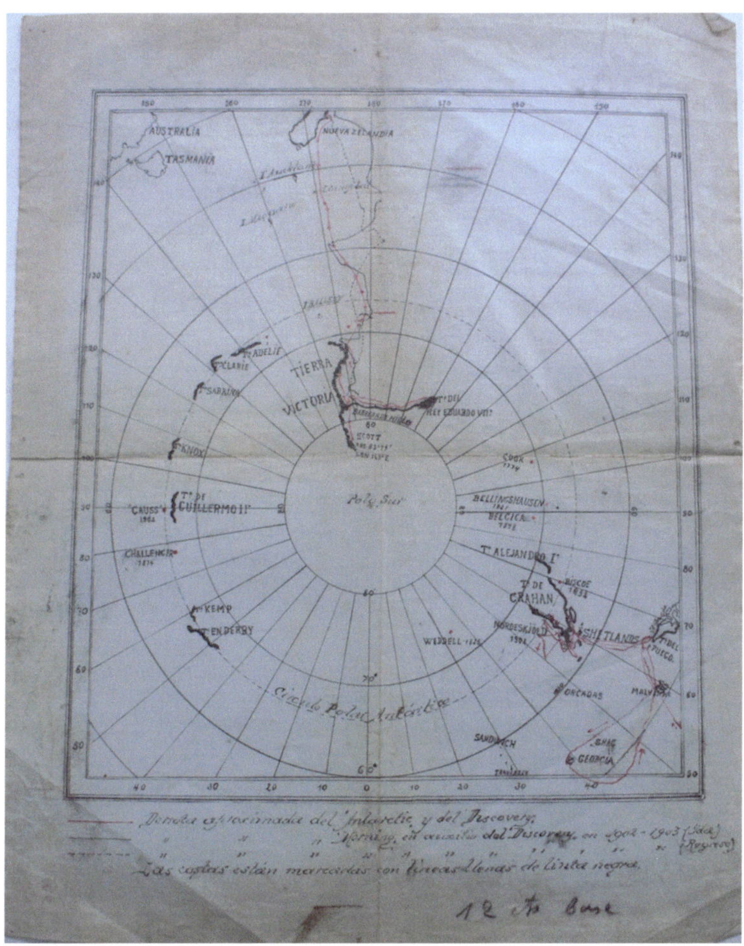

Map 1 Original map created by José María Sobral after the Swedish-Argentine Antarctic Expedition of 1901–1903, and reproduced in his book *Dos Años Entre los Hielos, 1901–1903* in 1904, showing the South Polar region and "the defeat of the *Antarctic*". The route of the ship *Antarctic* prior to its sinking is marked in red lines in the bottom portion of the map, indicating the Antarctic Peninsula region near the tip of South America. (Departamento de Estudios Históricos Navales, Archivo Histórico, Colección Sobral, Box 2, Envelope 5, Map of the South Pole with the *Antarctic* defeat.)

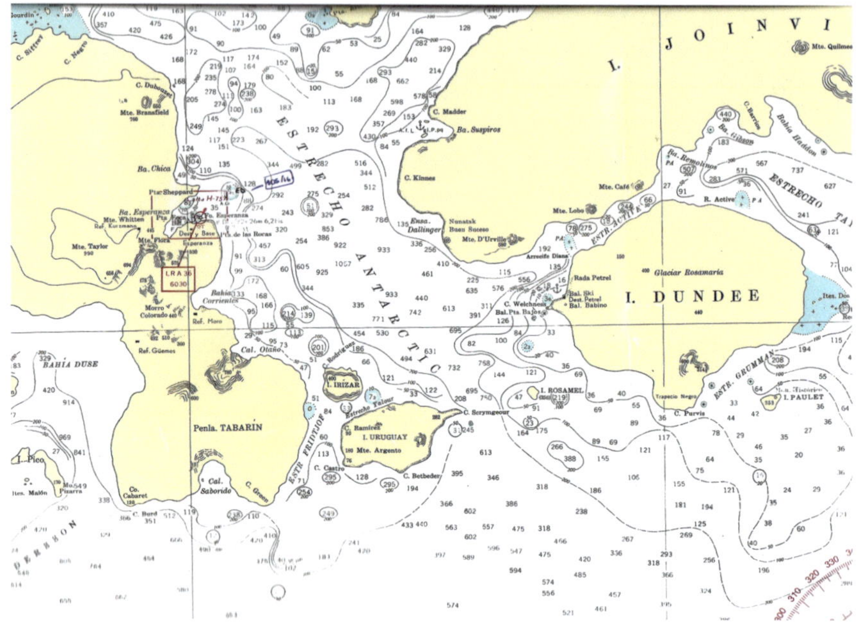

Map 2 Nautical chart showing Hope Bay (Bahía Esperanza) and Paulet Island. (Prepared by SIHN CN (RE) Della Rodolfa Daniel Rubén. Courtesy of the Naval Hydrography Service of Argentina. Carta H-713.   Escala 1: 300.000. Impresión: 10/1969; Reimpresión Julio 2010.)

Map 3 Nautical chart showing Snow Hill Island (Isla Cerro Nevado) and Seymour Island (Isla Marambio). (Prepared by SIHN CN (RE) Della Rodolfa Daniel Rubén. Courtesy of the Naval Hydrography Service of Argentina. Carta H-713.   Escala 1: 300.000. Impresión: 10/1969; Reimpresión Julio 2010.)

Dramatis Personae

Expedition Members, Mentors, and Friends

Argentine Navy
Dr. José María Sobral—Argentine Naval Under-Lieutenant, and later, Scientist and Geologist; worked in the Swedish-Argentine Antarctic Expedition in meteorological, magnetic, astronomical, and hydrographical research

Swedish Scientists
Dr. Nils Otto Gustaf Nordenskjöld—Expedition Leader, Geology Professor
Dr. Axel Ohlin—Professor of Zoology and Geology
Karl Andreas Andersson—Zoologist
Samuel A. Duse—Cartographer
Gösta Bodman—In charge of hydrographical tests and magnetic & meteorological observations
Carl Skottsberg—Botanist
Dr. Erik Ekelöf—Bacteriologist and Physician
Dr. Johan Gunnar Andersson—Lecturer and Geologist

Norwegian Crew
Captain Carl Anton Larsen—Captain of the *Antarctic*
F.L. Andreassen—First Mate
H.J. Haslum—Second Mate
Anders Karlsen—First Engineer
George Karlsen—Second Engineer
Axel R. Reinholdz—Third Mate
G.F. Schönbäck—Steward
Anton Olsen Ula—Boatswain
Ole Johnsen Björnerud—Smith
Ole Jonassen—Dog Sledge Driver
Toralf Grunden—Sailor
Ole Olansen—Sailor

Ole Christian Wennersgaard—Navigator and Sailor
Gustaf Åkerlundh—Sailor
Axel Andersson—Cook
Carl Johanson—Coalman
Wilhelm Holmberg—Coalman

American Artist
Frank Wilbert Stokes—Landscape Painter

Argentine Scientists & Mentors
Francisco P. "Perito" Moreno—Scientist, Explorer, and Collector
Lieutenant Commander Horacio Ballvé—Director of Observatory Island
Captain Onofre Betbeder—Minister of the Navy

Argentine Corvette *Uruguay* Crew
Lieutenant Commander Julián Irízar, Captain and Expedition Leader
Lieutenant Jorge Yalour, Officer
Lieutenant Felipe Fliess, Officer
Lieutenant Ricardo J. Hermelo, Second-in-Command
Lieutenant Alberto Chandler Bannen, Representative of Chili
Dr. José Gorrochategui, Medical Officer
Juan L. Bertodano, Chief Engineer
Gualterio Carminatti, Engineer
And the remainder of the crew

Sledge Dogs on the Expedition
Jim (Malvinas)
Cain (Malvinas)
Curri [also spelled Curry and Kurre] (Malvinas)
Peridota (Greenlander)
Elenita (Malvinas)
Fía (Greenlander)
Amager (Greenlander)
Abel (Malvinas)
Matilda (Malvinas)
Skottsberg (Malvinas)
Kalle (Greenlander)
Kvik (Greenlander)
Baskin (Greenlander)
Suggen (Greenlander)
Castor
Nemö
And other adults and puppies

Chapter 1
Historical Background: A Truly Unbelievable Story that Actually Happened

Abstract At a time when few had ventured into the Antarctic, José María Sobral was selected as a fresh, 21-year-old Argentine Navy Under-Lieutenant to accompany a world-renowned scientist to Antarctica. Otto Nordenskjöld had already conducted geological exploration in Patagonia and Tierra del Fuego in 1895–1897 and now had his sights set on Antarctica. His Swedish Antarctic Expedition would be one of a handful of high-profile expeditions inspired by the Sixth International Geographical Congress of 1895 and the Seventh International Geographical Congress of 1899, whose mandate had set into motion an international quest for the South Polar regions. Otto Nordenskjöld's expedition set off from Gothenburg, Sweden, on October 16, 1901, and stopped in Buenos Aires on December 16 for fuel and collaboration with the Argentines. Previously, he had requested assistance from the Argentine Observatory at the Isla de Los Estados and had asked for an Argentine Naval officer to be a part of his expedition. Sobral was selected and championed by the explorer Francisco P. Moreno, Observatory Chief Horacio Ballvé, and Minister of the Navy Captain Onofre Betbeder. A brief meeting with Sobral persuaded Nordenskjöld to invite him on the expedition, which would be one of the first few expeditions to land on the continent of Antarctica and the first expedition to intentionally overwinter in Antarctica for purely scientific research—later becoming the first also to spend two consecutive winters in Antarctica. A stranger to cold climes, and not sufficiently equipped, Sobral had dreamt of exploring the ice and embraced the challenge. The already audacious trip took on new depths of challenges when, after taking Sobral and his colleagues to Snow Hill in February 1902, the ship *Antarctic* sank, crushed by the ice near the Erebus and Terror Gulf in February 1903, leaving the three parties of the expedition stranded in three different locations on and around the Antarctic Peninsula. They each worked to survive and to continue their scientific research and discoveries over a second winter. The shipwreck and the second winter's survival were followed by nearly miraculous reunions of the three parties in mid-October 1903 and a daring rescue by the Argentine corvette *Uruguay* in early November 1903. All but one of the expeditioners survived (a sailor tragically died of heart disease), and the scientific samples that had been collected in Antarctica, for which the scientists and crewmembers had risked their lives, were successfully brought back to Buenos Aires. Sobral's life and identity would be

© Springer International Publishing AG 2018

M.R. Tahan, *The Life of José María Sobral*, Springer Biographies,

https://doi.org/10.1007/978-3-319-67268-7_1

forever intertwined with this expedition and its historic accomplishments, and he would go on to dedicate himself to geology and scientific exploration.

To understand the magnitude of the accomplishments of José María Sobral, one must know a little bit about the history that surrounded him. He was selected as a fresh, 21-year-old Navy Under-Lieutenant to accompany a world-renowned scientist to the Antarctic. Not only was Nils Otto Gustaf Nordenskjöld world renowned, but so was his uncle—Nils Adolf Erik Nordenskjöld, who had been the first to traverse the Northeast Passage across the Arctic in 1878–1879. Otto Nordenskjöld himself had already conducted geological exploration in Patagonia and Tierra del Fuego in 1895–1897, and now had his sights set on Antarctica. His Swedish Antarctic Expedition would be one of a handful of high profile expeditions inspired by the Sixth International Geographical Congress held in London, England, in July-August 1895, and the Seventh International Geographical Congress held in Berlin, Germany, in September, 1899, whose mandate was Antarctic exploration, and whose convening leaders had set into motion a quest for the South Polar regions by countries including England, Germany, Japan, and Sweden (and Norway, although no one knew this at the time—Roald Amundsen would famously and secretly join the race in 1909 for his 1910–1912 Antarctic Expedition, during which he and his sledge dogs reached the coveted South Pole in 1911). Otto Nordenskjöld's expedition had set off from Gothenburg, Sweden, on October 16, 1901, and had stopped in Buenos Aires on December 16 for fuel and collaboration with the Argentines.

And so Sobral found himself in heady company and in heady circumstances, indeed. Just as the up-and-coming Navy officer Robert Falcon Scott had been championed by Sir Clements R. Markham, President of the Royal Geographical Society, to head the British Antarctic Expedition of 1901–1904 (and again later the tragic 1910–1913 expedition), Sobral had been championed by no other than Francisco P. Moreno, the conservationist, explorer, scientist, and geographical enthusiast whose accomplishments had already made him a minor folk hero in Argentina. Moreno had been to Patagonia and studied its lakes and glaciers; he had established a natural history museum in La Plata; and now he was eager to have Argentina play a key role in the Antarctic scene. Nicknamed "Perito," meaning "Expert," and visually epitomizing the look of a bespectacled academic, the 49-year-old Moreno was a friend to the younger Nordenskjöld, whose barely 32 years belied his scholarly achievements. Perito Moreno put in a good word for the young Sobral, whom he viewed as exceptionally bright and talented with the potential to learn the necessary science quickly. Sobral had already sailed around the world, serving as a cadet on the ARA *Presidente Sarmiento* frigate ship from 1899–1900, and travelling to such faraway places as Asia and the USA. He was viewed by Perito Moreno—the man who would later have a famous glacier named after him—as just the right combination of mental prowess, physical strength, and determined discipline to represent his country in the Antarctic. And so he was recommend by Moreno to Nordenskjöld when the two great men met in Buenos Aires on December 16, 1901.

Five months prior, Nordenskjöld had requested assistance from the Argentine Observatory at the Isla de los Estados—Island of the States or Staaten Island—and had specifically asked for an Argentine Naval officer to be a part of his expedition. This request was a result of his correspondence with the Chief of the Observatory on Staaten Island, Lieutenant Commander Horacio Ballvé, who was collaborating with the Swedes on their expedition, and who had duly inquired if Nordenskjöld would also allow an Argentine Naval officer to accompany him. Nordenskjöld was fine with having someone representing the Argentine government being associated with his expedition, although his initial thinking was that the officer would remain on the ship. When he was asked by Moreno and the Minister of the Navy, Captain Onofre Betbeder, to allow the officer to overwinter at the station in Antarctica with him, the good professor nursed some serious doubts. A military officer on a geological surveying expedition at first seemed inappropriate. How would this young officer assimilate into the party and acclimate himself to the conditions? And what of the science—how would he understand it with no training, no education? In his book, *Antarctica or Two Years Amongst the Ice of the South Pole*, Nordenskjöld describes the moment (Nordenskjöld et al. 1905: 12):

> … I was fully sensible of the difficulties that could arise from introducing into our circle one who was a perfect stranger, and one who had no previous knowledge of what a wintering amid Polar ice involved. Still, I was most unwilling to answer in the negative, and delayed my reply until I had seen the young officer, Señor José M. Sobral, then sub-lieutenant in the Argentine Navy, who had been proposed to me by the Minister. Señor Sobral came on board the *Antarctic* early next morning, and he appeared so unaffected and affable, so interested in the question and so intrepid, that I cast all doubts to the winds and determined to run what risks there might be in accepting him, and the matter was definitely decided the very same day.

The brief meeting with Sobral had persuaded Nordenskjöld that this young man was up to the task. He would not leave Buenos Aires without him. Sobral's future was sealed.

Nordenskjöld's ship the *Antarctic* had traveled from the Atlantic Ocean along the muddy Rio de la Plata River to the Buenos Aires Port, where the former whaling ship had docked near the industrial, immigrant-filled, tango-minded La Boca district. The city's pulsing streets of workers, leisure seekers, entrepreneurs, and aristocrats had seemed liked an oasis to the Antarctic-bound ship's inhabitants. And the large three-masted vessel along with its inhabitants, in turn, had caught the eye of the elegant city's residents. They knew the ship was bound for the ice, and, in the sultry breezes of sunny Buenos Aires, they wondered how these men would feel, and they wished them well.

It must be noted that, up until this moment, precious few had ventured onto the Antarctic continent itself, let alone spent a winter there. In fact, one could count those expeditions on one hand. Ever since the discoveries of Captain James Cook, Captain William Smith, and James Clark Ross, only a few individuals had set foot on the Antarctic continent. Captain James Cook, with the ships *Resolution* and *Adventure*, had first crossed the Antarctic Circle during his expedition of 1772–1775, proving that Antarctica existed separately from South America, but failing to sight Antarctic land. Captain William Smith, along with naval officer Edward Bransfield, on the *Williams*, had first spied the Antarctic continent during their expedition of 1819–1820. Admiral Jules Dumont d'Urville, on the *Astrolabe* and *Zelee*, had located Louis Philippe Land (northern Graham Land) and Joinville

Island in 1837–1838. And James Clark Ross, on the *Erebus* and the *Terror,* had first penetrated the pack ice, discovering the Great Ice Barrier (later known as the Ross Ice Barrier and Ross Ice Shelf) during his expedition in 1839–1843.

But no one had actually walked on Antarctic land until Captain Leonard Kristensen arrived at Cape Adare in 1895, ironically on the *Antarctic*—the same ship that would now carry Nordenskjöld's expedition, who would become the first to spend two winters on the continent. (Kristensen's company of men had included Carsten Borchgrevink.) During that previous expedition, the *Antarctic* had come to hunt whales; now it would chase science.

Until this time, there had been only two expeditions to spend a winter in Antarctica; they were Commander Adrien de Gerlache's *Belgica* expedition of 1897–1899 and Carsten Egeberg Borchgrevink's *Southern Cross* expedition of 1898–1900.

The *Belgica* crew's overwintering was not by choice. The ship had become frozen into the ice off Graham Land, in the Bellingshausen Sea, south of Bransfield Strait. The captain suffered insanity from the never-ending darkness and lack of nutrition, and the crew fell into depression. The entire expedition grew sick and nearly died from scurvy, except for two members, who force-fed fresh seal and penguin meat to the rest of the crew. These two were the ship's doctor and second-in-command—the American Dr. Frederick Cook and the Norwegian Roald Amundsen. They nourished the men to recovery. After more than a year locked in the ice, from March 1898 to early 1899, the men dug trenches and dynamited the ice, several times, in order to free themselves. The ship was eventually freed and returned from the Antarctic. One member, the Belgian Lieutenant Emile Danco, was a casualty of the ice; he died a few months after the ship had been caught. But most of the crew returned home safely—physically speaking, at any rate.

Carsten Borchgrevink's overwintering was indeed intentional, and courageously calculated. He was the first to spend a winter on Antarctic land. His *Southern Cross* expedition made a landing at Cape Adare, off Ross Sea, in February 1899, and stayed on land until January 1900. The Norwegian explorer was the first to hike the Great Ice Barrier, and the first to build a prefabricated hut on Antarctic land—the very first human habitat on the continent. Moreover, he was the first to conduct a sledge trip across the Antarctic ice for exploration, employing the first sledge dogs on the southern continent. While he did take magnetic observations, his overwintering was more for the sake of exploration than science.

Nordenskjöld's expedition, then, was the first to intentionally spend a winter in Antarctica for purely scientific research—and the first to end up spending two consecutive winters at that. The esteemed Swedish scientist, learned professor, and famous geologist was on a mission of science—he was about to embark on an audacious trip. And he was to be accompanied by a certain young Under-Lieutenant, José María Sobral, a native son of Argentina, an Entre Ríos-born military man whose lifelong dream was to explore the ice about which he had read so much.

The ice, however, was a stranger to Sobral. The reality of Buenos Aires at that time of his departure was a sweltering average of 34 °C. The warm summer nights did not usually reach lower than 17 °C. The young Alférez de Navío Sobral was not sufficiently prepared, like the Scandinavians or the British were. He had hardly ever seen snow. But he was game to embrace it.

No one fathomed at that time that the expedition would dwell in the depth of the ice for two long years. No one foresaw the amazing adventure that was yet to greet them: The research, the wreckage, the reunion, and the rescue.

The story unfolds in this way: The Swedish-Argentine Antarctic Expedition of 1901–1903, supported in part by the Argentine government, who provided Nordenskjöld with food and coal, received supplies at Buenos Aires, and set sail from that port in December 1901. The expedition was comprised of 29 members who formed three parties to study the Antarctic Peninsula. Nordenskjöld and Sobral's party of six worked and lived at Snow Hill for 21 months, beginning in February 1902, making long, exhausting sledge journeys to conduct scientific experiments and collect geological material, and making important discoveries about climate, glaciology, palaeontology, and continental drift. When the *Antarctic* found it difficult to traverse the ice in order to return to Snow Hill, it set ashore Johan Gunnar Andersson's party of three at Hope Bay, in December 1902, with 25 days' provisions for three men to enable them to go on a sledge journey and a depot of provisions for nine men over two months. Captain Carl Larsen basically said to Andersson and his two companions, "Wait here, I'll be back," but never did return to pick them up. The three castaways survived for nearly ten months in a stone hut they built around their tent, feeding on penguins and seals, while continuing to conduct scientific studies and collect fossil samples. Captain Larsen's contingent of 20, including the scientist Carl Skottsberg, escaped their sinking ship the *Antarctic* as it was crushed by the ice near the Erebus and Terror Gulf in February 1903 and managed to exist on the volcanic Paulet Island for over eight months. Only one crewmember died, due to heart illness. In mid October 1903, the first of three amazing reunions occurred. Andersson and his two mates, who had begun walking from Hope Bay toward Snow Hill in late September, miraculously met up with Nordenskjöld while the professor was conducting a scientific sledging journey on Vega Island. In early November 1903, the second reunion occurred. The corvette *Uruguay*, which had been dispatched by the Argentine government—at the urging of Perito Moreno—to search for the tardy expedition, arrived in the vicinity. Its Captain Julián Irízar discovered a sign that Sobral and Andersson had left on Seymour Island, which led him to find the two parties at Snow Hill on the following day. That very evening, the third miraculous reunion occurred. Captain Larsen and several members of his stranded crew showed up at the door of Snow Hill, having crossed over the waters from Paulet Island just in time to board the ship that had come to rescue them all. As if saving all of them was not enough, the *Uruguay*, at the insistence of the scientists, retrieved the fossil collections left at Seymour Island and Hope Bay, in addition to picking up the remaining crewmembers on Paulet Island. The little Argentine corvette returned to Buenos Aires in December 1903 with its full complement—the *Uruguay* crew, the entire Swedish-Argentine Expedition (less the sailor who had perished), and the scientific samples collected in Antarctica, for which the scientists and crewmembers had risked their lives.

In other words, storytellers and dramatists could not have created a more perfect ending.

Looking at the expeditioners' experience in hindsight, and analyzing their situation, even the famous (and, to some, infamous) Roald Amundsen was in awe of the achievements and endurance of this Swedish-Argentine Expedition. In his book on his own South Pole conquest, *The South Pole: An Account of the Norwegian Antarctic Expedition in the "Fram", 1910–1912*, written in 1912, Amundsen found it necessary to point out the amazing details of the Swedish-Argentine expedition, calling it "one of the most interesting .. in the Polar regions" and nearly gushing that "it sounds almost incredible"; this from the man who calmly and methodically reached the southern axis of the world employing meticulous strategy, steely nerve, and, sometimes, cold-hearted calculation. He marveled at Sobral and his companions, going on to say (Amundsen 2002: 17):

> Both from a scientific and from a popular point of view this expedition may be considered one of the most interesting the South Polar regions have to show.

Amundsen was not alone in his admiration of the Sobral expedition. The great patriarch of Arctic exploration himself, Norway's Professor Fridtjof Nansen, who had brilliantly invented the ice drifting method, and who had designed the ship *Fram* in which Amundsen had made his victorious journey to Antarctica, also marveled at the Swedish-Argentine Expedition's accomplishments, and included them in his lectures that he gave worldwide. His handwritten notes and lantern photos outline for his lecture at the Geographical Society of Kristiania (Oslo) in March of 1912 include the Nordenskjöld expedition and the sinking of the Antarctic (Nansen notes 1912).

That sinking, and the chain of events that it unleashed, would forever be intertwined with Sobral's identity and his life. Even without the shipwreck and the second overwintering, counting only the voyage to Antarctica and the first winter spent there, Sobral had accomplished much in this expedition, and the Antarctic experience whetted his appetite for geology and further Antarctic exploration. Through his own initiative, commitment, and hard work, Sobral spent a lifetime excelling in geological science—in Sweden, in Norway, and in Argentina. He was sought after internationally and pursued his passion in his home country. But his great love—Antarctica, to whose cold embrace he wished to return, was out of reach to him forever more; he was never allowed to return to the continent that he had so boldly explored.

References

Amundsen Roald (2002) The south pole: an account of the Norwegian Antarctic expedition in the "Fram", 1910–1912 (translator A.G. Chater). Birlinn Limited (reprint of original book published in 1912), Edinburgh

Nansen Fridtjof (1912) Handwritten notes and outline—Roald Amundsens Færd 1910–1912, lecture held at the geographical society, Kristiania, 3 March 1912. Oslo, Nasjonalbiblioteket, National Library of Norway, NB Ms. Fol. 1924:5:3

Nordenskjöld N. Otto G, Johan Gunnar Andersson, Carl Skottsberg, Carl Anton Larsen (1905) Antarctica: or two years amongst the ice of the South Pole. The Macmillan Co, New York

Chapter 2
Born to Sail, with a Spirit for Science

Abstract An account of José María Sobral's life prior to his being selected to participate in the Swedish-Argentine Antarctic Expedition of 1901–1903, including his experiences in the Naval Academy and the Argentine Navy.

He was born on April 14, 1880, in the province of Entre Rios, in the northeastern part of Argentina, just south of Santa Fe where Roald Amundsen would later write his book about reaching the coveted South Pole.

A fitting birthplace, this physical convergence of Antarctic history, as Jose Maria Sobral would himself aspire to, and achieve, Antarctic exploits of historic proportion.

His birth home was on Calle San Martín 633, in the city of Gualeguaychú, a city named after the river that bordered it, which in turn originated from the waters of the Uruguay River.

Sobral was born near the water that he would later aspire to sail.

The son of a notary public and a homemaker, Sobral was one of eight children. His father Enrique Sobral and mother Maria Luisa Iturrioz were loving parents whom Sobral adored and of whom he would write and think incessantly during his lonely years in Antarctica (Sobral 1898–1961).

Sobral was the oldest of the children. From the sense of responsibility that he would later exhibit in his working years, it would only seem logical to think that he took responsibility for his seven siblings.

Moving to Buenos Aires during his primary schooling, Sobral graduated from high school at Colegio Nacional Norte (Northern National College) (Fundación 2017; Marenssi 2007).

At the age of 14, Sobral enrolled in the Escuela Naval Militar (Naval Academy). The date was October 26, 1894, when the young Sobral sat for his admission examination. By December, he was officially in the Navy.

The day he entered the Navy, his father Enrique Sobral wrote to him a lengthy and emotional letter in which he urged his son to uphold the national honor in times of war, to be a guardian for peace, and to seek science among its followers, so that "your name is repeated by your contemporaries and proudly, fondly remembered by future generations" (Sobral 2016). Sobral would take his father's advice to heart.

© Springer International Publishing AG 2018

M.R. Tahan, *The Life of José María Sobral*, Springer Biographies,
https://doi.org/10.1007/978-3-319-67268-7_2

The following year, on September 7, 1895, the Naval Academy discharged the young Sobral, citing reasons of health. He adamantly re-enrolled a few months later, on December 5 (Rabassa 2003). He was now at the threshold of taking the path to become a Navy officer.

At the naval school, he met and befriended Jorge Yalour, who he would again encounter in Antarctica as a lieutenant on the Argentine ship sent to rescue him.

He completed his studies at the Naval Academy and was graduated as part of Class Number 24, with a ranking of 19—at the bottom of the class. He had successfully fast-tracked his education, however, completing the required courses in only 3 years. And so, on August 8, 1898, he was graduated as Midshipman.

Shortly thereafter, Sobral found himself sailing around the world.

He was assigned to the training frigate ARA *Presidente Sarmiento* on its first voyage, which lasted from 1899 to 1900, visiting tropical ports-of-call and exotic locations. It was an ideal assignment for an ambitious novice. Sobral was able to see the world, sailing to Japan, China, Hawaii, San Francisco, New York, and other locales (Sobral 1899–1900). On that extensive voyage, Sobral met Julián Irízar, who he would later meet in Antarctica as the commander of the ship sent to rescue him.

On the *Sarmiento*, Sobral was under the command of Onofre Betbeder, who, a year later, in 1901, would become Argentina's Minister of the Navy and who would usher in Sobral as the Argentine member selected to participate in the Swedish Antarctic Expedition.

Sobral also served for a year on the gunboat cruiser ARA *Patria*, where again he met Irízar, this time serving under his command.

During his schooling, training, and assignments to naval ships, Sobral learned and adapted quickly. He exhibited an interest for technical and scientific subjects. In addition to his training in all things involving the maritime vocation, Sobral was also proficient in languages. He learned to speak and write the English language to an impressive degree.

As of 1901, Sobral was back in Buenos Aires. On October 28, he visited a Sanatorium Quirurgico (surgical clinic) on Avenida Montes de Oca where he had radiography done on his foot (Quirurgico 1901). Whether the foot problem was related to the health situation that had caused his early discharge when he first enrolled in naval school, or whether it was due to an isolated incident, is not certain. Interestingly, it foreshadowed problems he would later have in Antarctica. But certainly, the reason for the X-ray did not stand in the way of his resuming his immediate duties. For, a mere two months later, he would be called upon to take on the role of his lifetime.

Ever since his childhood, Sobral had harbored a dream to become a Polar explorer (Sobral 1904: 45–46):

> . . . it was always my wish to make a trip like the one I was going to undertake.
>
> I had read the account of many explorers, and of course I did not ignore the dangers and hardships that those who go to those regions must suffer, but I was completely sure of my physical resistance, and I did not doubt my morality.

I knew that the frozen masses had been the shroud covering the remains of many of those who had gone to the Polar regions, transmitting to posterity, from that solitary sepulcher, the respected names of martyrs who had succumbed for the sake of science, but also that more were the number of those who had returned to refer to humankind the mysteries that nature keeps there.

And since many have returned. Why should not I return?

And so, when the opportunity arose—through Argentina's noted scientist Francisco P. Moreno, and through the Minister of the Navy Onofre Betbeder, as well as through future director of Observatory Island, Lieutenant Commander Horacio Ballvé—for Sobral to join Otto Nordenskjöld on the Swedish Antarctic Expedition in December of 1901, Sobral jumped at the chance.

After meeting with Nordenskjöld on December 17, and gaining the great scientist's approval, Sobral received his official orders from the Navy. He was now a full member of the Swedish-Argentine Antarctic Expedition. There was only one thing left to do: go on a shopping spree.

Walking around Buenos Aires in the searing sun, and hurrying along the hot streets of the city, Sobral began looking for warm polar clothing. A more impossible task was difficult to imagine.

He tried on all manner of layers of clothing to create enough covering for the Antarctic weather, as the Buenos Aires stores did not carry parkas or reindeer boots or anoraks or snow gloves. Neither did the fashionable stores on Florida, nor did the midrange shops downtown carry even the lightest winter clothing.

And so he was inventive, putting together a winter outfit that could take on the Antarctic temperatures.

In store after store, he piled on shirts and sweaters, hoping they would work. When he had exhausted the stores, he proceeded to the Navy depot, frantically foraging through the warehouses.

An assortment of pants, shoes, shirts, sweaters, boots, hats, and jackets was the best he could do. It would have to do.

He was about to embark on his lifelong dream—to sail to the Antarctic.

On the morning of December 20, Sobral found himself in front of the *Antarctic* ship. The crew aboard was preparing to set sail in the damp warmth of the Buenos Aires summer. The Polar vessel was being seen off by hundreds of well-wishers gathered at the docks to convey their *buen viaje*. The hour was early, but the energy was high.

The ship stood over 33 m high above the water and over 4 m below the waterline. It was over 41 m in length, and its waistline was 9 m wide. The vessel ate coal like a humpback whale eats krill—insatiably, and in mass quantities. At least four tons of the pure carbon was poured down its throat a day, gobbled up quickly and *en masse* by the engine. The fossil fuel fueled the ship's speed—a maximum of six knots. And while the ship ate like a whale, its true vocation was that of a whale-eater. Everything about the *Antarctic* equipped it for the hunt: the hull with large bindings made of greenheart hardwood; the reinforced bow, with half-inch irons and side-by-side strappings on the outside, and metal and wood on the inside; the several whaling boats for one-on-one (or more accurately twenty-on-one) sea mammal retrieval; and the observation barrel strategically located on the main mast.

With this whaling pedigree inherent in its very being, and the additional retrofitting the ship had undergone in Sweden prior to the departure the previous autumn, the *Antarctic* was fitted out and fit for travel in Antarctica.

Standing on its deck, Sobral was outfitted in his uniform—his "winter" clothes stowed for the time being. Besides the winter clothing he had frantically tried to obtain, Sobral had armed himself with a diary journal—a large one, most likely meant to be used as a ship's log. Measuring approximately 27 cm across and 39.9 cm high, the diary had a dark, hard cover and a red label declaring "Diario" in gold embossed lettering. It would be his companion on this expedition. He would use this as his personal voyage diary.

At 21, and at an average height and physique, especially in comparison with some of the veteran explorers surrounding him, he looked very much the part of the Ensign—the junior, the symbol, the representative of the Navy and the Argentine flag. But he knew he had a voracious appetite for study, and a brain and frame fit for exploration.

This ship he stood upon had carried the very first humans to step foot on the Antarctic continent; now, it would carry the very first Argentine to explore the white continent.

Sobral may not have thought about the whaling legacy that had brought this ship to this port, but he was eager for the Polar adventure.

Figures 2.1, 2.2, 2.3, 2.4, and 2.5.

Fig. 2.1 José María Sobral, Navy ensign, Under-Lieutenant, and later Doctor of Geology, had dreamt of being a Polar explorer from a young age (Departamento de Estudios Históricos Navales, Archivo Histórico, Archivo Fotográfico, P-0140 c, Sobral, José María. Alférez.)

Fig. 2.2 The house where José María Sobral was born, in Gualeguaychú, in the province of Entre Rios (Departamento de Estudios Históricos Navales, Archivo Histórico, Archivo Fotográfico, P-0140 j, Sobral, José María. Casa de Gualeguaychu donde nació el Alférez.)

Fig. 2.3 Navy Midshipman José María Sobral (Departamento de Estudios Históricos Navales, Archivo Histórico, Archivo Fotográfico, P-0140 h, Sobral, José María. Guardiamarina.)

Fig. 2.4 The *Presidente Sarmiento* training ship upon which José María Sobral served (Departamento de Estudios Históricos Navales, Archivo Histórico, Colección Sobral, Box 2, Envelope 1, newspaper cutting.)

Fig. 2.5 José María Sobral, Alférez [Departamento de Estudios Históricos Navales, Archivo Histórico, Archivo Fotográfico, P-0140 b, Sobral, José María. Alférez (Sector A-planera verde).]

References

Fundación Marambio Alférez de Navío José María Sobral (2017) http://www.marambio.aq/alferezsobral.html. Accessed 14 June 2017

Marenssi Sergio (2007) Doctor José María Sobral (1880–1961) de los hielos antárticos al olvido argentino. Revista de la Asociación Geológica Argentina, versión On-line ISSN 1851-8249. Rev Asoc Geol Argent 62(4) Buenos Aires oct./dic 2007. http://www.scielo.org.ar/scielo.php?pid=S0004-48222007000400002&script=sci_arttext. Accessed 6 Apr 2015

Rabassa Jorge (2003) Estudio preliminar. In *Dos años entre los hielos, 1901–1903,* José María Sobral (reprint of original book published in 1904), 11–46. Eudeba / Universidad de Buenos Aires, Colección Reservada del Museo del Fin del Mundo, Buenos Aires

Sanatorium Quirurgico (1901) Radiography receipt for José María Sobral. Departamento de Estudios Históricos Navales, Archivo Histórico, Colección Sobral, Box 2, Envelope 1. Receipt for foot X-ray, dated 28 Oct 1901. Archivo D.E.H.N.–A.R.A., Buenos Aires

Sobral Guillermo José (2016) Letter from Enrique Sobral to his son José María Sobral on the day of his entering the Naval Academy [1894]. Communication sent from Guillermo José Sobral to the author, received 9 Nov 2016

Sobral José María (1899–1900) Hotel menus, purchase receipts, and letters of correspondence collected by José María Sobral. Departamento de Estudios Históricos Navales, Archivo

Histórico, Colección Sobral, Box 2, Envelope 1; Box 3, Envelopes 17 and 19. Sarmiento programs and banquet menus 1899–1900. Archivo D.E.H.N.–A.R.A., Buenos Aires

Sobral José María (1898–1961) Original personal notes, manuscripts, letters, diaries, and photographs. Departamento de Estudios Históricos Navales, Archivo Histórico, Colección Sobral. Archivo D.E.H.N.–A.R.A., Buenos Aires

Sobral José María (1904) Dos años entre los hielos, 1901–1903 (Two years amidst the ice, 1901–1903). Imprenta de J. Tragant y Cia., Bolivar 319, Buenos Aires

Chapter 3
Expedition Diary: The Voyage to Antarctica

Abstract The journey of the Swedish-Argentine Antarctic Expedition to Antarctica, December 21, 1901, through February 12, 1902 is documented in José María Sobral's diary entries on board the ship the *Antarctic* during the voyage from Buenos Aires to Antarctica. He writes of the following: Making preparations for departure, December 17 through 21; sailing, observations on whale hunting, oceanographic testing, and acclimation to Swedish language and customs, from Buenos Aires to the Malvinas (Falklands), December 21–December 31; Visiting Port Stanley, bringing aboard Malvinas dogs to supplement the Greenland dogs, and meeting with officials, December 31, 1901 through January 1, 1902; Traveling from Malvinas to Isla de los Estados for magnetic meteorological observations, and to Observatorio Island, location of Argentina's Observatory and new Lighthouse, January 1–January 6; magnetic work performed and translation of information between Observatorio Island's Argentine personnel and the Swedish expedition, January 6; Bird specimen collecting, hydrographic research, first sighting of icebergs, King George Island and South Shetland Islands (Pleistocene and Mesozoic comparison studies), seeing the first penguins, Louis Phillippe Land, Joinville Island, Hope Bay (Esperanza), Paulet Island, Seymour Island (Marambio), and fossil collecting, January 6–16; the provisions deposited at Seymour Island, the lack of surface snow, an attempt to sail further south of 66° that is thwarted by pack ice, and the retreat made northerly, January 17–20; oceanographic studies in the Weddell Sea, the voyage through pack ice back to Seymour Island and then to Snow Hill, Cockburn Island, Erebus and Terror Gulf (Sidney Herbert Bay), Sobral's concerns regarding Argentina's geopolitical standing, and the first ski journey on snow, January 20 to February 12.

The following are the daily journal entries from the expedition diary written by Under-Lieutenant José María Sobral, of the Swedish-Argentine Antarctic Expedition, on board the ship *Antarctic*, during the voyage from Buenos Aires to Antarctica, from December 21, 1901, through February 12, 1902 (Sobral 1901–1903, Diary pages 7 through 47).

M.R. Tahan, *The Life of José María Sobral*, Springer Biographies,
https://doi.org/10.1007/978-3-319-67268-7_3

3.1   There Is No Anesthesia for the Heart

3.1.1   December 21, 1901, Saturday

On Monday the 16th in the afternoon, the Swedish ship "Antarctic" arrived; I was expected to board it to set off on a scientific commission in the Antarctic regions.

I didn't have anything prepared, but such was my excitement to go on that expedition that, on the 17th, 18th, and 19th, after consulting with Dr. [Otto] Nordenskjöld about the things I should obtain, I finally got everything ready, and on the 20th in the morning, I had my equipment packed, and I was ready to set sail. On the 17th, 18th, 19th, and 20th, I was in a continuous state of overexcitement. The men who were in charge of loading the coal were on strike, but since we were urged to depart, the strikers appointed a commission for the shipping of such fuel.

On the 20th, after boarding with my luggage at 10 am, I asked the commander of the ship, Mr. [Carl Anton] Larsen, what time we would be leaving. He told me it would be between 2 and 5 pm, but that departure could be put off until Saturday. I went home and had lunch, and the time to say good-bye finally came. I hugged everyone. My mother broke into tears and made a thousand recommendations to me. These are the moments when one realizes that there is no anesthesia for the heart. Tenderness flows like the water of a wellspring. Everything is small compared to a mother's pain.

For many reasons, there have been times when I would have retracted my decision if I'd been able to. But the reaction would immediately come, and the enthusiasm would return even more strongly.

I boarded accompanied by [my friend] Esquivel and my brother Enrique, and I started to settle in, but I had to suspend the task, because the great number of visitors was making it impossible. It was tremendously hot.

My father was aboard, and around 4:30 or 5, he said good-bye because the ship was about to leave, and I promised that I would come home if the ship didn't depart.

The "Antarctic" moved, aided by tugboats, and was moored to the side of the south dock in order to continue loading coal.

Captain Larsen told me we wouldn't leave until past 4, on the 21st, so I went to give a second good-bye to my parents and siblings. At 9:30, I was departing for the last time and, accompanied by my friend Arturo Esquivel, we headed toward the south dock. We said our last good-bye, and then, I tried to get some sleep [on the ship].

When I awoke, the "Antarctic" was moving away from the wall, it was 6:11 am.

At 9:30 am, they called me for *breakfast*, because they eat more or less like the British and the Americans do: 9:30 *breakfast* (in Swedish it is called *frukost*); *lunch* was at 2:30 pm; and *Dinner* at 8:30 pm (*Middag*).

There is a dog that is very sick; his suffering seems to be a consequence of the heat. Mr. [Frank Wilbert] Stokes worries about him. He has a book that talks about dogs' illnesses. He had it [the dog] bathed well with soap and warm water, and after that, the dog was dried and left to rest in the shade. He gave him milk to drink. It looks like he is getting better. At 4:30 or 5 pm, we saw some lightning in the southeast, and the breeze blowing in that direction started to escalate. It looked like we were going to have a southeast strong wind. A stormy cumulonimbus formed and started moving to the north. Some raindrops began to fall on us, and we could see two concentric rainbows. At first, we could only see the fourth part of the most defined one, which was the most interior, and an eighth part of the exterior one. And as clouds were growing larger, the visible part of the rainbow was more visible. The shape of the cloud was as follows: The base of the cumulonimbus was completely horizontal, and its edges were very well defined, while the lower side of the cloud was shadowy. The upper part was white, and its edges were also perfectly defined. The wind wasn't strong and the night was clear. Sails were raised during the night.

Sobral found the Antarctic *docked at the Buenos Aires port near the La Boca barrio, where the Riachuelo River flows into the Rio de la Plata. There, a mass of movement could be felt, and a sea of people seen, where the sounds of Italian and Spanish conversations mixed with the breezes through the palm trees and the water lapping against the dock.*

He was under specific orders, issued by the Minister of the Navy, assigning him to participate in the Swedish Antarctic Expedition led by Dr. Otto Nordenskjöld and to represent Argentina in this scientific mission.

At the time that Sobral boarded the ship, he knew only one soul—the highly respected geologist Otto Nordenskjöld, whom he had just met a few days ago, on December 17th, when the professor had interviewed the young Under-Lieutenant for acceptance into the expedition. It was immediately after that interview that Sobral received his official orders.

The Antarctic, *which would take him to the South and into the ice, was a storied ship. Built as a Norwegian sealer in 1871, and later used as a whaler, she had carried the first human expedition to step onto Antarctic land in 1895, under the command of Captain Leonard Kristensen. Carsten Borchgrevink, also, was on that expedition, and he returned a few years later to the very same spot—at Cape Adare —to spend the first winter on the Antarctic continent. Nordenskjöld purchased the* Antarctic *in 1900 and equipped her for his scientific expedition to Antarctica.*

Like his ship, Captain Carl Anton Larsen was a Norwegian hunter, too—a sealer and a whaler by trade. The 41-year-old veteran had made his mark on the North before venturing South, sailing to Seymour Island on the Jason *in 1892–1893 and 1894, and, as a byproduct of his hunting, becoming the first person to bring back fossils from the Antarctic continent. He also had the distinction of naming some of the islands he found in the vicinity, and discovered King Oscar II Land.*

The dog who is ill on board is one of the fourteen Greenland dogs brought by Nordenskjöld to serve as transportation on the ice. These northern dogs suffered greatly traveling through the heat of the equator and the tropical zones, and the heat in Buenos Aires at the time was quite severe. Nine had already died from the heat. Sobral paid attention to the dogs' plight from the very start and demonstrated true empathy for the dogs (Tahan 2016a, b). He would later disclose, in his personal notes written at the end of 1903, that the dogs were usually allowed to roam free on the deck and that they were fed a ration of dried fish twice daily (Sobral 1903).

On board the ship, Sobral befriends the American artist Frank Wilbert Stokes, who joined the expedition in Buenos Aires. The celebrated painter had corresponded with Nordenskjöld and requested to be allowed to accompany the expedition, an idea that Nordenskjöld liked very much in that the artist would be able to visually represent the Antarctic landscape and wildlife for public consumption. He probably viewed him as a visual documentarian, a creative ambassador. F.W. Stokes would become one of the first artists to paint the Antarctic. Although Stokes paid his own way, he, too, was a veteran of Polar exploration, having painted the Arctic region when he accompanied Robert Peary's expedition in 1892, and having travelled to North Greenland with Admiral Peary in 1893–1894. His oil paintings of the Inuit people and of icebergs in the North Polar region were shown at the Art Institute of Chicago (Staff Writer 1901; Stokes 1903).

Whereas Stokes would paint the Antarctic landscape with his brushes, Sobral would attempt to paint it with his words. This entry is the first of many in which Sobral describes the sky and clouds and weather phenomena in what he attempted to be observational yet poetic language. It gives a clue as to his future persistence in keeping an accurate ongoing record of meteorological developments.

At the beginning of the entry, Sobral questions his commitment to go to Antarctica when faced with his mother's grief at the possibility of losing her son. But his fervor to explore burns even more intensely each time he reconsiders his decision, and the questioning of himself only further reaffirms his desire to depart to the white continent.

3.1.2 December 22, 1901

I woke up at 7 am. There was a small breeze and they took advantage of it with all the sails. The poor dog howled all night. Who knows what he has. It would be a pity if he died, because in that case we would only have four left. It's very hot, but the cloudiness is increasing, so we should expect a cooler afternoon. I have started to study Swedish, I think I'll be able to learn it. The wind is blowing NE and diminishing.

Sobral was already feeling alone on the ship. Other than Stokes—with whom he could relate—he could not easily speak with the rest of the expedition members. The Swedish scientists and Norwegian crew spoke their own language. Sobral was able to use the English language to communicate with them, but it was not the same as speaking to his colleagues in their own language. This was one of the reasons he became determined to learn the Swedish language, fluently, and he set this goal for himself on the second day of his voyage.

As for the dogs, whose contingent had already been reduced by nine, and one of whom now continued to howl piteously, perhaps Sobral felt that he, like them, was not understood. He was the alien. The lone wolf. The underdog.

3.1.3 December 23, 1901, Monday

I woke up and went on deck, and the captain was performing a maneuver of throwing a [trawl] net into the sea to drag along the bottom and collect fish, shellfish, and seaweed. These nets are of different sizes, and they have an iron ring in the mouth, in order to keep the net completely open; a weight, according to the depth, takes the net down to the bottom. A wire cable coiled to a reel is rolled up to a winch drum; then, it passes through an anchor base, then through an engine hanging from the mainsail, and finally to the net.

When Misters Ohlin, Scottberg, and Anderson [Axel Ohlin, Carl Skottsberg, and K.A. Andersson] took the net out and had it with them, they classified all the different shellfish and fish that had been in the net and placed them into jars.

The wind is not favoring us right now, and the knife sail [fore-and-aft sails] can barely be held up in the wind. The morning was very cloudy, but the afternoon is now beautiful. The wind is from the SE. The poor dog is still howling, and he's completely separated from the rest. There is some rough water due to the wind blowing from the SE, with a strength of 4 or 5 [on the Beaufort scale]. It is a pity we don't have it on the beam. I have to point out something pretty weird for my tradition, that is, the kind of food. If they have analogies with the American and the British regarding the times of the day they take meals, it is not the same with the kind of food. At 8:30 am or 9 am, they have tea, coffee, or cocoa, cold cuts, butter, and bread (there is this bread made of wheat that is very nice). At 2:30, they eat *lunch* and, besides cold cuts, bread, and butter, there's wine, beer, tea, coffee with milk, and one or two [main] dishes. Today's was in the form of soup. We ate a plate of fruit, very tasty, very good, and they say it's anti-scorbutic. At night, cold cuts, butter, beer, bread, one or two more dishes. This night, I ate something for the first time in my life. Fried oatmeal flour, and after that, milk. It's very nice. And I am very satisfied.

In the early entries of his diary, Sobral, naturally, misspells some of the names of his ship's Swedish contingent, spelling some of them phonetically. He has just been introduced to these new colleagues and tries to familiarize himself with them. As the journey progresses, he corrects these names in his diary.

In his handwritten notes that he later prepared for his speeches and for his Naval report at the end of 1903, Sobral expanded on this diary entry, specifying his observations about the work on the ship (Sobral 1903). He wrote:

> *The time during the day is employed by the crew to load the coal that still remains on the deck, and by the scientific personnel to continue their work; meteorological observations are made three times a day, at 7 am, 2 pm, and 9 pm. The zoologists devote their time to fixing the specimens collected in the last "catch", etc.*

Sobral was very preoccupied with food—and rightly so—and this is the first taste of his obsession with nutrition. His discovery of certain new culinary experiences delighted him, including his first nibble of an oatmeal biscuit with milk. The anti-scorbutic food, including fresh meat and fruit, was a necessary precaution to stave off scurvy—the bane of any Antarctic expeditioner.

3.1.4   December 24, 1901

When I woke up, I went up to the deck, it was 8 am, and the morning was beautiful. At some point the engine stopped and we continued navigating by sail because we had a breeze that kept us moving for three miles. At 12 they started the engine again because the breeze had ceased. According to the point calculated by Captain Larsen, today at 12 we will be 960 miles away from Isla de Año Nuevo [New Year's Island].

The poor dog still is whining in the night. The doctor gave him some injections of morphine. At 3:00 in the afternoon the Swedish flag and the North American flag were put on top of the same pole, and the Argentine flag on the foremast, and we started decorating the table to celebrate Christmas because the Swedish have parties beginning from the 24th. They put all kinds of gifts that are numbered, cigarettes, sweets, magazines, etc., and it was like a raffle. A card was pulled out, and each one got a present according to the number on it. So all the crewmembers except for the one who was on guard duty were in the room that was adorned with Swedish flags and the lights of candles of many colors. They made numerous toasts, both for the union between Sweden and Norway, and for personal reasons. Old Haslum [H.J. Haslum] played the Swedish and the Norwegian national anthems on the accordion, as well as some other musical pieces. Accompanied by the old man, Scottsberg [Skottsberg], Botsman [Gösta Bodman], and one of the engineers sang songs from their country. Old Haslum is an interesting guy. He's spent around twenty summers in the Ices of the North, and the corresponding winters in Norway, and he embarked when he was in the tropics, so he didn't feel the heat much.

Scottsberg [Skottsberg], Duse, and Botsman [Bodman] read some verses as jokes for some of those present. We drank port wine and Swedish mineral water, we ate fruit, sweets (this started at around 6 pm), and around 11 everyone went to eat. We ate around 12 pm. And we went to bed around 1 am.

Sobral describes the joy of the holiday celebration and the pain of the dog's suffering, all in the same breath. He is attuned to the behavior of the men and the animals that surround him on the ship. In his 1903 notes, in which he references this diary entry, he mentions that the sick dog's constant whimpering had kept the men up all night and that the morphine was intended to calm him down; meanwhile, the men were celebrating "Jul afton"—Christmas Eve (Sobral 1903). At this time, as apparent from this diary entry, Sobral is also beginning to familiarize himself with the scientists, and—name spelling errors notwithstanding (for he is still learning the language)—he is observing and finding out more about them.

3.1.5   December 25, 1901

Tailwind all day; in the afternoon, we stopped the engine and navigated only with the sails. People on the deck are dancing and playing. They're throwing arrows against a piece of wood divided into circles like a target. It is remarkable how well sailors are treated on this ship.

Today there was a session of phonograph and gramophone in the chamber, and flags were raised the same as yesterday. So far, the weather has been magnificent. The dog has cried all night. Depending on the wind we get, we will see Las Malvinas [the Falklands] before spotting Isla de los Estados [Staaten Island].

Sobral further detailed this day of festivities in his 1903 notes, saying that "The party that had begun yesterday continued all day," and noting that the meals were special for the holiday, and that "The Swedish punch *circulates with profusion" (Sobral 1903). Sobral himself did not drink alcohol, but he appreciated and partook in the spirit of the drinking of libations.*

In those same notes, he also confides his disgust at the lack of cleanliness on the ship. A fastidious individual, Sobral did not appreciate the dirt that surrounded him. He wrote (Sobral 1903):

> *What a difference I feel coming from one of our clean ships to this one, so dirty! One cannot hold on to any part of the ship without getting their hands dirty. There has barely been one washing-down, done with buckets, since we took on coal in Buenos Aires. Being in the cabins or in the chamber is unbearable. If it is hot on the deck, the reader can imagine what it is like on the second deck, where the chamber and the cabinets are – places where there are no portholes, and where the ventilation is created only through a skylight and the entrance hatchway.*

3.1.6   December 26, 1901

The day is a little cloudy, wind from the N, the engine is off. Navigating with the sail at a speed of 6 miles. The day is getting cloudier.

Today I started reading [Fridtjof] Nansen, and I had to leave the book because I, like him, have left behind everything that I love, and reading his lamentations makes me sad.

I sometimes experience periods of discouragement and sadness, and so I go to the deck and walk around until I can become distracted by something and so that I can forget my sorrow.

There are four healthy dogs: two couples; one of the female dogs is very skinny. Sometimes, they give her such a shake that I think they're going to end up killing her. The engine hasn't worked today. The wind persists in the N.

Sobral's expressions of his loneliness are touchingly poignant. Even turning to the writings of his heroes, like the Arctic explorer Fridtjof Nansen, does not give him comfort, but actually accentuates his isolation.

It is not clear who was "shaking" the dogs on the ship, but, again, Sobral identifies with the dogs and their plight.

3.1.7 December 27, 1901

The morning is presenting itself quite nicely. Luckily, the wind from the N persists. Now it is said that we're going to Las Malvinas [the Falkland Islands] before going to Isla de los Estados [Staaten Island]. The wind tends to move to the NW. The barometer is going down, and you can see a belt of dark clouds to the south and SW. Are we going to have bad weather? Today they [the crew members] have been cleaning these little canons that shoot Remington bullets; these will be placed on the boats for the fishing of whales. Happiness reigns on board. The saddest one is me, because I don't have a friend to communicate with. They treat me very nicely, but not with the confidence of a friend. And then they speak a different language [from me]. Many times they may even be talking about me without my understanding. Aside from the meals, I either stay in my cabin or walk on the deck; the one I talk to the most is Mr. Stokes; he is always cheerful, making jokes, and he seems to be a good person. If I don't become closer with the rest of the members of the expedition, my life will be frankly very weary. At night, almost everyone plays cards. 11 pm the breeze has increased, we have all the sails on wind, and we sail approximately 7 miles.

Sobral uses the innocent term "pesca," which means "fishing," to describe what Larsen et al . are doing to the whales—which would be more accurately described as hunting. Larsen would later go on to found the Compañia Argentina de Pesca— Argentine Fishing Company—in 1904, after his return from this expedition. This would warp-speed-kick-off the whaling industry of the Antarctic waters, for which he also built Grytviken, the whaling station on South Georgia Island. (Grytviken was the first whaling station on land, and the place where Ernest Shackleton sought help for his stranded crew in 1915 after the wreck of his ship Endurance. *Shackleton was buried there in 1922. One of Larsen's partners in the Pesca whaling company was Don Pedro Christophersen, the Norwegian-born Argentine who would later finance Roald Amundsen's Antarctic Expedition, allowing his ship the* Fram *to return to Antarctica to pick up Amundsen and his expedition members.)*

As reflected in this entry, Sobral's feelings of isolation and lack of communication continue, and his friendship with Stokes strengthens. He seems to appreciate the American Artist's humor and sensibilities. It is ironic that his best friend on the expedition, Stokes, is the one member who later decided to turn back rather than spend the winter in Antarctica. He had been very prolific in his art during the voyage and remained on the ship when Sobral and his companions were set ashore at Snow Hill. Stokes then disembarked at the Malvinas (Falkland) Islands, depriving his friend Sobral of his companionship, but saving himself two winters in the Antarctic.

3.1.8   December 28, 1901, Saturday

Last night the poor sick dog died, and they threw him overboard this morning.

Today from 6:30 am till 9 am, they took soundings and the temperature and samples of the water at different levels of depth. With a 1000-meter sounding line, they didn't touch the ocean floor; this is used for soundings and measuring the temperature of the water, a device invented by [Fridtjof] Nansen for the storms and water. And also the thermometer of the invention of Negretti and Zambra [Enrico Negretti and Joseph Zambra]. Temperatures have been measured at 1000, 700, 500, and between 100 and 500. The temperature on the surface has diminished a lot. Today at 4 pm it was 9.5°, while at 2 pm it was 10.35. According to Captain Larsen, our location now is 275 miles from Port Stanley. Very big albatross can be seen. Larsen and [Samuel A.] Duse shot at them with no result. Samples of animal life were taken at the depths mentioned above. The wind moved to the west during the night. Today the breeze has increased, and we are doing 8 miles at times.

3.1.9   December 29, 1901, Sunday

The engine has been working since the early morning, because the breeze has decreased a lot. However, it [the wind] started moving toward the north again, and today at 7 am, we had it in the stern; therefore, we had the rigging in cross. The barometer is lowering. The wind has increased in intensity, and so the engine has been turned off. While we were finishing our lunch from 3:30 to 4:30 pm, a hailstorm occurred and hail fell, and the stones were very small. And from that time on it just continued raining. The wind increased at times, and it kept moving in circles, clockwise. Between 12 at night and 1 am, the barometer went down to 742, and the wind acquired its maximum strength. And then the barometer started to rise. At that same time, the direction of the wind was SE and, furling the square sails, we rode out the storm with the engine and the knife sail [fore-and-aft sails]. Some petrels were seen.

3.1.10   December 30, 1901, Monday

The ship is heaving to, the barometer keeps rising. Some blue patches can be seen in the sky. The strength of the wind has diminished and the surge is quite intense, the ship looks very________ [unreadable]. The barometer has been rising all day. At 4 pm, we saw land from the starboard side. At that time, the wind had calmed and a heavy sea was the only thing left from yesterday's rough weather. Some sea wolves were seen during the afternoon. The albatross were flying in circles over the ship, with their black wings steady, barely moving their bodies; it looks like an unknown force propels their fast flight.

3.2 To Think that I Am a Foreigner in My Land!

3.2.1 *December 31, 1901, Tuesday*

Today at 6 am, we anchored at Port Stanley. To think that I am a foreigner in my land! It's a beautiful bay of more than four miles long and half a mile of width. It is oriented from east to west; its depth with low tide varies between 3 and 4 arms; population is situated in the southern part; from what I see, it must not even have 1000 inhabitants. The entrance of the bay has about 300 m in its wider part. There are two merchant steamboats; one British canon ship; two sailboats; four or five pontoons; and us. An officer from the British ship came to greet the commander. A doctor also came to ensure that there were no plagues on board. At 9 am, everyone went ashore, only Mr. Stokes and I stayed on board. None of the crew has gone off the ship yet.

Today I realized that it was not certain if, when we returned, the ship would stop in Buenos Aires, or if maybe I would have to stay in the Malvinas [Falklands] to take the transport that goes to Montevideo, or if the ship would leave me in Montevideo, so I went to talk to the North American consul Mr. Rowe [John E. Rowen]. Mr. Stokes introduced me, and he received me very well and offered me everything I needed, and agreed to store my clothes and the money that my father will send to me at Port Stanley.

Last night, Captain Larsen and Dr. Nordenskjöld were invited to have dinner on shore. There was a small party aboard, too, and the gramophone played an important role.

Last night, something very *particular* happened to me with Dr. [Erik] Ekelöf. We began talking about the things that were happening between Chile and Argentina, and he said, among other things, that people from Chile were much better than people from Argentina. That the population of Argentina was made up of an infinity of races, and that the Chilean population was not, and he said other things to this respect.

At last I said: Doctor, I think you don't like my people. It's true, he replied. Your people, I don't like. Then you must not think I'm nice, I said. It's true, he replied. I became astonished with such frankness. And I replied that I preferred his talking to me like this, and that I liked his being sincere. The lieutenant Duse, who had heard this, started to talk to him in Swedish, and even though I don't know the language, I came to understand that he was saying that it was necessary to be a little diplomatic. I asked Dr. Ekelöf if this was what Duse had told him, and he replied yes, but he hadn't been diplomatic, and couldn't really be, because he said what he felt. This has created for me a very bad impression, because it makes me think that everyone must have the same opinion. At least I know for sure that I'm *persona non grata* to two of them: Duse and Ekelöf.

After speaking with Mr. Stokes about this, I wrote a little and I went to bed at 1:00 am.

The year is over. What does the new year hold for us? Where will we be a year from now? This and a thousand other questions are what, to those like me, who don't have a friend aboard, think about; those who, like me, do not get to hear their own language.

The visit to Port Stanley in the Falkland Islands truly reflects the "foreign-ness" that Sobral feels—as a stranger in what he views as his own Argentine land occupied by English. It also unleashes the problematic relationship between Sobral and most of his fellow expeditioners.

Sobral gets his first taste of bald-faced prejudice, perhaps even racism, when he hears the words from Ekelöf regarding the good doctor's disapproval of the mixed race that was the Argentine population. To make matters worse, Ekelöf confirms Sobral's fear that he is not regarded as a sympathetic person, as "simpatico." When he catches on that Duse, speaking in Swedish so Sobral won't understand, probably agrees with the doctor but nonetheless chides him for being so blunt, Sobral's worst fears are confirmed. He handles the moment calmly, however, complimenting Ekelöf for his candor. But he now believes that the sentiment which Ekelöf opined about him is shared by all the others. And this only deepens his sense of isolation.

It is ironic that the topic which brings Ekelöf's hurtful sentiments to the fore is already a painful one for Sobral—the fact that he believed Argentina and Chile were on the verge of war. This potential conflict was due to the two countries' boundary disputes, and the build-up of armaments by the two countries' respective navies. It is a problem about which Sobral obsesses. And it is the subject matter with which he had engaged Ekelöf.

Sobral's uncertainties include how he will get home after the expedition— whether the ship will drop him off in the Malvinas [Falklands] or Montevideo or Buenos Aires, and again he finds a friend in Stokes, who takes him to the American consulate and helps him arrange to have his things, which his father will send to him, stored there at Port Stanley awaiting his return. (Port Stanley is where Stokes returned from the Antarctic to set sail back to the USA.)

In notes that he would later write for his speeches and for his Naval report in 1903, Sobral would say that he did feel that Port Stanley was "excellent as a military port," and that, while it now served all the ships that traveled to the Pacific, he hoped that one day the Argentine ports would serve that purpose (Sobral 1903).

He would also write about the fur seals that had been hunted to near extinction, specifically the South American fur seals, which he states had been "completely exterminated" (Sobral 1903):

> *Port Stanley is frequented by South American fur seal hunters; when they are questioned, their answer is that they hunt them at South Georgia, but because there, as well as on every*

other island and land of the south, there aren't any [fur seals], one can almost be certain that they come from our coasts.

And so, the year closes for Sobral with a painful episode that will stay with him throughout the expedition, and that defines what he views as his precarious standing with these learned men from Sweden.

3.2.2 January 1, 1902, Wednesday

Today, in my land, perhaps all will be rejoicing, everyone will receive the greetings from their relatives and friends, while I receive the courteous but cold greetings from people who have known me for not even two weeks and who have everything different from me.

Today in the morning, after *frukost*, everybody left to go ashore (to the town), except for Scottsberg [Skottsberg], who went to the north side of the bay with the objective of collecting the largest number of samples of the flora possible. Around 12 pm, some British officials from the cannon ship "Ninphe" [possibly *Nymphe* man of war] came to visit our ship, and then they left.

Nordenskjöld came, and Captain Larsen arrived shortly after. This morning, 8 dogs were brought onto the ship, they were purchased in Port Stanley, and they cost eighteen pounds. They are very expensive, the vendors who sold them took advantage of us because we needed them urgently. It looks like the dogs are going to endure the cold well, but they're not as strong as the dogs they brought from the North.

About 2:30 in the afternoon, the governor [William Grey-Wilson] and the consul of the USA, Mr. Rowen, came to visit the ship. Mr. Rowen and a couple of families visited the ship. We shared a glass of champagne, and we had a very nice time. At 5:30 pm, they left. Botman [Bodman] and I went on a small boat to pick up Scottsberg [Skottsberg]. The ship whistled several times so that the lagging ones would get on board. We were about to finish lifting anchors when Stokes, Duse, Ekelöf, and Anderson [K.A. Andersson] got on board. The British canon ship made a signal wishing us *bon voyage*, and all the ships were waving at us with their flags. It was 6 pm. Around 7 pm, we had room to maneuver, and a heavy sea swell made us remember that we were no longer anchored in the warm and nice bay of Port Stanley. In this port, we were told that 2 North A. [American] ships are sailing around the "South Shetlands," hunting fur seals, so we will likely encounter them. We have the wind in a good direction, but it increases in intensity, and the sea is getting heavier and heavier.

In his notes written later in 1903, Sobral mentions that the ship Antarctic *was, as usual, "filthy," and that the English women at Port Stanley would exclaim that it was "So dirty!" (Sobral 1903). He also laments that, despite his reminding Captain Larsen, the* Antarctic *crew never thanked the English gunship that sig-naled to them to wish them a good trip; likewise, they did not reciprocate the visit paid to them by the commander and officer of the English ship. To Sobral, etiquette and proper conduct were very important.*

The two ships he mentions were US sealers, hunting seals in the South Shetlands region. Sobral refers to fur seals as "lobos" or "lobos de dos pelos," killed for their pelts, as opposed to the "focas," which are the seals he hunted for food in Antarctica.

As he bids farewell to the old year and welcomes the new, Sobral very much realizes the reality of his situation. Later he would write that "everyone keeps in their hearts new hopes, new longings which the new year . . . will either destroy or encourage."

3.2.3   January 2, 1902, Thursday

The sea is still heavy, and the wind has a power of 6. We are rolling a lot. The barometer, which had decreased in Port Stanley, is now rising quite a bit, and at 10 pm, we changed course and the engine was turned off. The wind has the same velocity, and the sea is still very heavy, making it impossible to write.

The poor dogs are dizzy (the new ones). They have put them under the derrick.

At 6 or 7 pm, the wind abated, and the reefs of the topsails were released, and the topgallants were tautened. The average temperature of the air is 6 °C (the air), and the temperature of the ocean is 9, a very cloudy day.

The dogs who had been brought on board the previous day did not understand their new home. Sobral later describes them in his personal notes as "running from one place to another on the ship" and "staring in fright at the large waves" (Sobral 1903). Those who received a bath from the heavy sea, he said, would "run away and curl up behind the upper deck." Nordenskjöld described these dogs as nothing more than sheep dogs, which, of course, was the primary vocation of dogs on the Falkland Islands (Nordenskjöld et al. 1905). In addition to the sea waves, the Falkland dogs also found enemies in the Greenlanders. Sobral would later write that they would growl at each other whenever they were in close proximity and that "it wasn't long before there was a clash between the two races," which would be alleviated through the cooperation of the men on deck (Sobral 1903). Sobral paid special attention to the dogs, both those from the Malvinas/Falklands and those from Greenland.

3.2.4   January 3, 1902, Friday

Today the sea has diminished greatly, and the wind is from the N.

3.2.5 January 4, 1902, Saturday

We are moving smoothly, the wind is to the NW, and the day has cleared up. The temperature is very nice. Because the breeze is weak, we have been steam-navigating since last night, but with all the sails in the wind.

This morning, all the new dogs were taken to the deck; they are very docile. Around 4 pm, a ship was spotted on the starboard side.

The wind changed completely. This calm predicts a change of direction. I was right. After a while, we ended up feeling the breeze from the NW, and it started increasing little by little. At 5 pm, the barometer started to go down rapidly and the wind became colder. Twilights are already very long. At 10 pm, land was sighted from the port side of the ship.

In his notes written for his speeches in 1903, Sobral further expanded on the state of the dogs at this time; he wrote (Sobral 1903):

> *To prevent the new dogs from going to the sea, yesterday afternoon they were locked up in the upper deck, and today, with the calm, they have been let out to the deck; it seems that the Greenlanders don't look at them with friendly faces.*

3.2.6 January 5, 1902, Sunday

We have spent all day riding out the waves in front of Islas de Ano Nuevo [New Year Islands]. Every now and then, _________ [unreadable] changes, keeping us noticeably in the same place.

The wind has blown from the SW and NW. At times, the veil of clouds that was interposing itself between us and the island would disappear, and we could see the island's high lands from a 20-mile distance. But then the curtain would fall again and in a forceful manner, sending us downpours.

The wind has been blowing strongly, and in a cascading-descending fashion, sometimes accompanied by rain which, when hitting one's face, feels like pins. A lot of albatross can be seen.

At 6 pm, the wind began to calm, and so did the sea, forming a SW that allowed us to cross with fore-and-aft sails and navigate at full steam, permitting us to go to the Isla Observatorio [Observatory Island], which is the new name of the most eastern of the New Year Islands. We spent the night in front of the island.

3.3   The Best Meteorological Observatory in the Southern Hemisphere

3.3.1   January 6, 1902, Monday

At 5 in the morning, the ship came close to the small cove located in the south side of the Observatory Island, and we got into a boat to go to land. Nordenskjöld, Duse, Botman [Bodman], Scottsberg [Skottsberg], Stokes, three sailors, and myself. We were taking the magnetic apparatus and someone to accompany us so that we could make comparisons [of the instruments]. Once we hit land, with some difficulties, we were finally able to secure the boat, and then at that moment, a soldier and a prisoner appeared and helped us with the task.

We hiked the side of the island through a very narrow and steep path [trail] covered with vegetation; once we were up there, we saw a tent and a little wooden house, which is where the prisoners who work on the island live. And they indicated to us the way to go to the observatory. Around 200 m ahead, the path divided into two. The one on the left took us to the observatory, located at the center of the island, a little closer to the N coast. And the one on the right led to the lighthouse under construction, close to the N coast. I went to the kitchen (everyone was sleeping, it was 5 am), and I woke up a man who was sleeping there. I asked him for Lieutenant Ballvé [Lieutenant Commander Horacio Ballvé], and he said he is occupied in Puerto Cook, but that Under-Lieutenant Plate and the Commander are here. I go to where he indicated to me, and, after knocking on the door, I yell: Plate! Plate! He opened the door. He was puzzled to be called in such a manner, and he thought he recognized my voice, so at last he opened the door, and we embraced.

I immediately went to the next room and met with Ballvé, knowing that Nordenskjöld and the others who were accompanying me were there [waiting]; they [Ballvé and his men] got dressed in five minutes, and I introduced everyone.

We began walking through the facilities. The meteorological observatory was ready. But the observations hadn't started yet because they were still installing the magnetic [portion], and they'd had little time. The wooden armor part of the magnetic observatory is not completed yet. A lining of cork and tar will be placed on it; I think this layer will be 10 cm thick. Inside the little house, there is a series of masonry pillars above which they are going to install the magnetic instruments. Everything is completed with an intent to be permanent; there is a heater in the middle, surrounded by asbestos, which enables the exit of hot air from the top, and the entrance of cold air from the bottom; thus, in this way, a circulation of the air is generated, which ensures that the temperature in the compartment remains constant. It has a double ceiling, too, so that the air trapped in between them also contributes to maintaining a constant temperature.

The first thing you see when you get close to the island is the _______ [unreadable—possibly wind station] where there are installed the anemometers, wind vanes, heliographs, etc. Nearby, there are the pluviometers, data recorders, totalizers, and the one from the Bureau Central M de P. Close to that place, there is a stall with electrical installations of the _________ [unreadable] anemographs and thermometric casings. There is another stall for barometers, heliographs, etc. Anyway, one day, this will be the best observatory of its kind in the southern hemisphere.

According to what they told me, the lighthouse will also be ready in a month. The house of the officers hasn't been built yet; the botanist has gathered some samples of plants. Because the magnetic observatory was not ready, we did not compare the instruments of this type. We only compared the chronometer. Dr. Nordenskjöld arranged with Ballvé that, in addition to the observations that they make simultaneously between the seasons and the expeditions, they also will make others simultaneously between this expedition and the Argentine observatory.

Ballvé and Plate accompanied us to the beach. It was 8 am when we left the landing. With all of us on board, they have thrown into the sea a net that has a kind of razor in its mouth, which was dragged along the seabed at the speed of the ship; a moment later, it was pulled out, and it was full of seaweed, rocks, snails, spiders, etc.

It must have been 8:30 when, after moving the ship, we set course to the east to turn at Punta San Juan, and then at 11 we were in front of that point, from where we moved to the SE toward "Shetlands." The British navigation chart says it is very dangerous to get closer than 10 miles to Punta San Juan when the current is against the wind, because of the strong overfalls. We passed by this point at a distance of 3 miles, with wind from the NW against the current, and we didn't observe or see absolutely anything at all.

The wind from the NW was blowing quite hard, but it started to diminish at around 7–8 pm and moved to the W until it reached SW. At 9, a downpour fell, water and wind.

Now 9 days!—the last civilized ones; now it is directly to the ice.

This morning I saw a "Nacion" [*La Nacion* newspaper] of the 22nd where they were giving the news of the Minister's retirement, showing the gravity of the circumstances. If I would have followed my impulse, I would have stayed. Mine is a very critical situation.

This was the chance for Sobral to prove his indispensable assistance to the expedition and to Nordenskjöld, who had previously made an agreement with Lieutenant Commander Horacio Ballvé to collaborate together on the expedition. Ballvé and the Argentine Observatory, as part of the International Geographical Congress agreement, were to assist the international expeditions traveling to the Antarctic, including the Swedish Expedition. Part of the original plan had been for Nordenskjöld to calibrate his instruments with those at the Observatory, but because the Observatory was not yet completed, that goal was not completely fulfilled at this time (Rabassa 2003). Sobral, however, fulfilled his role, translating for the Swedish scientists, and touring the facility with them. And Ballvé was able to provide other types of helpful cooperation. The observatory would be completed around the time the expedition returned, and, true to Sobral's words, it was indeed one of the best meteorological observatories of its time.

In addition to the magnetic observatory, a new lighthouse was also being constructed, utilizing the labor of prisoners who resided on this tiny island located just several miles north of Staaten Island.

Sobral's relationship with Ballvé was friendly, as can be seen. And Plate was an old shipmate with whom he had served.

Meanwhile, the newspaper that Sobral picks up on the island again stirs up his fears regarding the state of affairs in his home country, and the possibility of war with Chile on the horizon. His elation at helping his expedition on Observatory Island is transformed once more to doubt about leaving Argentina at such a crucial time.

3.3.2   January 7, 1902, Tuesday

All day we've had a fresh wind from the SW. The ship is navigating with the sails filled and the ________ [unreadable], with a handful of curls. We rolled a lot.

In the afternoon, the wind started to calm, it rained, and they turned on the engine. The temperature of the water is above 6°.

This diary entry is written around the time of the entrance into the Drake Passage, the body of water that separates Cape Horn, South America, from the South Shetland Islands; it is notorious for its winds and waves and rough seas.

3.3.3   January 8, 1902, Wednesday

It has been a splendid day today. Average temperature was +5° but without any wind, so we didn't feel any cold. Yesterday, with +6°, it was very uncomfortable to be on the bridge because of the wind. The breeze is weak and from the W. We are engine-navigating [sailing with the engine] but with all the cloth caught [sails unfurled and filled]. You can see some albatross and other birds (white and black spots, petrels). This morning, Duse, Ekelöf, Stokes, and I had a conversation, in which Duse and Ekelöf appeared to be very upset with Nordenskjöld. Ekelöf said that if he [Nordenskjöld] could not be a good boss, then at least he should be a good gentleman; to me, he was; Duse even insinuated he was a coward.

The thing in question is that they say that Nordenskjöld wants to waste two or more months hunting whales in the Shetlands, and then after that set up the winter station. They say, and I actually think this is most reasonable, that we should establish the station immediately, as far south as possible. These two men seem to have very revolutionary ideas. Who knows whether this will bring more serious consequences.

After the *Middag*, since it was such a nice day, we went to have coffee on the poop deck; the sailor who was on duty navigating was told to step aside and Botman [Bodman] took the helm and began piloting. Photographic views were taken.

Water was at 6° at 10 pm last night. And today, it was at 3.2° and it started to rise; at 10 pm, it was 4.16°. Lat [latitude] = 58°. Sunset was at 9 pm, and twilights are very long. Today, we had a beautiful sunset with a very nice afternoon. The pale sun (not red like many draw the sun in these high latitudes) partially covered by clouds; the rest of the sky was completely clear, and the temperature was extremely nice.

At 12, there was a weak breeze from the north, the temperature of the water was −4.10. In order to preserve samples of sea water in bottles, they put vapor inside them for one or two minutes so the glass doesn't leak. Some of them broke.

Sobral's talk about whaling is telling. (This time he uses the term "whaling"—"Cazando"—rather than "fishing.") The idea to hunt whales in the waters surrounding the South Shetlands may or may not have been an effort by Nordenskjöld to placate Larsen, who was an avid commercial whale hunter, or an attempt by him to seemingly accomplish both things at once—whaling and science. Always chivalrous, Sobral defends Nordenskjöld as a gentleman, although he does agree with the wisdom of establishing the winter station first, and going as far south as possible now, before engaging in any other extra-curricular activities.

His attention to record taking and keeping of meteorological conditions, location, and scientific sampling begins to be evident here.

3.3.4 January 9, 1902, Thursday

The morning has been misty. At 10 am, the temperature is 2.5° (air) and water temperature is 2.9°. Bar [barometer] = 738.5. The barometer has gone down a lot. I see petrels and other white birds in flocks. Wind is from N although weak. At 10 am, the engine was turned off, and we continued navigating with the sails. The fog is about 800 m, probably due to the fact that the N wind has put into contact layers of both cold and warm air together.

A hook baited with meat was thrown into the water to catch birds. One got tangled so it was brought on board. It has a white chest and blue back. They killed him by giving him chloroform. Some photographs of the bird were taken. Hundreds of birds, of different species, fly around the ship. The fog dissipated at 3 pm, but the sky was still very cloudy. Some whales were seen. One came as close as 300 m from the ship. At 8 pm, we had fog again; at 9:30, it dissipated; the wind is moving to the SW, so all square rigs were rolled up, leaving only the fore-and-aft sails, and the engine was turned on, clear sky. Today, I note that faces look very happy in general. I think they have resolved whatever was happening. Water temperature at 10 pm is 3.20°.

3.4 Icebergs . . . Each One of Them More Whimsical Than the Other

3.4.1 January 10, 1902, Friday

Yesterday, some snow fell. The morning has presented itself nicely. The temperature of the air was +0.7°, and water temperature was +1.4° at 8 am. At 10 am, water temperature was +1.2°.

We have wind from the west again. We are navigating with all the sails, always with SSE course.

Everyone has covered themselves with many heavy coats. I am wearing 1 knitted shirt, 1 English shirt, and a coat; because everyone is telling me that this is going to be bad for my health, I have decided to augment my warm clothes with a vest. Captain Larsen hunts with a hook the pigeons from the cape. Several pictures were taken. At 1:07, Capt. Larsen sighted land from the portside of the ship. The bird was killed with chloroform. The other was already embalmed, and this one will have the same end. The first land that we have sighted is an island in front of King George Island. Once we reconnoitered the land, we started to coast alongside of it because we were going to go by the Nelson Strait between Roberts Island and Nelson Island. The land looks white and very even (no doubt it's because of the snow). When you see that land from far away, it looks like a stack of clouds. At 3:15, the first *iceberg* was seen.

Along the coast, a lot of icebergs were sighted.

We continued until 1 am, when we came into a little bay on Nelson Island. [The previous sentence is crossed out.] It is impossible to accurately describe the effects of the light on the ice, both on the land and on the icebergs, but especially on the icebergs. It is true that there are no flowers, no plants, no trees here, but there is ice, and it has more beautiful colors than flowers, from the red, orange, green, violet, blue, white; they appear in turns as if each iceberg were a prism that one rotates in front of the sun; then there are the shapes, each one of them more whimsical than the other.

Sobral's lack of proper attire for the Polar regions is now felt for the first time. This is yet another challenge he will have to endure.

The young Under-Lieutenant and his expedition have arrived at the South Shetland Islands, where they spy King George Island. It is the first land sighted. More to Sobral's delight is the first iceberg sighted, followed by many more, which he describes in terms of their beauty and magnificent colors and all the imagery that they conjure.

3.4.2 January 11, 1902, Saturday

I woke up at 7 am. I had gone to bed at 12 pm. And I felt the ship wasn't moving. Having a strange feeling, I dressed quickly and went to the deck. I checked the ship's cabins, and there was no one inside them, except for Ohlin's. So I get to the deck and I see Haslum on the bridge: Good Morning, Haslum, where are the rest of the people? They left for land, at 3 am, in two boats. Great was my surprise. I immediately went to W. Stokes's cabin to see whether he also had left, and I found him asleep. And I told him what had happened, and it made a bad impression on him. After that, Ohlin came out, and when he realized that we had not been awakened, he said that this had been very wrongly done.

When we were having coffee, the cook's assistant told me, on behalf of Botman [Bodman], that we had to make the observations. So it seems that they did not forget about us, but they didn't want us to go with them.

Afterward, Ohlin spoke to Nordenskjöld about this matter, and Nordenskjöld excused himself with Stokes and myself, saying that he didn't know they had not woken us up. Ohlin seems to be disgusted with his procedure. They came back from the Island around nine with 100 seals, penguins, albatross, and other types of birds. Their skins and skeletons were pulled off, and others were embalmed, and only the skeleton was kept from the ones that had their skin broken or cracked or looking bad. The only plant that there is here is lichen. At around 8:30, the E wind began to blow, which is actually getting fresher. At 10 am, after they had brought the seals on board, we went toward Orleans Inlet. The wind is turning into evil weather, and shortly after we heaved to portside ________ [unreadable] the engine is working. The wind is blowing at 19 m per second, which corresponds to 9 on the Beaufort scale.

Little by little, we're losing sight of Nelson and Robert islands. By 3 pm, we were passing by Middle Island. Some icebergs are seen, and whales. Water temperature is going higher by 1°. And it has been even higher during the day. During the night (which is not night), you can see the twilight light. At 10 pm, the Astrolabe Island is sighted. Despite the intense twilight light, some stars can be seen.

It is austral summer, and the beginning of the midnight sun, hence the long twilights that Sobral is observing. True to character, he records this, along with all the other measurements.

3.4.3 January 12, 1902, Sunday

At 7 am, we were in front of Louis Felippe Land [Land of Louis Philippe]. We have left little Astrolabe Island on the port quarter, and you can see Trinity Land perfectly. The land of Louis Felippe seems to be much more to the west than the navigation chart shows. We're luffing head to the wind as much as possible, to take the right course later on and so get to Orleans Inlet, which is what they're trying to explore. The wind that was blowing very strongly from the SSE has suddenly calmed. At 11 am, we are at a part of the sea with many small pieces of the ice, which would break into smaller pieces, crunching continuously. The sound you hear is analogous to the one you hear when something ________ [unreadable]. All the land in our sight is high and steep. Projected in the middle of the entrance, you can see a land that seems to be very vast because it extends to the SW as far as the eye can see. This one is not on the navigation chart.

Trinity Island is an island completely different from what is marked on the navigation chart. On its opposite NE, it has around seven small islands, and to the W, a pretty big one. At 12, I observed the meridian and latitude, which was 63° 39′ 15″. We have no longitude observed, but, guided by the navigation chart, we must be at 59°. We are still hugging the coast of Louis Filippe Land, and a narrow lake appeared in front of us, whose end we could not see: Louis Felippe extended to the left, and to the right the other land that wasn't marked on the navigation chart. Maybe this land is Danco Land. The course that we're taking is approximately WSW. At four, I went to observe these heights to calculate some longitudes, and I found that I did not have a horizon. Making the best of it, I observed some of them, calculating the height that the horizon would have been. I calculated three longitudes 60° 20° W average point and the extreme latitude from 12 m 60° 43′ 21″. This point, according to the British navigation chart of the South Shetlands (3205), is located at the entrance of Davis Gilbert Inlet, and according to the course, we were taking, we were going to navigate in the "Belgica" Strait.

I calculated a straight line of height and an average point very close to the one before. We kept navigating, and our strait became larger to the left, the land was one continuous Louis Filippe; to the right it looked like big islands, but in the straits that separated them, other very tall lands would appear in such a way that everything could be either one same land joined by the coast that we could not see very well, or one archipelago of big islands.

On board, Lieutenant Duse took sights, and soon, there was talk that we were in Belgica Strait. At 7:25 pm, we lost sight of Trinidad Island. At 9, Nordenskjöld, Stokes, Botman [Bodman], Skottsberg, Anderson [Andersson], and I got off the ship at a group of four rocky islands, two-thirds of which were not covered in snow; these little lands were just nests of ________ [unreadable], penguins, etc. When we got off, the penguins received us with fear, because they saw that their dirty, but for them splendid, mansion was being invaded. Almost all of them had little chicks and eggs; their nests were as poor and wild as their owners. And the place where they made them was composed of a little of guano with some rocks, and the rest of the area had the skeletons of some other penguins. ________ and various other birds were killed for the collection onboard.

When we went back to the boat, we encountered a seal (Weddel Seal) [Weddell] on the shore. It looked at us with concern but without the willingness to run away. I was going to shoot at it with my gun, but since we couldn't take it with us, I left it. Some samples of rock and moss were taken from this island. We had the intention of observing on this island _______ [unreadable] with the difficult _______ [un-readable] but we couldn't hold the sun.

At 10:30, as soon as we returned, the ship set off again.

We sailed all night. Duse was making a sketch and taking pictures. The island where we had gotten out of the boat was covered with moss in the part where there wasn't any ice. A very nice day.

This was a momentous occasion for Sobral and his companions, who were seeing the true land of Antarctica for the first time. Approaching from the north, they crossed the strait where the Belgica *expedition had overwintered a few years prior. (Danco Isand was named after the lieutenant who died on the* Belgica *expedition while the ship was locked in the ice.)*

Trinity Land and Louis Philippe Land are the tip of the Antarctic Peninsula— Trinity Peninsula. It is the first Antarctic land ever sighted, first seen by Edward Bransfield in 1820. The Swedish-Argentine expedition was now determining if this area was a continuous formation of land or a united group of islands, eventually discovering that it was, indeed, continuous land.

Sobral and the scientists collected flora and fauna from this area, and so they killed samples of the wildlife to study, but, from his decision not to shoot the seal, it seems that Sobral did not wish to kill these animals unnecessarily.

3.4.4   January 13, 1902, Monday

Comparing some of Dr. Kook's [Frederick Cook] sightings of the land with the sites that one finds, one concludes that they are just the same, so that I start doubting that this is any other strait than the Strait of "Belgica." We arrived at an almost round bay in the morning. It was reached through Wilhelmina Bay; being between the islands of Nansen, Brooklyn, Pilsener, it could have an almost round shape. After circumnavigating it, we continued to the Strait. A perspective appeared as if the channel were divided in two. After that, it was seen that the one on the left was no more than a bay. While this was happening, and we couldn't lose more time, because it looked like whatever it was, the exit of that strait was the W, and what we needed to do was to go to the E. So it was decided to go back through where we had come from, to repeat the same route. Along the entire strait we traversed, an infinite number of whales were seen. In general, in the channel, we've never seen more than one or two icebergs together, lots of currents from the NS and SN. Another thing that currents make noticeable is the sudden changes in the temperature of the water. At 12, when we were going out of the rounded bay, the temperature of the water was +2.2° (this was near the east coast and at 2 pm, when we were almost in the middle of the channel.) The temperature was +0.3, and at 4 pm, it was +1.3. It is noticeable the little saltiness of the water. Samples of this water were taken. Between 1 and 2, some snow fell, and the wind was NE. In various parts, the similarities between the pictures of Dr. Cook and the land of the Strait were seen. Maybe they weren't exactly the same because of different perspectives (between our observations and those of Dr. Cook), but when we saw Cape Murray and ____ [blank] and the bay, it became evident this was the Belgica Strait.

Besides being in a hurry to take the picture, and being nervous to take the sight of the SW from the east coast, which is from where Cook had taken it, I took three pictures in the same plain, when I realized it wasn't exactly the same. The bay ____ [blank] between Cape Murray and ____ [blank] is very similar to what the Gerlache [Adrien de Gerlache] navigation chart indicates. At 11 or 12, we have in the diagonal an island with two *Hummocks*, and it couldn't be any other than the one on the navigation chart of G.S., because, relating it to Cape Murray and to the bay ____ [blank], it was exactly the [accurate] situation. From cape ____ [blank] in the western coast and from cape ____ [blank] in the eastern, to the ENE, which is to where the strait continues, there's no navigation chart. But what seems inexplicable is that the island situated at the ENE of the cape wasn't seen by the "Belgica," because there is no other explanation for its absence on the navigation chart. It could only be that they entered the strait from the north of the cape and during heavy fog, which prevented them from seeing the island and the lands of the ENE. In the space between these islands or lands, far away on the horizon, some lands can be seen; undoubtedly, the western part of the strait must be a large archipelago. If this strait is Belgica, the land of Louis Filippe and Danco is one and the same. The entrance to the strait is in front of Trinity Island exactly where _______ [unreadable] Orleans Inlet. Since we left Astrolabe Island, we have been sailing along the coast of Louis Filippe the entire time. If this strait is not "Belgica," the Orleans Inlet is nothing but the entrance to a strait that runs parallel with the "Belgica" Strait. We can only confirm the certainty of this situation in an approximate way, by estimating, and by taking Astrolabe Island as the starting point. A very nice day.

One can almost sense the excitement and envision the flurry of activity on board the ship Antarctic, *as Sobral and the scientists attempt to ascertain their position on the map, and compare the land that they are now sighting to the findings of the previous expedition, the* Belgica. *Sobral is flustered and all aflutter as he runs around the deck trying to take photographs from the same perspective as Dr. Frederick Cook's (note that he corrects the spelling of the name and spells it accurately thereafter). This is another important day for the expedition, and Sobral hits the mark correctly regarding the significance of the land sightings, although he does not yet know the names of the capes and bays that he is seeing. His penchant for accuracy shows through his words and through his struggles to be certain of the discoveries they are making. He was learning very quickly.*

3.4.5   *January 14, 1902, Tuesday*

At 6 am, we had Trinity Land in our sight, and we were navigating in the ENE direction, more or less. And then at 8:30, we headed toward Louis Filippe Land in order to reconnoiter a bay (2). At 10:30, after circumnavigating it, we made sketches and took photographs. I took various pictures of the cape (1). This is the most

N point of Louis Felippe Land. The sea is completely calm and there is no wind. To the E of the cape (2), there is another cape almost completely covered in snow. It is quite high, and it has the shape of a cone. This cape (cone) seems to be within the bay delimited by the cape (1). At 10:05, we went directly to Astrolabe Island, and, every so often, we went off course in order to avoid the icebergs that were appearing along our way in a large number. These enormous chunks of ice adopt very curious shapes, and the effects of the light above them are beautiful. Some have enormous crevices in their insides, where waves enter, and you can hear the same sound as when water hits a rock. Sometimes an enormous iceberg, eroded by the sea waves, gets divided into two, but is united by a little bridge. When the bridge breaks, and if its height is more than the respective base, you can see it [the iceberg] fall to different sides with a loud sound. Livingston and Middle Island were seen. At 2 pm, we headed toward Cape Rogue and _______ [unreadable] with the idea of getting off at cape (1). Nordenskjöld, Skottsberg, Ekelöf, Stokes, a sailor, and I were in the boat. The sailor and I rowed, because I really like that exercise.

Nordenskjöld collected geological samples; Skottsberg [collected] moss, seaweed, and lichen. And I, after taking some pictures, started shooting the target with my revolver at the penguins and other birds, many penguins (Pygoscelis Antarctica with white neck collar) [Pygoscelis antarcticus]. There were three seals around 50 m away from us when we arrived inland. They were on top of the snow, sleeping. I took pictures of them from an approximately 5 m distance.

After coming back and using the boat around 5 pm, we continued navigating, hugging the coast.

A great number of whales come near the ship in pairs. They're not of a very large size, but they spout great jets of water. We were on the bridge when we were surprised by a loud noise. There were avalanches off the glaciers. Enormous chunks of ice came loose from the glacier and, falling onto the water, formed icebergs. From Cape Roquemaurel up to as far as the eye can see, toward the E, it is all a huge glacier. Temperature of the water is $-0.3°$ at 10 pm. It's the first time it goes below 0. At 9 pm, we had Astrolabe Island to the stern of the ship, and we moved away from the coast because small rocky islands were seen from the bow of the ship that could break off and form dangerous reefs. A large amount of unknown reefs were seen. During the night (it's not night), some icebergs passed, and some of them, the ones that weren't that tall, were carrying seals and penguins on them. Beautiful day.

While we were on land, some nets were thrown into the deep sea, pulling a good quantity of seafood, seaweed, etc. Water temperatures and samples were taken at depths _______ [unreadable] 250 m; at 200 m, temp = -0.4.

The expedition is now approaching Trinity Peninsula as Sobral sets off on a mini-excursion in a boat to observe—and shoot at—the abundant wildlife. Back on the ship, he watches whales spouting water and icebergs calving off of glaciers. As the sun does not set during the Antarctic summer, there is no night.

3.4.6 *January 15, 1902, Wednesday*

All night we sailed, leaving behind us a lot of islets and reefs between the coast and us. At 6 pm, we were turning at cape ____ [crossed out]. On the port side, we have a very big, but not very tall, island, covered in ice. The wind is from the S. We were sailing along the coast with a SE direction, more or less. It looks like there is a very narrow strait between that not-so-tall island covered in snow, and the Land of Joinville that we are seeing. The navigation chart is very erroneous in this part.

Between Joinville and Louis Felippe Land, there is a big island (X). Between this island and Louis Felippe, there is a very narrow strait that leads to the gulf of the "Terror and Erebus." And between Joinville and island X, there is another strait with a rocky island in the middle, very tall (see booklet). This strait leads to the same gulf. The exit of the strait is the entrance to the gulf. Around 10–15 miles from the cape, there is a nice little bay. The land that surrounds it is very high, but there is an exception, one at the North, which is relatively low, but only accessible from a rocky point that isn't covered with snow. The one in the south is high, but it ends in a lower land. It is rock with no snow. It looks like a good landing. A ship could be in the bay, but I don't know the depth that it has; I mention this with such detail because it's very rare to find a place to dock in this land. Between the cape and the bay, there is an enormous glacier. You see chunks of ice floating, full of penguins.

After identifying that the land that we could see to the E of "Louis Felippe" was an island, and seeing that the strait that separates such land is full of floating chunks of ice, it was decided to pass between island X and Joinville; besides, this route was shorter to go to Paulet Island, where some observations would be made. Island X seems to be divided into two by a strait. "Paulet" is 15 or 20 miles from island X.

On the N side of Paulet Island, there is a good landing. The beach is rocky. At 5 pm, 3 boats left the ship, two of them took the sailors to hunt seals. Haslum was in one of these boats, and Cristaisen [Ole Christian Wennersgaard] was in the other one. The first boat brought 27 seals, and the second 21. Nordenskjöld, Bodman, Stokes, Ohlin, Duse, Scottsberg [Skottsberg], 2 sailors, and I were on the third boat. Nordenskjöld picked geological samples. Scottsberg [Skottsberg] collected botanical samples. Ohlin collected some Magellanic penguins. And Duse made some solar observations with the theodolite. And Bodman was carrying the Neumeyar device to observe the declination, deviation, and horizontal power of the terrestrial magnetism. Ekelöf went to land in a little canoe that he has. At 7 pm, we came back on board and, after hoisting the boats and the seals, we went in the direction of Cape Seymour. Paulet Island is an extinct volcano, and it's inhabited by millions of Pygoscelis Adeliae [Pygoscelis adeliae] penguins, of the kind that has a black head and chin. The seals are of the Weddel [Weddell] kind (Leptonychotes Weddeli) [Leptonychotes weddellii]. Almost all the penguins were with chicks. And the ones that weren't, were about to have. They live stacked up [on top of each other]. From far away, one can see where they have their nests, even though they're not there, because their guano is red. It stinks. The guano is completely red, and it emanates a very disgusting odor. On the island, there are some puddles of freshwater [lakes], but there lie a lot of penguins and all kinds of birds there. The beach is of gravel. The barometer went up to 748, and it remained like this until today, when it increased to 751.5; at this moment, it is at that height. Before we boarded [the ship again] at Paulet Island, a Weddell seal was flipping around [in the water] and, after observing us for a while and seeing that we didn't have any hostile intentions, it came on land about 2 m away from us, and we were petting it, and it didn't do anything at all, and it looked like it was the first time it had ever seen men. A precious day.

3.5 Cape Seymour and the Pack Ice

3.5.1 January 16, 1902, Thursday

At 8 am, we were in the southern part of Cape Seymour. A N breeze was beginning to grow stronger. Around the cape, there were about 4 or 3 miles of large chunks of ice (*ice pack*), and very close one to the other, and some separated, and some 2, 3, 100 m, which made it very difficult for the ship to reach land. The helm had to be maneuvered continuously, the engine was working very slowly, and many times the ship would crash against the ice. The front made a lot of pieces of ice fly. We tried to get closer to land from the other side, and it was worse. I think we could have reached land in the boats.

After the last attempt to go there [to land], we came out of the ice and headed toward the south. What for? I asked. And they said, to go and reconnoiter the *pack ice*. We could see Robertson Island and ____[blank]; we navigated until 1:15 pm, and then we went back to Cape Seymour. Around 10 miles to the south of the cape, we saw enormous icebergs, and it can be guessed that this pack ice extends all the way, because, in the far distance, we can see clear sky. When the boat went from Cape Seymour to reconnoiter the ice, the [ship's] engine was turned off, and we raised the topsail and a ________ [unreadable]. The breeze from the N is 5.

Captain Larsen did not make another attempt, foreseeing that this wind of the north would bring fog. And indeed, at 9 pm, it suddenly came upon us, completely covering Cape Seymour, which is in front of us. It's remarkable that this cape doesn't have any snow, except for some in the crevices of the rocks. Maybe it's because of the way the wind comes, but there are other lands that are exposed in every possible way to the wind, and they do have meters of snow. It may be that the wind shapes the rock. At 7 pm, a seabed net was thrown twice into the sea at a depth of 140 m, pulling out beautiful samples that are going to enrich the collections. The temperature of the water at 10 pm is −1.5.

Tomorrow, if possible, 3 or 4 men will disembark in a boat to place a letter containing our news, and providing the course we are taking, so that, if we cannot return, they will know where to look for us. As soon as this is done, we'll go to the

South. There was an idea to leave some food provisions here, but this is not for sure, we probably won't do it.

At 11 pm, we set forward in the SE direction, getting away from the coast; wind is NW, fog is increasing. We can't see land.

The floes that had accumulated along the coast prevented the ship from making its first entrance into Cape Seymour. In his 1903 notes, Sobral would later write that "This was the first time in our journey that the ice declared itself our enemy, and since then, it never stopped being hostile, hindering our advance towards the south" (Sobral 1903). He describes the battle with the ice as such:

At first, floes were separated by 50 or 100 meters, but these distances diminished as we approached, and shortly after, they had reduced to 3 or 4 meters. It is then when the ship starts its hard work; from the crow's nest, a small channel is spotted, that is, a place where floes are more separated; we must get there, and we move forward by means of the engine and hits of the prow. At a convenient distance, we speed up, and when the ship has enough push, the engine is turned off, directing the ship, so that, with the clash, the floe moves in a direction that clears the way; the whole ship shakes, and masts bend over to as far as their elasticity allows them, causing whoever is in the crow's nest to feel a strange sensation of being rocked so roughly at that height.

And those clashes occur repeatedly for hours, and sometimes for days, and they are only interrupted when the space in between floes is so small that it is not possible to maneuver; in that case, the ship remains immobile, and so, if it is cold, it is exposed to getting stuck and also to suffering the effect of pressures. When the ship is trapped by the floes, making it impossible to advance or to move backwards, and young ice forms in between the floes, one has to force the ship to roll, making all the crew move from side to side with the aim of avoiding the adherence of new ice; this is what the English call "sallying a ship".

Getting in between those floes so close to land would have been dangerous, in that we could have gotten stuck or suffered a storm, not only because of the pressures which, being against the land, would be strong, but also because, swept by ice, we could become stranded, and it is clear that in those conditions the ship would be in great danger.

3.5.2   *January 17, 1902, Friday*

During the night, the ship has stayed in front of the cape, maneuvering so as not to hit the ice. At 8 am, the fog continues. Sometimes it dissipates and allows us to see the cape and the adjacent land. At 9:15 am, we slowly set in motion in the direction of the ice pack in a new attempt. At 11:30, we went to land in two boats. The wind from the NW, which began to blow harder, dissipated all the fog, and it was a splendid day. Nordenskjöld, Stokes, Bodman, Skottsberg, Duse, Ekelöf, and I went to land. A post was secured with rocks and 3 _______ [unreadable] with a large knot. Tied to the post there is a bottle, and inside this bottle, there is a letter to the Scottish expedition that is coming next summer. In this letter, they are told the direction we are going, so that, if, for any reason, we cannot come back, they can come and rescue us. The post is located on a fairly high hill to the south of the cape, but, seen from the sea, it projects on the land, Seymour, so I think it's very difficult for a ship to see it, as it won't be able to approach the land due to the ice pack that is surrounding the cape.

According to Dr. Nordenskjöld, Cape Seymour (and the adjacent capes) is very interesting for geologists.

It's made up not of a compact rock but of separated rocks, and on the rocks, there is a layer of sand that in many parts is very thick.[1] There is a very curious thing, which is the little snow that there is on it. This could be explained by saying that the cause is the wind, if only the parts that are protected against it had snow; but no, there are parts protected from it that don't have any snow, and there are other parts completely exposed to all kinds of winds, from all sides, that do have snow.

There are small streams that come down from where there's snow, down to the sea, during days of relative heat; therefore, they must have some water.

So the post and the provisions are in a small cape located in the south of Cape Seymour. The provisions are in a crate placed at the base of the hill where the post is. This crate is covered with an impermeable canvas, with rocks on top of it and on the sides. All this land is full of penguins (Pygoscelis Adeliae), the kind that have a black neck. They're excellent to eat in stew. Another good thing to eat that I haven't seen in Cape Seymour but that can be found in the rest of these lands is a white bird (____) [blank]. Some seals were hunted. In one of the boats, an insect was found ________ [unreadable] to the lobster of the kind . . .

Botanical and geological samples were collected—lichens and moss. At 3 pm, after being on land for 3 h, we went back to the ship. At 4 pm, after pulling up all the boats, and tightening the topsails and the foresail, we set off to the south, and the engine was stopped. Wind NNW.

At 9 pm, the breeze decreased, and the fog came upon us. Enormous icebergs are seen, some of them are many miles long. A lot of whales are seen, too. The engine was started again. The sky has splendid colors at 11 pm. Again, snow petrels (Pagodroma nivea) were seen.

---

[1]A lot of fossil wood has been found.

The Scottish expedition to whom Sobral refers is the Scottish National Antarctic Expedition of 1902–1904, led by Dr. William Speirs Bruce, and planned to begin the year following the Swedish Expedition's departure. Bruce had met with Nordenskjöld in London during Nordenskjöld's departure south, traveling there from the Scottish Geographical Society in order to see him off. The two had agreed on a rescue pact, wherein Bruce would rescue Nordenskjöld, if necessary, when he followed him—a year later—into the same area of Antarctica into which the Swedish Expedition had ventured, in the Weddell Sea. Sailing on the Scotia, *Bruce's successes included the establishing of a base on Laurie Island in the South Orkney Islands in 1903, and, on that base, a weather station, which, in an agreement between him and the government of Argentina, was acquired the following year by Argentina. That base became Base Orcadas, and houses the oldest functioning Antarctic weather station operating to this day, with meteorological records that date back to 1903. It is for Bruce that Nordenskjöld left a letter, detailing the Swedish Expedition's course, at Cape Seymour.*

Rich in fossils and ancient soil, Seymour Island is basically a barren-looking black rock, mostly snowless, yet bountiful in its geological and paleontological treasures. Carl Anton Larsen had been the first to land at Seymour during his whaling voyages in 1892–1893, and had found the first Antarctic fossils here. Now Nordenskjöld could not wait to plumb its depths for geological discoveries.

3.5.3   January 18, 1902, Saturday

The wind began to blow from the SW, clearing the sky a little. We have the ice pack in front of us. The temperature of the water is 0°. It has been snowing almost all day. We went into the ice pack at 12:45, and various seals of the Weddell, Grabier [sic] type were seen sleeping on the ice. White petrels were seen in flocks, *Cape Pigeons* [Cape petrels]. Port side approximately 5 and 8 miles from starboard, we have pack ice, and open sea to the bow. At 4:30, the ship stopped, and water temperatures and samples were taken between the surface and 300 m [meters]. Then, the net was thrown into the sea, and little fish from that depth were pulled out. Surface temperature was −1° and at 300 m was ____ [blank]; the air was −0.8°. At 5 pm, *ice blink* was seen in front of the bow, which indicates the presence of ice. At 9 pm, we were back in the ice pack. This ice pack is the most sad that one can imagine. It's the gathering of several pieces of ice, more or less large flat chunks, separated by 5, 10, 50, 100 m. Every once in a while, some seals are seen, and the "Pagodroma Nieva" [snow petrel], as white as the snow, flies around without even acknowledging the ship; no penguins are seen. The breeze continues to blow SW. At 12 pm, the ice pack was still present. Land will be sighted any time now. The temperature of the water is +0.05°. At 11 pm, we spotted the largest iceberg that we have seen thus far; it looked to be several miles long.

3.6 The Furthest South We Have Reached Is 66° 5′

3.6.1 January 19, 1902, Sunday

We've been in the ice pack all night, trying to get as close as possible to the land. In the morning, many crashes woke me up. Our route was closed to the SSW and W because of the ice pack, so it was decided to go back to the North and try to get near the land of Rey Oscar [King Oscar Land] in between the lands of Robertson Island, etc. The furthest south we have reached is 66° 5′, in front of Cape Framnäs. At 1 pm, we anchored the ship to the ice field that separates us from King Oscar Land. Magnetic observations were made (of inclination and horizontal power). Dr. Nordenskjöld made an excursion of many miles of distance on *skis*. Jonasen [Ole Jonassen] began training the Greenlander dogs in pulling the sledge. For this being their first time, they did very well on the ice. Here, there are around 20 cm of snow. Before the dogs were tied to the sledge, they were released, with great content for them, as they rolled around in the snow. A royal penguin [Emperor Penguin] was hunted. During all this time, the ship was anchored to the *ice field* from the stern with an anchor and _______ [unreadable], the anchor in the ice. There was a moment when pieces of ice started to stack up around the ship, threatening to amass pressure on it. This was at approximately 3 pm. So we all went aboard except for Nordenskjöld, who was pretty far away, and Captain Larsen decided to lift the anchor, and the "Antarctic" was surrounded from everywhere by pieces of ice separated by little distance, so that the ship advanced at half power against the ice; the ice would crunch in front of the "Antarctic," but it wouldn't break, so they went back for a moment and again tried to go through, and this time the ice separated, letting us go through. And we didn't stop until the sea was free of ice, at around two miles from the ice field. The land is over 20 miles away. Pygoscelis Adeliae (white neck) [penguins] were seen again. At 6 pm, Nordenskjöld returned. They lifted the boat that he was in, and we continued sailing along the edge of the ice field. A lot of whales are seen, and seals. Seals sleep on the ice, and sometimes not even the loudest screams can wake them. At night, the gramophone was played. A curious thing happened to me when the ship was anchored to the ice field. I was walking along the ice and saw that the ice in a particular place was greenish yellow. So I said to myself that this must have been caused by some seaweed. So I started collecting the snow, and then, Bodman asked me what I was doing, and I said that I was collecting this for the botanical [collection]. When he heard me, he burst out laughing and told me that Duse had urinated there; that was the seaweed that I was collecting. Temperature in the air is −1°, and water is −0.2°. The barometer has been at 755 since yesterday. Cloudy sky. Despite being cloudy, the brightness of the snow is very intense. I am so happy to have the goggles that my dad bought for me! They seem to be excellent.

This was the furthest south that the expedition could reach, encountering ice that would not allow them to penetrate further. Sobral reports it as 66° 5′ in his diary, and Nordenskjöld would later report it as 66° 10 to 15′ S in his book (Nordenskjöld et al. 1905). It was the first time the dogs were able to leave the ship and frolick on the ice. And it was the first time that an Emperor Penguin (Aptenodytes Forsteri) was caught.

This was also another close call with the pack ice, which had become a closed pack and was closing in around the ship. Sobral would later write in his speech notes that it would have been illusory to establish the winter station here, as a solid field of ice separated them from any land, and the difficulties of transportation would have been impossible to overcome.

In the midst of the ice, Sobral is learning the scientific names of the wildlife he is seeing for the very first time; he is learning about the flora, and learning about pack ice and snow—even going so far as to collect yellow snow, thinking it contains the traces of some strange seaweed. The earnest young ensign is following the trail of his colleague's urinary tracks. It is a revelation of his personality, and his humor, that he recounts this episode with such matter-of-fact self-deprecation.

3.6.2   January 20, 1902, Monday

At 8 am, we were sailing in a path formed by the ice field on the W and an ice pack on the E. Approximated direction NW. We had land in our sight, and it looks like it's Robertson Island et [etc.]. The ice field is full of lakes in many parts, and the layer of ice in the channels is very thin, but it would be very dangerous to risk navigating there to reach land, because the ship could get stuck there, who knows for how long, and the expedition would then fail.

Whales, seals, and Pygoscelis Adeliae [penguins] are seen.

At 3 pm, we could finally head to the W, directly to land. This seems to be part of the Louis Filippe. The SE part.

At 6, we made it to the ice field that separates us from the land. A hill completely covered in snow reaches the sea, and from there, the land takes a SE to NW direction, and the edge of the ice field, which is basically at the entrance of the hill, is oriented NS. Some parts of the lands in the SW don't have any snow, and they're 15 miles from the water, and far away we can see a series of peaks. If this white hill is part of Mount Haddington, or of *Snow Hill*, those peaks we are seeing must belong to the strait of "Belgica" (or to ours).

At 6:30 pm we—Nordenskjöld, Bodman, Duse, Ekelöf, Skottsberg, and myself —went down to the snow. I was going to start skiing for the first time. At first, it was very difficult for me. But after an hour, I could walk somewhat, but with some bumps. We killed a Grabier seal.

Duse and I were together because I couldn't catch up with the others, who had gone very far.

At 8:30 pm, we came back on board with a great appetite. Ice here doesn't have a layer of snow. Lakes are seen on many parts of the ice field, and holes made by the seals.

At two in the afternoon, the barometer was at 750 and was decreasing rapidly. At 12 pm, it was at 742, and at 10 pm strong winds began to blow from the NNW, leeward of the land, so we kept sailing with the engine.

The barometer continues decreasing.

During the afternoon, a complete calm reigned. The silence in the ice is extraordinary, every once in a while disturbed by the spouting of whales or the crying of penguins. Beautiful colors were seen in the sky tonight, E direction.

According to Sobral's 1903 notes, that night, "a strong wind began to blow; we left the engine on all night and navigated leeward of 'Snow Hill'" (Sobral 1903).

3.7 The Idea of Suicide Comes to My Mind

3.7.1 *January 21, 1900 [sic], Tuesday*

The barometer has gone down to 707. It has a tendency to rise after the wind blows a lot at night. The day was beautiful. At 3:30 pm, we started moving with the intent to get out of the ice pack and set off to the SE and navigate to the south as much as possible, like we had done once before when we moved a little bit away from the coast and visited Cape Seymour and the Island of Cockburn, which are the characteristic spots to reconnoiter that part of the coast. This confirms our belief that the location where we had been was a part of "Snow Hill" or Mount Haddington. Very early in the morning, the ship was adorned with buntings to celebrate the anniversary of King Oscar's birthday. The Argentine flag was on the largest mast, and the North American [USA] was on the foremast. At 10:30, I observed some sun heights, with which I calculated some straight lines, and with the meridian I situated at 9 = 64° 50′ and W = _____ [blank]. At 12 pm, a toast was made to King Oscar, the *Kroner Prince* [crown prince]. The day was beautiful, a soft breeze from the NW was helping the engine that, at ¼ of its power, was taking us to the south. We had a little swell, remains of last night's weather. At 5 pm, food was served in the chamber, and it was a big banquet. It finished at 7:30. Dr. Nordenskjöld gave a speech. During the day, the water temperature rose up to +2°, but it started decreasing at night, and at 10 pm, it was +1.45. Barometer increased until 11 pm, when it was 739, and then it started to go down; at that time, some fog came from the W and the breeze has eased a lot. Sunset was at 9:30. The thought that my country could be at war obsesses me. Every moment I remember that, and I regret this trip a thousand times. I think that at this specific time my companions, maybe my brothers, are dying for our homeland, while I am here. I think that when I return, everyone will see me as a deserter, and for the first time in my life, the idea of suicide comes to my mind.

Sometimes, in my crazy delirium, I come up with the idea that the governor has to send a ship to come and get me, because it is only less than 1000 miles away, or that my father should make a sacrifice and hire a ship. Anyway, I've been kind of crazy, especially today. I think that I have a double duty to fulfill with my country: As an Argentine, and as a soldier; these reproaches that I make to myself make me sad, very sad. These ideas and these thoughts are a consequence of the news received on Isla de los Estatdos. Oh, I would really go back on the ship, but, if there's no war, I'm going to look like a coward who comes back because he is afraid. Today it's been a month since we left B.A. [Buenos Aires].

The distance that separates I de E from Cape Seymour (1). (1) Cape Seymour is the likely place for the station if we don't find any other land where it's possible to establish it.

From the erroneously written date at the top of the entry (1900 rather than 1902), to the thoughts of suicide at the end of his stream of conscience, this journal entry is quite extraordinary and reveals much about Sobral. The joyous celebration of the birthday of Sweden's King Oscar must have set off his thinking about his own country and brought these intense feelings—thoughts that he had kept deep inside —up to the surface, so much so that he reveals these feelings to his diary. He is torn between his role as an Antarctic explorer and an Argentine patriot. He fears that his country is at war, a thought that he cannot discard, and that continues to haunt him on his journey. In Hamlet-esque indecisiveness, he contemplates which of the two may be the most wise course of action—stay the course, or return home. And yet he knows that these are "crazy" ideas (he uses the word loco*), because the reality is that he must continue with this expedition to Antarctica. Sobral had intense feelings of loyalty, and these entries reflect the mind of an individual who always tried to see a situation from many different perspectives. He analyzed and dissected observable situations and events, and theorized about them, while drawing the most reasonable conclusion. And so, in this diary entry, he operates on several planes at once—the existential and the practical. Meanwhile he continues to faithfully record the temperatures, the barometers readings, the sun observations, the ship's positioning, and the activities of the ever-present ice.*

The final statement in this entry is especially painful. Sobral was measuring the distance from his soon-to-be home (Seymour Island) to his actual home land (as represented by the Island of the States—Isla de los Estados). Seymour was indeed Nordenskjöld's first choice as a wintering site, and it was his original plan to establish his base there, a plan he impulsively changed. Sobral could not know at this time that the plan would be altered, and that Nordenskjöld would later regret this.

3.7.2 *January 22, 1902, Wednesday*

At 6 am, the wind was blowing from the SE and it started to snow, which sticks to the cables and deck, everything on the deck had a nice appearance, covered with snow windward and with no snow leeward. Water temperature is −0.20° at 10 am. It's a bad sign. This sudden drop in temperature means that the ice must not be very far away. In the surroundings, no iceberg is seen, not even a little piece of ice. The barometer decreased to 732. This decrease coincides with the abundant snowfall. The snow is not dry. It's a snow in flakes, ________ [unreadable] mixed with a bit of water, so you can't see its shape. At 12, the water was −0.5° and the air was −0.5°. Very weak wind from the SE. SE direction. Bar = ____ [blank]. We must be at around 66° latitude and ____ [blank] longitude. At 1 pm, we entered a sea full of little pieces of ice, and after a while, we had the pack ice in front of us. At 2 pm, NE direction and changing continuously to avoid the big chunks of ice. At 2 pm, the water temperature was −1.5° and air temperature +0.5°. Today the heater was lit because of the humidity. Temperature in the chamber is + 16°. Sounding was begun at 4 pm; at 4:30, the sounding line was pulled out, it reached the seabed at a depth of 1050 m [meters], but it didn't bring samples of ________ [unreadable]. Temperatures and samples of water were taken between 0 and 700 m. At 700 m it was −0.8; at 500 m. it was +0.11° and −1.40° on the surface. This finished at 6 pm; we pulled up plankton from a 700-m depth. At 6 pm, a large bottom net and a smaller surface one were both thrown into the water. At 9:30, they were pulled out, but nothing came out. Only the small one brought samples from the seabed, some rocks. But the big net broke and nothing came out but its iron mouth. The net was secured to the mouth by copper wires, all of which broke; maybe a big rock got into the net and that's why it broke. Who knows!

At 9:45, we set the ship into motion again, sailing along the ice pack, which is very thick, and we took a N course, but it looks like we will soon be able to head toward the south again, because the color of the ice and the direction where we are going are very dark, which indicates lack of ice [i.e., a water sky]. On the ice pack, you can see a lot of penguins. I couldn't distinguish what kind they were. Some whales have been seen today. The barometer started to go up and it stopped snowing at 4 pm. Sky completely overcast. At 12 pm, a soft breeze from the SW.

I broke a thermometer.

3.7.3   January 23, 1902, Thursday

At 10 am, we are heading N with the idea of avoiding the ice pack and going to the South. Barometer still going up. A fresh wind from the SW. All day we've been navigating to the N, and we were able to fall a little to the NE in the afternoon. It's been snowing on and off, but very little. We're at 65° latitude.

3.7.4   January 24, 1902, Friday

All morning we were navigating along the edge of the ice pack until 12:30, when it was decided to penetrate it; in general, the way between the *floes* presented no difficulties. S ____ [blank], W ____ [blank], min temp = −1.8, max = −1. Bar 745, wind SE and weak. The sails that were tightened were rolled up when we entered into the ice pack, because the continuous going around in circles, which had to be done to avoid the floes, didn't allow us to use any sails.

It is seen that some of the floes are full of Pygoscelis Adeliae (black collar) and some seals.

Three seals were killed—Leopard (1). Three meters long on average. They will be used to show the different colors that they have, even though they are of the same type. Two of them had black backs, and the other was of a very light gray. In the afternoon, we repeatedly crashed against the floes, which made us note that this wasn't open sea. We had to turn on the engine to propel the ship through the floes. The "Antarctic" proved to be a strong ship. With each [violent] thrust of the bow, all the ship would shudder. The smallest interstice, the smallest opening that presented itself, was taken advantage of by the "Antarctic." If it was very small, with two or three bow thrusts, the gap would become larger, and then the ship would move forward.

One of the three dead seals had fish in its stomach, the other two didn't.

Between SE and E, there is a black line on the horizon that indicates to us that there is open sea. But, with the engine at half speed, it took us nearly two hours to go one mile. At night, the wind from EN started blowing gently.

(1) Today we ate a Leopard [seal] stew. It happened to be very tasty, much more tasty than the meat of a penguin or a white bird. Probably this is partly due to the ability of our cook.

3.7.5 *January 25, 1902, Saturday*

All night we have been opening our way by thrusting with the bow. Some penguins and seals have been seen. The situation at the meridian indicated $S = 65°\ 00'\ 00''$, $W = 50°\ 43'$—by my reading of the sun's passage through the meridian.

They realized that it was useless to be working so hard with the ship, we were barely advancing, and we were wasting a lot of coal. So, we decided to go out a little to the north and see whether the ice pack could be more easily attacked from another side. At 6 pm, we navigated through a great bay formed by the pack, in the direction of an enormous iceberg that could be seen from NNE. At 7 pm, it began to get fresher from ENE. A little snow was falling, and some fog was coming.

Since nothing could be seen, they resolved to stay at the cape all night until the wind calmed a little, and the sky was clear. In order to not waste more time, at 9:27, a sounding was begun, and they began also collecting samples of the water and measurements of temperature at different depths. This was completed at 11:45.

At 400 m temp = ____ [blank]; 300 m +0.31; 200 m + 0.22; 150 m temp = −0.34; 100 m temp = −1.6; 70 m − 1.63; 40 m temp = −0.77; 20 m temp = −0.62; 10 m temp = −1.13; 0 m − 0.90.

The temperature of the depth was ____ [blank] and the background temperature was ____ [blank]. S = ____ [blank], W = ____ [blank].

The samples of the temperature and the water were taken with the Nansen device, and the soundings with the Negretti and Zambra thermometer.

At 10 pm, the wind went to the NE.

3.7.6 January 26, 1902, Sunday

During all of the night, the wind has been blowing quite strongly, with a force of 5 (on the Beaufort scale). The wind, more or less, between NE and ENE. It has snowed, and we've had fog at intervals. In the morning, there was very thick fog, and we're surrounded by floes; the ENE still continues.

At 2 pm, wind ENE is 6. Sky completely overcast. Maximum air temperature is $-1°$; _______ [unreadable] minimum $= -2.8$, bar $= 743$. The barometer has been decreasing since yesterday, when it marked 749. Today at 2 pm, a net was thrown into between 400 and 500 m of depth, and only very little plankton was pulled out. At 11 pm, a net was dropped between 200 and 300 m, bringing enough plankton. The wind keeps blowing strongly and in a squall-like manner. But the sea is not violent, because of the many floes that surround us. During the day, petrels have been seen, and ____ [blank].

A seal appeared sleeping on a floe, and when it heard our yells and our whistles, it woke up (which is very rare, because they generally have a very heavy sleep). It looked kind of nervous, but not that much; only after a piece of wood was thrown at its head, did the seal make as if to attempt to bite, and then went into the water.

3.7.7 January 27, 1902, Monday

During the night, the wind was SE, and it is still blowing strongly. We're almost in the same place, surrounded by floes. It looks like these strong winds have separated the floes a lot, leaving the ice pack in better condition to be traversed. Will we make it very far to the south? Will we find the land where we can establish the station? I believe it's very difficult. I think that the station will be set up in Cape Seymour. At 2 pm, barometer 735; temp $= -1.8°$, wind from the SE at 9 m per second. Overcast sky. Water temperature $= -1.35$. At 5 pm, the sky cleared up and the sun shone in all its glory, but this did not last very long, because, only 2 h later, the sky was overcast with clouds. The sunset was beautiful. The sun, covered by a thick curtain of strange clouds, gave these a purple brightness. It looked as if, far away, in the mysterious south, in the unknown regions of the Pole, there were an enormous brazier, reflecting the sea in its shining brilliant colors; on that side, it looked like fire, while the north, full of nimbuses, gave the sea a greenish black color. Some seals were seen, and many petrels and cape pigeons flew around the boat.

Some snow has fallen, and the barometer is going up.

3.8   I Think We're not Even Going to Reach 70°

3.8.1   *January [sic] 28, 1902*

The wind has completely ceased, and we are navigating between N and NE in the ice pack. Seals were hunted. In the afternoon, the wind started to blow from the NE. We are in the middle of a great bay formed by the pack. A lot of birds: snow petrels, storm petrels, cape petrels, and many other types. Barometer hit 743 at 12 am, and then, it started to go down. In the afternoon, the wind is still blowing from the NE. Several icebergs were seen. Air temperature is $-2°$, and water temperature has varied between $-0.7°$ and $+0.30°$, and water temperature increases as we move away from the ice pack. At 8 pm, while some of us were walking along the deck, a big whale dove less than 50 m away from us.

We are navigating in an approximately NE direction. The sea is becoming populated by more floes, although we can still sail with all freedom in this direction. Sea temperature is decreasing; at night, it is $-0.8°$. At this time of the day, you can't read on the deck, which means that the summer is going away. The barometer marks 739. There is some heavy sea; I think we're not even going to reach 70° of latitude at this pace. I think we're wasting coal.

At this hour, while I'm writing, my people are probably sleeping. My mother may be thinking of me. A long year still separates us!

The last few entries reflect Sobral's growing sophistication with record-keeping, and his growing concern over the ship's seeming treadmill progress, which is burning up fuel and staying in the same place. Despite navigating and maneuvering, they have not made much more progress further south. They are still at 65° (they have actually backtracked from 66°) and Sobral doubts that they will reach 70°. He still believes Cape Seymour will be their winter station location. And he keeps thinking of his family, whom he believes he will not see for yet another year. Sobral was close to his mother, as these diary entries reflect. He did not realize that it would actually be another two years before he would see her again.

3.8.2   *January 29, 1902, Wednesday*

The wind is between NNE and SE, blowing at 10 m per second. It snowed copiously all day. Since 6 o'clock, we have barely moved.

At 8 pm, the wind eased significantly and the barometer had gone down to 728 (at 0°). It now has a tendency to remain at this level. There are a lot of floes around us, a lot of birds. Our life is always the same. There is nothing in particular on board. Everyone enjoys great health. At 12 at night, we are in a free area in the sea, but surrounded by ice pack.

At 2 pm, the air temperature was $-1°$. Bar = 732. Overcast sky, wind ENE. The lowest water temperature was $-0.60°$ and the highest $-0.20°$.

At 12 at night, it has stopped snowing, but a light rain is falling. Plankton was pulled out from the surface twice. Barometer = 727. Sea and wind calm. Only a weak breeze from ENE. Every night when I come to write the last paragraphs of my poor notes, I remember my dear mother. When will I see her again?

What happens to me is really strange. Sometimes, chased by the obsession that my country is at war, thinking that maybe my people are suffering, I start being sad, very sad, and I curse the day that I craved coming on this expedition, and I even dream that I have returned to Buenos Aires, and that my peers treat me as a deserter; other times, I think that it is impossible that we are at war, and I calm myself, although I am then ashamed of my tranquility, and I try to mortify myself; I believe that there [in Buenos Aires], they are in peace, and that it can't be any other way. It is God's will!

3.8.3 *January 30, 1902, Thursday*

At 6 am, we began moving again after being practically immobile all night. On the deck, every so often, the overcast sky opened its black curtain, the smiling face of the sun appears, so desired in these places where everything is ice. Sun rays not only come out to melt the ice, but also, even though they are enemies to the death, they end up adorning the ice with their very beautiful and delicate colors. Breeze from the W is blowing, and there is some heavy sea that, with its clean, effortless waves, rhythmically moves the huge icebergs that we have in our sight.

The barometer has gone up one millimeter since 2 in the morning, and it tends to continue in that direction. Today, besides the various types of petrels of this region, we have seen an "ocifraga gigantea" [Ossifraga gigantea, or Giant Petrel] flying around, something very rare, because it's the first time we have seen this bird far from land, although it might be that there is land nearby. In the afternoon, the sky cleared up, the speed was 5, thermometer $-0.5°$. At 9 pm, something with the appearance of land was seen from the SE.

Regarding our approximated position, it is not in the same marking as the land that Ross [James Clark Ross] thought he saw in the year ____ [blank] at S = ____ [blank] W = ____ [blank], but the appearance of land (Ross's) is 70 miles away from us.

Closer, some icebergs can be seen that have a very similar manner of shading [as that of the land], so it wouldn't be strange if our land was nothing more than an enormous iceberg.

At 9:15, the prow was set [in the direction of the sighted land].

All kinds of comments were made in regard to this respect; every so often, there was a change of mind, but, in general, no one would dare to formulate a conclusive opinion. The sky has been completely cloudy since 6 pm.

Some seals and whales were seen.

At 12:45, we were a mile away from the thing, and we still couldn't confirm whether it was land or an iceberg. It was perfectly seen on the left side, on top of it there was a black part that looked like rock without snow, and there was a glacier in the middle part, but only when we were a 1/4 mile away were we able to see that it was an iceberg. That black part was shadow. This didn't make us lose our way; on the contrary, we went in an excellent direction.

At 12, we passed by it and continued our course E.

Sobral was impressed with the performance of the Antarctic. *In his notes that he wrote in 1903, when referring to this diary entry, he stated that "In this short time, the ship has behaved very well; it seems to be an excellent sailor" (Sobral 1903).*

3.8.4   *January 31, 1902 [originally wrote 1901], Friday*

This morning, the pack was so compact that we had to go back N for several hours and then navigate NE. The sky cleared up a little, and then it got cloudy again, and it snowed copiously all afternoon. A lot of seals are seen sleeping calmly on the floes. It is incredible what a heavy sleep these animals have.

We have passed 50 m away from some of them, and, despite our yells and whistles, they did not move. Some whales were seen also. Today, the boats have been prepared for hunting whales.

Water temperature has varied between $-1.05°$ and $-1.40°$.

The barometer, which had climbed up to 738 at 6 in the afternoon, has now started to fall rapidly. At 2 am, and due to a thick fog, the engine was stopped, and we just remained there in the same spot.

3.9   We Have Barely Made It to the Polar Circle

3.9.1   *February 1, 1902, Saturday*

The day is dark and sad. A thick fog has extended its ashen attire over us, impeding our ability to see beyond 100 m. The ship is stopped, and it's swaying in the midst of the floes, at the impulse of a slight sea swell.

The snow that fell yesterday is now felt on the heads of all of those who are walking on the deck, falling in the form of rain, pieces of _______ [unreadable—presumably "slush"], because the temperature has increased a lot.

The first month of the year is gone, and, in general, it has been good, but it hasn't let us reach the south. We have barely made it to the Polar Circle, and we haven't been able to reach the so desired Antarctic land.

Between 1 and 3 pm, plankton was pulled out from 900 m of depth, and samples and temperatures were taken of the water at very diverse depths: 300 m, +0.34; 200 m, +0.20; 150 m, −0.29; 100 m, −1.3; 20 m, −1.57. At 10 pm, the barometer dropped to 724 at 0° and it continued like that until 2 pm, when it started to go up again. At 6 pm, when the barometer marked 726, wind from W started to blow and it became stronger, causing quite an ocean swell. At 8 pm, we were heaving to. Water temperature has varied between −1.25° and −1.40°. Enormous icebergs are seen. In the afternoon, it snowed a little, and at night, we were in an open space with no floes.

The first of February was a crucial day for the expedition. According to Nordenskjöld, they were now at 63° 30' S and 45° 7' W, having backtracked on their course (Nordenskjöld et al. 1905). They were having a difficult time continuing their ice-blocked track to Weddell Sea, and they knew they would have to establish their winter station soon. It was also during these first several days of February that the final group of overwintering party members was selected and that Stokes made the fateful decision to not remain with the wintering party but to return on the Antarctic *to the Malvinas (Falkland Islands). Sobral, of course, had already been confirmed as an overwintering member, and his fate was sealed. His mind must have been extremely preoccupied with these prospects, as can be seen by his diary entries during these days of February.*

3.9.2 *February 2, 1902, Sunday [corrected from "January"]*

The wind has decreased a lot, and the barometer still rises. There's a little of the swelling sea, and there are a lot of birds flying around the ship. They especially notice one another and this is why they are never seen. A specimen of (*Larus domicanus?*) [Kelp Gull] and a lead-colored albatross.

At 10:30, a boat with Nordenskjöld and Anderson [K.A. Andersson] left, and they hunted: 1 lead-colored albatross, some cape pigeons, blue petrels, and storm petrels.

At 12, the boat returned.

It's snowing a little.

At 12 m [noon], the barometer reached 736, with a tendency to decrease.

At 2 pm, sounding was begun. ____ [blank] was found at a depth of ____ [blank]. Plankton was pulled out from a depth of ____ [blank]. Samples of water and temperature were taken. The sounding brought samples from the ocean floor.

At 6:30, we headed toward the N. The wind has moved to the N. The barometer continues to drop rapidly. Large icebergs are seen, as are many whales. At 12 pm at night, the wind started to blow from the N with a lot of strength. One of the female dogs, the ones that were brought from the Malvinas [Falklands], had 8 puppies, but 3 were killed because they say that she can't take care of so many.

As the ship sways with the slight sea swells and slowly makes its way among the entourage of ice floes surrounding it, the expedition members make the best of the "down time" by collecting samples of water, flora, and fauna. Sobral meticulously keeps track of all the activities while learning the scientific aspects and terminology of the work. He also keeps an eye on the dogs, especially with the birth of the puppies, noting the killing of three of the newborns because "they say" that the mother cannot care for that many offspring. Killing puppies was unfortunately a common practice among some of the Polar explorers—Roald Amundsen killed all the female puppies born on the Fram *on his way to Antarctica (Tahan 2016c).*

3.9.3 February 3, 1902 Monday [corrected from "January"]

Barometer dropped down to 726 at 2 am, and at 4 am it started to rise again.

The day is gray. What an ugly day! At night, the wind moved W, blowing between WSW and WNW with much force.

The sudden highs and lows that the barometer has had these days. Today, a lot of whales and birds have been seen.

Big icebergs in our path. Water temperature has varied between −1.2 and 0.5.

In the afternoon, the wind began to diminish, and therefore, the sea became more calm as well. We now have real nights, quite dark ones. A lot of whales and birds. At 10 pm, the barometer marked 743.

3.9.4 February 4, 1902, Tuesday
[corrected from "January"]

The day has no sun, but it is calm. As we were entering the pack ice again, many enormous icebergs were seen. Plankton was pulled from a depth of ____ [blank], and water temperature was taken.

In the afternoon, we saw an iceberg of a very curious color and shape: It had the shape of a ship wrecked on the rocks, and the color of the rocks and that of the ship's hull were exactly the same as the one you would see on the line of floatation [waterline] of the ship when it's dirty; that color is probably produced by seaweed. The ship [*Antarctic*] was stopped between two and half past five.

Another iceberg was seen, a piece of which was completely blue, as if it had been painted.

Sobral further commented on his observations of these icebergs in his 1903 notes, recalling that "some of them were colored yellow on the flotation line due to the adherence of diatoms" and that most of them "were tabular, rectangular," while "the regularity of their stratifications showed that their origin was some snowdrift" (Sobral 1903).

3.9.5 February 5, 1902, Wednesday
[corrected from "January"]

The barometer had maintained the same height since the day before yesterday. Today, at 12 at night, it went up to 745.5. Today was the coldest day that we've had. The temperature was as low as −5°. Wind from ____ [blank].

Large icebergs are seen, a great quantity of whales, some penguins, cape pigeons, and other petrels. The day has been cloudy. The dogs fight; today, there was a great battle between Jim and the black [one].

The weather and ice conditions that had greeted the Antarctic *that year were worse than usual, and definitely worse than had been expected. By this time, according to his personal notes written in 1903, Sobral surmised that "our advance toward the south has concluded, and we have probably had a bad summer" (Sobral 1903).*

3.9.6   February 6, 1902, Thursday [corrected from "January"]

A splendid day. We've had sun. The thermometer has remained at around −5° centigrade. In the morning, the wind from the S has been fresh, but it eased in the afternoon. It was a splendid sunset. At 5:30, the ship was stopped and a sounding of the depth was undertaken. Some samples of the sea floor were collected. Zoological specimens were pulled out from a depth of ____ [blank], and samples and temperatures of the water were taken at different levels of depth.

At 9 pm, the engine was started, but it was stopped at 11 pm. Today, for the first time after many days, we have stars. It's a shame that there are some clouds that cover a great part of the sky. At 12 at night, the barometer marked 748 and it started to go down.

3.10   Cape Seymour

3.10.1   February 7, Friday [1902]

In the morning, the sun came out a little, but it began to get cloudy shortly after, and we had another gray day. At 3 pm, the fog came suddenly, and we were navigating between floes. According to the estimated location at midday, we are about 150 w from Cape Seymour [52° W]. At 4 pm, the sea was seen covered by a great quantity of dead little fish. Maybe some volcanic shock. Many were taken on board. A lot of penguins and seals are seen on the floes, and the spouting of whales can be heard in the distance. At 4:30, they attacked a big floe with the aim of making drinking water. This floe was full of *Hummocks* and crevices presenting the most beautiful tones of blue. On the ice, there was a thick layer of snow. This ice was cut into pieces, and it was taken in baskets to the tanks. At 7:30, the work was completed, but because the fog was persistent, we stayed anchored to the floe. The sea is in a state of complete calm, and there is no wind blowing at all. The barometer at 12 at night marks 735.73 and is still decreasing. Only the murmur of the water hitting against the great floes is heard outside; everything else is in silence. At 8 pm, we saw an enormous blue whale coming in the direction of the ship. We all ran to the photographic machine, but, as we were getting there, the whale passed under the ship and under the floe that we were anchored to, and we no longer saw it.

3.10.2 February 8, Saturday [1902]

After we left the floe, we stayed the entire night almost in the same place.

At 6:30 am, we moved, stopping the ship several times to hunt seals. Most of them were leopards. The barometer reached 733, and at 8 am, it started to rise. Some snow fell from 8 to 10 am. Some birds were hunted, including a Pagodroma Nivea [snow petrel], that we didn't have [previously]. At 1 pm, we left the pack and headed toward Cape Seymour. At 12 pm, Bar = 736.

3.10.3 January 9 [sic—should be February 9], 1902 Sunday

At 4 am, we sighted Joinville Strait and the adjacent lands, and at 6 am, Cockburn Island and Cape Seymour were sighted. Quite strong wind from SSE. Surrounding Cape Seymour there is very little sea floor, and it is of sand and rocks.

The wind is moving to the S, and it is blowing with great force. At 4:30, it had a velocity of 16 m per second. At this same time, it started to snow. The barometer has been at 736.5 from 12 at night yesterday until 10 am today. Since then, until 6 pm, it has risen to 738.5.

Cockburn Island is an inaccessible volcanic island of horizontal elliptical section and oriented NS. Its height, according to the British navigation chart, is 2760 ft. It has very little snow.

The wind has moved to the SW, and at 7:30 pm, it blew reaching a speed of 21 m per second. But the sea did not swell much, because we are leeward of the coast.

The rocks at Cape Seymour are later described by Sobral as being "a huge number of erratic stones that probably were transported by snowdrifts in former times." The 21-meter-per-second winds that he and his mates were experiencing are described by him as being "the strongest storm that we have borne since we came to the south" (Sobral 1903).

3.10.4 February 10 Monday [1902]
[corrected from "January"]

The wind has taken us 15 miles to the North. We are in front of Sidney Herbert Bay [Erebus and Terror Gulf] at 8 am. In general, all the land to the W of Cape Seymour is covered with ice and snow. And in the depth of the aforementioned bay, there are high lands completely free of snow, which, according to Nordenskjöld, are made of basalt.

The glaciers reach the sea. At 9 am, we pointed the prow toward the end of the bay to reconnoiter it. In the back part of it, there seems to be a narrow channel that continues toward the W. After reconnoitering the bay, at 1 pm, we started to navigate along the coast in the direction of Cape Gage.

At 9 am, the wind calmed down a great deal, moving to the W. At 3 pm, it switched to NNW of Cape Gage; [we have hit a] sand bank. The wind begins to blow with strength again, this time from the S. With the engine at full power and the sails unfurled, we were able to make it off [the sand bottom and back into the water again], sailing to the NE with the objective of going into Admiralty Inlet, closer to Cockburn than to Gage. We went through a belt of ice pack. Large icebergs can be seen. The temperature was −7° last night. The coast is white with frost, and the ________ [unreadable—land formations] 10 m from the sea are also all completely white, and look as if they were great altar candles. At 2 pm, the barometer started to go up very slowly. At 12 at night, the barometer marks 740.58. Wind as strong as yesterday's. We are heaving to.

The sightings of new land and the near-miss with the sandbank keep the crew busy on this day. Sobral faithfully interprets Nordenskjöld's reading of the land— red basalt at Sidney Herbert Bay, and black basalt at Cape Gage—in his 1903 notes. In those, he describes Seymour Island and the northeast part of Snow Hill as being the only lands not covered with ice and snow. And Cape Gordon as having "dark parts devoid of snow." From the crow's nest, Captain Larsen spots a strait that follows a westerly direction. But in the midst of these sightings, the ship runs aground on a sandbank, and the crew works feverishly to release it, setting it free and into deeper water again. The ship itself becomes encrusted in ice, the waves from the stormy sea soaking it with water that promptly freezes in the −7° temperature. Sobral reports that "Several sailors got frostbitten fingers during the maneuvers . . . because of their hands being permanently wet" and were "in bad condition" (Sobral 1903).

3.10.5   *January [sic] 11, 1902, Tuesday [this should be February]*

[This diary page is left blank; however, as previously stated, Sobral prepared handwritten notes for his speeches in Buenos Aires, and for his official report to the Minister of the Navy, in December 1903. In these notes, he refers to his diary, and copies many of the entries, as well as further details them. He also adds the two dates of February 11 and 12, which are missing from this voyage diary. Those two entries follow in their entirety.]

We are to the NNW of Cockburn Island and around the latitude of Cape Gordon. The vast floe barrier extends from Cape Gage to Cape Seymour, surrounding

Cockburn Island and hindering our entrance to Admiralty Inlet. Yesterday's wind has accumulated a lot of ice in a few hours, and this shows how ice conditions can change greatly in a short period of time.

Yesterday there was only a belt of floes that was not difficult to pass through between Cape Gage and Cockburn Island; and the one that is there today, despite all the efforts that we have made, is not allowing us to force our way through it. Paulet Island, and all the islands in the vicinity of Joinville, can be seen. At 11:30, a seafloor net was thrown; at 1:30, after pulling up the net with excellent results for zoologists, we moved toward the pack that we had in front of us, which was thick and difficult. At 9 pm, after a lot of work, we came out of it and headed for Cockburn Island.

3.10.6 February 12, 1902 [Wednesday]

During the night, we navigated very slowly toward Admiralty Inlet. At 7 am, we were in front of an area of "Snow Hill" completely devoid of snow, which, from onboard, looked good for establishing the winter station.

At 7:30, a boat was sent with the chief [Nordenskjöld] and some [science and crew] members of the expedition, with the aim of exploring the surroundings to see whether this was indeed an appropriate location for the installation of winter headquarters.

The place was considered to be good, and in the afternoon, the disembarkation of everything pertaining to the station began.

The disembarkation of provisions and of our things was as fast as you can imagine.

The first thing that was put on land was the wood for the building of the house, and immediately the carpenter and a sailor went to work.

But the building of our house was [ultimately] left to us, for, in the early morning of the 14th, along with the current of the flow, a little bit of ice began to enter Admiralty Inlet, and we deemed it not prudent to keep the ship waiting, and so it set sail. It had as much to do as possible to go to the south to place a deposit of provisions on the Seal Islands that we could use as a base.

And so ends the voyage portion of the diary, describing the journey to Antarctica on the ship Antarctic. *The expedition has backtracked from 66° 30' to establish their base at 64° 22' S and 57° 10' W, according to what Sobral would later write in his book (Sobral 1904).*

Sobral has learned much, seen much, and endured a doubtful situation. But he and the expedition members are now at Snow Hill, Nordenskjöld having rejected Seymour Island as his wintering station, which had been his initial plan. Sobral and his companions would all have to come to terms with the implications of that decision. But for now, Sobral is hoping that this is a good place to study Antarctica.

Figures 3.1, 3.2, 3.3, 3.4, 3.5, and 3.6.

Fig. 3.1 Alférez José María Sobral, the first Argentine to embark on an Antarctic expedition. (Departamento de Estudios Históricos Navales, Archivo Histórico, Archivo Fotográfico, P-0140 m, Sobral, José María. Alférez.)

Fig. 3.2 The Antarctic expedition diary of José María Sobral. (Departamento de Estudios Históricos Navales, Archivo Histórico, El Diario de Alférez de Navío José María Sobral.)

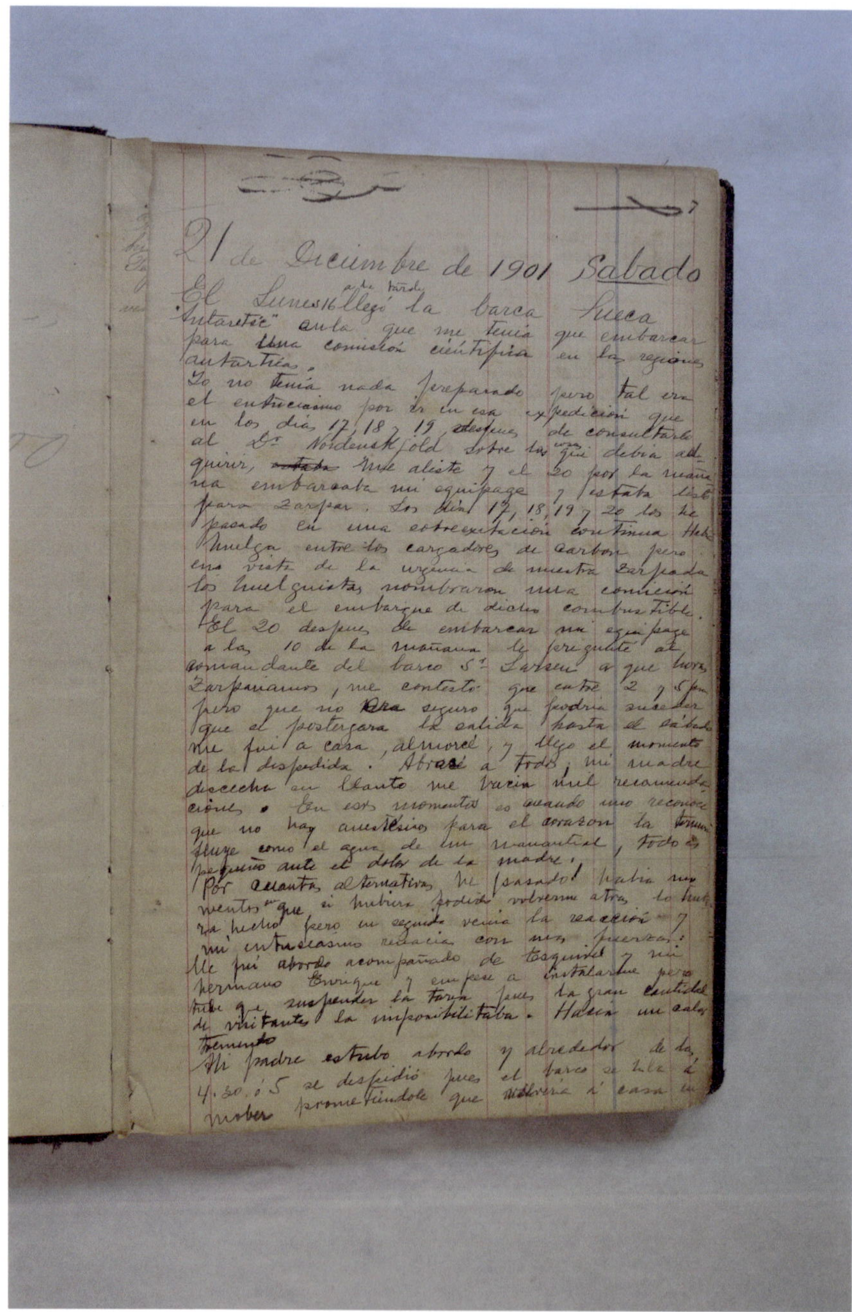

Fig. 3.3 The beginning of the diary, and the beginning of José María Sobral's journey. (Departamento de Estudios Históricos Navales, Archivo Histórico, El Diario de Alférez de Navío José María Sobral.)

Fig. 3.4 The ship *Antarctic* departed from Buenos Aires on December 21, 1901. (Photo from *Dos Años Entre los Hielos, 1901–1903*, by José María Sobral, 1904, p 49.)

Fig. 3.5 Soundings, water sampling, and oceanographic studies were performed aboard the *Antarctic* during the Swedish Antarctic Expedition. Photographic image by José María Sobral. (Departamento de Estudios Históricos Navales, Archivo Histórico, Archivo Fotográfico, Ex-0040 o, Expedición Sueca - Dibujos de Sobral.)

Fig. 3.6 One of the tabular icebergs sighted during the voyage of the *Antarctic*. Photographic image by José María Sobral. (Departamento de Estudios Históricos Navales, Archivo Histórico, Archivo Fotográfico, Ex-0039 z, Expedición Sueca - Dibujos de Sobral.)

References

Nordenskjöld N Otto G, Andersson JG, Skottsberg C, Larsen CA (1905) Antarctica: or two years amongst the ice of the South Pole. The Macmillan Co, New York

Rabassa J (2003) Estudio preliminare. In *Dos años entre los hielos, 1901-1903,* José María Sobral (reprint of original book published in 1904), 1–22. Eudeba, Colección Reservada del Museo del Fin del Mundo, Buenos Aires

Sobral JM (1901–1903) Expedition diary of José María Sobral, Swedish Antarctic Expedition of 1901-1903. Departamento de Estudios Históricos Navales, Archivo Histórico. El Diario de Alférez de Navío José María Sobral. Archivo D.E.H.N.—A.R.A, Buenos Aires

Sobral JM (1903) Handwritten notes for speeches and for official report to the Minister of the Navy. Departamento de Estudios Históricos Navales, Archivo Histórico, Colección Sobral. Box 2, Envelope 4, Manuscrito referido al viaje a la Antartida y Conferencia en el Centro Naval 1903. Archivo D.E.H.N.—A.R.A, Buenos Aires

Sobral JM (1904) *Dos años entre los hielos, 1901-1903 (Two years amidst the ice, 1901-1903).* Imprenta de J. Tragant y Cia., Bolivar 319, Buenos Aires

Staff Writer (1901) Artist in the Antarctic: Frank Wilbert Stokes to join Swedish Expedition. Los Angeles Herald, October 2. (California Digital Newspaper Collection, https://cdnc.ucr.edu/cgi-bin/cdnc?a=d&d=LAH19011002.2.19. Accessed 2 May 2017)

Stokes FW (1903) An artist in the Antarctic. The Century Magazine August: 521–528. http://www.antarctic-circle.org/stokes.pdf. Accessed 18 April 2015

Tahan MR (2016a). Paw prints in human Antarctic history. Scientific Committee on Antarctic Research (SCAR). The 34th SCAR Biennial Meetings and 2016 Open Science Conference in Kuala Lumpur, Malaysia. Abstract and poster. Abstract Book, 20–30 August 2016, page 1017. ISBN 978-0-948277-32-0. http://www.scar.org/scar_media/documents/outreach/communications/SCAR_OSC_2016_Abstracts.pdf

Tahan MR (2016b) Paw prints in human Antarctic history: the use of sled dogs in Antarctic exploration. Latin American Antarctic Historians. The XVI Encuentro de Historiadores Antarticos Latinoamericanos in Buenos Aires, Argentina. 28–29 October 2016. Paper and presentation. http://dna.gob.ar/xvi-encuentro-de-historiadores-antarticos-latinoamericanos

Tahan MR (2016c) The canine explorers: the sled dogs who helped Roald Amundsen reach the South Pole. J Antarct Aff. Agenda Antártica, Buenos Aires, Argentina, and the Antarctic and Southern Ocean Coalition (ASOC), Washington, DC. Volume III–July 2016, pp 31–50. ISSN 2451-7755. http://www.agendaantartica.org/agendaen/july2016.pdf

Chapter 4
The First Winter in Antarctica

Abstract The first winter at Snow Hill, February 12, 1902, through January 24, 1903, was productive for the Swedish Antarctic Expedition, whose activities included: The initial finding of marine fossils at Snow Hill; the unloading of food and supplies; construction of the prefabricated house; departure of the *Antarctic*; daily meteorological and astronomical observations; extensive scientific and bacteriological studies; paleontological discoveries on Seymour Island, including the finding of the first fossil vertebrates from a giant penguin species; lengthy camping trips and plateau hikes; boat trips to Snow Hill Glacier and Lockyer Island; and fossil-collecting sledge journeys of the six-member party: José María Sobral, Nils Otto Nordenskjöld, Gösta Bodman, Erik Ekelöf, Gustaf Åkerlundh, and Ole Jonassen.

4.1   The Hellos and Goodbyes of February

José María Sobral had finally realized his childhood dream. He was now an inhabitant of Antarctica, and he would remain so for what he thought would be the duration of a year.

Sobral did not write about this first year at Snow Hill in the same expedition diary in which he recounted the ship's voyage to Antarctica and the second year spent at Snow Hill.

But he did write about this first winter in his book *Dos Años Entre los Hielos, 1901–1903*, although, given the vast amount of information and innermost private thoughts that are found in his diary and not found in his official book, one can surmise that that is the "public" version of his experience.

On February 12, 1902, the *Antarctic* anchored off of Snow Hill, and the Swedish-Argentine Expedition began offloading the supplies, habitat components, and scientific equipment needed from the ship.

The expedition leader had selected a picturesque location that was also protected from the southern winds. The winter station house would be positioned in the snow-free lee of the hill, which itself was snow-covered. To the south-southwest was a glacier. To the north-northwest was the sea. And supposedly, to the east-northeast was terrain made of sand and rocks and fertile with fossils. (Sobral 1904).

© Springer International Publishing AG 2018   69
M.R. Tahan, *The Life of José María Sobral*, Springer Biographies,
https://doi.org/10.1007/978-3-319-67268-7_4

For two days, the crew unloaded provisions and began building the little pre-fabricated house that would serve as the home of the six expeditioners selected for overwintering: Nils Otto Nordenskjöld, Gösta Bodman, Erik Ekelöf, Gustaf Åkerlundh, Ole Jonassen, and, of course José María Sobral.

Sobral assisted in bringing the scientific instruments and equipment from the ship to shore and setting them up for installation. A small hut was constructed to house the magnetic instruments.

The artist F.W. Stokes took the opportunity to complete more painting sketches, finding a vantage point and creating drawings depicting the snowy hill, the ice, the waters, and the colorful skies. But he decided that he had completed enough paintings and that Snow Hill was not far south enough to enable him to paint the aurora australis, and so he begged to return north with the ship (Nordenskjöld et al. 1905). And so Sobral would be deprived of the one friend he had made on the ship during the voyage south.

By February 14, the ice began to build up in the water, and so Captain Carl Larsen made the call to depart before the ship became locked in. They would return later with additional provisions. In the meantime, homebuilding was turned over to the six remaining members at Snow Hill.

Two small boats were rowed to shore. They would serve as the men's transportation over water. Not to be forgotten, the dogs were also brought on shore to be part of the wintering party. They were the men's primary means of transportation, across the land and sea ice. They were the original Greenlanders, the Malvinas/Falkland Island dogs, and the puppies born en route.

As Sobral and Nordenskjöld carried the building material from the shore to the hill, the rest of the team worked on assembling the house (Nordenskjöld et al. 1905).

The six expeditioners proceeded to build their outpost on this alien world. It was almost as though they were the sole inhabitants on an icy planet.

Meanwhile, the ship *Antarctic* attempted to make its way south to drop off a depot of provisions to be used during the expeditioners' future sledging excursions. On board, F.W. Stokes reminisced about the temporary departure from Snow Hill (Stokes 1903: 521–528):

> The Swedish flag was dipped, while the little whale-boat of the winter party grew smaller as our vessel threaded its way between the floes, out of the inlet, during which time many themes for the brush appeared. At supper we numbered six instead of ten, and the absent ones were missed.

In its attempt to go further south to establish the depot, the ship was stopped by ice. It was forced to go northwards again.

The *Antarctic* returned on February 21 and was greeted by the six expeditioners as though they had not seen it for years. Its brief absence had given them a little taste of the separation they would soon feel from the rest of the world (Nordenskjöld et al. 1905). Because now it would leave for the entire winter, seeking more temperate waters and climate closer to Argentina.

Disembarking for a final farewell, Larsen brought to Nordenskjöld additional cases of coal for the winter. F.W. Stokes presented his new-found friend Sobral with a small gift before departing. Final letters of correspondence were given by the six men to the *Antarctic* crew to take back with them to be sent to family and loved ones.

And then they watched the ship disappear.

4.1.1 Thoughts on Colonization

Meanwhile, on this very same day of February 21, in the Malvinas/Falkland Islands, a future member of the expedition arrived from Sweden at Port Stanley. His name was J.G. Andersson. He had been assigned by Nordenskjöld as the second expedition leader, and he awaited his rendezvous with the *Antarctic*, scheduled to take place upon the ship's arrival almost a month hence.

In its journey north, the *Antarctic* stopped to load coal for fuel at the southernmost city on earth, Ushuaia, Argentina, in the tip of Tierra del Fuego. Arriving at Port Stanley on March 27, the ship allowed the artist F.W. Stokes to disembark. He promptly took a steamer to the United States via Montevideo. The *Antarctic* remained in the Malvinas/Falklands until November 5th, when it would begin its journey back to Antarctica.

Meanwhile, back on the continent of Antarctica, the inhabitants were taking stock of their new kingdom. They were now completely alone.

The first month passed on the white continent. Sobral mused about his reign on this frozen realm. He took a metaphorical moment to reflect on his place there in the larger scheme of things, as he saw them. He provides a revealing look into these musings, which are his own thoughts regarding the harnessing of Antarctica and its resources, and the domination of the lands of certain "savage" people by "civilized" colonialists—a startling and shocking view held by this amiable and equality-minded individual. To this end, in his book, he paraphrases a passage from his diary written during this month at Snow Hill (Sobral 1904: 106–107):

> … I will make a few comments to remember what I had written in February of 1902 from Snow Hill. I have been of the belief, and continue to believe, that the Argentine Republic should have taken possession of these southern lands. (This is now also what many politicians believe.)

> Historical antecedents give credence to these opinions. In general, the possession that European nations took of the Asiatic and African continents began as explorations for study and ended up as armed expeditions.

> In the Antarctic regions, life is very hard; this is true, but these difficulties are due to the lack of resources for living.

> In the warmer European colonies, life is also difficult, due to the heat which is unbearable, but this difficulty is neutralized with the resources from civilization; the same can be said for the glacier areas.

Many of the first explorers of Central Africa, India, and our own Chaco could have died of hunger and illnesses produced by the heat, like those who have died of hunger and cold in Siberia; but that happened when these inhospitable areas lacked the necessary resources. Today, they are very hospitable areas because those resources have been brought to those areas.

A few years ago, our own Tierra del Fuego was inhabited only by the savage; today, we have a national territory, a nascent town, and a regular government where our national sovereignty is respected.

Because of that, it is my belief that the same thing should be done with these sub polar lands, whose seas, with more or less some difficulties, could be navigated any time of the year, and that offer some benefits to the fisheries that could be established there.

4.2   Life and Work at Snow Hill

The inhabitants of Snow Hill quickly established a daily routine with work assignments and explorative activities.

Nordenskjöld was in charge of all things Geology. Ekelöf monitored and maintained everyone's health, as well as studied the bacteriology. Bodman and Sobral conducted and recorded continuous magnetic, meteorological, and astronomic observations. Jonassen trained and took care of the dogs, maintained the sledges and performed jack-of-all-trades tasks. Åkerlundh cooked and served meals.

Breakfast was served at 9:30 am, after which those who smoked tobacco would light up their pipes and head off to their tasks. Lunch was served at 2:30, coffee at 5:00 pm, and dinner at 9:00 pm. Following dinner each evening, the men would socialize until 11:00 pm, and read from 12:00 am to 1:00 am. (Sobral 1904). Given Sobral's place as the somewhat "odd man out," and his just beginning to learn the Swedish language, most likely he did not engage in as much socializing as the others.

Nordenskjöld, Ekelöf, Bodman, and Sobral performed guard work during the nights, rotating shifts, and diurnal sentinel work was done by Sobral and Bodman. These guard watches were necessary because the expedition did not have the equipment required to register certain atmospheric phenomena, and so they made manual observations on an hourly basis.

In terms of personal grooming, given the amount of fuel necessary to melt ice for water, the men bathed every 15 days.

For entertainment, they played a gramophone that regaled them with orchestral works and operas.

Their living arrangements were snug yet provided a sense of private space for each individual. The house was divided into five compartments that included a central dining space, a living space, and a workroom. At the corners of the house were the sleeping areas, with bunk beds for two people per corner. Sobral and

Nordenskjöld shared the northwest corner, Bodman and Ekelöf the northeast, and Jonassen and Åkerlundh the southwest. In the southeast corner was the kitchen. (Sobral 1904).

On some of the walls, they hung work charts and instruments. On others, they hung pin-up photos.

4.3   A Beautiful Day in the Neighborhood

Nordenskjöld immediately began making brief excursions to the surrounding area to gauge the terrain and look for fossils. He took Sobral with him on the lengthiest of these excursions on March 9th (Nordenskjöld et al. 1905). The two explorers traveled south of Snow Hill, walking over the ice, as their skis could not negotiate the hard ridges in the snow. Traveling for 10 h and reaching a height of 300 m, they discovered three nunataks that Ross had not indicated in his writings about the area. The two men left on their excursion while the sun was shining gloriously, but returned in a heavy fog that barely allowed them to find their way back. It was Nordenskjöld's way of getting to know the area, and Sobral was the companion he had chosen.

4.4   The First Sea Excursion

Once the winter station was built, and prior to winter's arrival, the men decided to go explore Lockyer Island by boat and to deposit a depot of provisions there in preparation for later excursions. This was especially necessary as the *Antarctic* had been unable to deposit an additional cache of provisions.

Sobral, Nordenskjöld, and Jonassen departed from the station on March 11, 1902, with five dogs, one sledge, and enough provisions for six days. They took with them a stove, a reindeer pelt for sleeping (for three persons), and a small tent (Sobral 1904). And they took their boat, which had been named *Tromsø* in honor of the little Arctic town in Norway.

The way to Lockyer Island presented several challenges, and the men were met with many difficulties. The water ways were an obstacle course of fast-moving icebergs that they constantly had to avoid. When the icebergs collided with one another in their path, they formed both a physical and an aural risk, as the noise of the collision was deafening.

When they had reached Lockyer, Nordenskjöld and Jonassen left Sobral in the boat so that they could climb an iceberg nearby to reconnoiter the immediate area. Sobral's only colleagues left on the boat with him were the dogs. As Nordenskjöld and Jonassen had taken with them the ice hooks to help them in their climbing, Sobral did not have these to use as ice anchors. His boat was wedged in between the iceberg that the two men were climbing, and the moving sea ice around him. The rush of the coming ice, and the force of the water, nearly crushed and carried the boat away, with everyone and everything in it—Sobral, the dogs, and the provisions. Attempting to keep the boat steady and in place with the oars, Sobral simultaneously called out to his colleagues for help, but to no avail, as they could not hear him. One of the dogs became so unsettled by the motion of the water and the sound and impact of the crashing ice that he jumped overboard into the icy water. Almost immediately, he swam back toward the boat, unable to withstand the coldness of the water. In the midst of trying to save the boat, and having his hands full with rowing and keeping the craft steady, Sobral reached out and helped the dog, viewing him as his equal partner in this critical situation. He valiantly pulled the dog back into the boat, then redoubled his efforts with the oars. Just at this point, Jonassen heard the distress call from Sobral, and ran toward the boat, leaping over treacherous bergs of ice along the water, and reaching Sobral just in time to jump on board and help Sobral row the boat to safety. (Sobral 1904; Sobral Boletin 1903).

After the close call with the boat, and because the way to Lockyer seemed impossible to negotiate, it was decided to deposit the provisions at Haddington (James Ross Island), in a protected cape, which they named Cape Depósito.

The three men then camped on the frozen field of ice, approximately 50 m from the water. They settled in for a much-needed good night's sleep. At around 7:00 am, they had a rude awakening when Jonassen noticed that the tent had been invaded by water. He yelled out a warning that woke up the other two men. The ice upon which they lay, which they had thought was sufficiently removed from the water, had been quickly eroded by the wind, which had pushed huge chunks of the ice out into the water, so that, one moment more, and the men and tent would have become submerged into the open water. All three scrambled to put on their boots and strike the tent with all their equipment, quickly moving away from the water. Jonassen's observational abilities and quick reaction had been fortunate for them, and their fast moving had saved their lives.

Following the tent incident, while camping to wait out hurricane winds, and contending with frozen clothing and sleeping gear, Sobral barely escaped getting frostbitten, his finger freezing overnight, but recovering sufficiently.

The return home was just as treacherous, with Sobral and Jonassen rowing frantically to prevent the strong current from driving them into a submerged glacier.

After four weary days, they thankfully returned home to the house at Snow Hill, safe and sound, with all limbs attached. Sobral maintained that Ekelöf had been preparing himself to amputate several of their fingers upon their homecoming, assuming that they would return with frostbite; but luckily, they did not require his services.

During their absence, Bodman had been able to install the magnetic instruments at the station.

Besides the magnetic and meteorological work that he began, Sobral was learning the ways and the properties of the different types of ice, and recording his observations about sea ice and land ice.

A kennel was built for the sledge dogs, but the dogs were showing signs of inability to co-exist. Specifically, the Greenland dogs began attacking the Malvinas dogs. In order to stop their continuous fighting, the men were impelled to create two different spaces within the kennel, so as to separate the two groups. Though peace would never be fulfilled between the two, at least this precaution prevented further fighting.

4.5   Winter Excursions

April brought cold to the wintering party at Snow Hill, and with it a lengthening of the night. Sobral measured the freezing temperatures and observed the austere landscape that presented a contrast of black (from the basalt) against white (from the snow and ice). These were the colors of life in this Polar region.

Along with the disappearance of the relatively warmer temperatures was the disappearance of the wildlife, specifically the birds.

Sobral's impressions and writings at this time centered on the prevalence of storms and the intensification of the cold. And yet life could be endured in this region, he asserted, if one were properly attired. Alas, he was not so, as his garments from Buenos Aires barely covered him adequately enough to fight the coldness or the icy dampness.

The magnetic and meteorological installations were completed in April, and Sobral began to use the instruments to steadily make magnetic and meteorological observations and faithfully record them. He occupied himself with these magnetic measurements, which began on April 1st and continued until the very end of the nearly two-year stay at Snow Hill. This was an historic achievement and represented the very first continuous recording and documentation of meteorological data in Antarctica. And it kept Sobral busy.

In addition, Sobral assisted Nordenskjöld in another first: installing a series of monitoring posts which he planted in the ice along the glacier to ascertain the speed and direction of the wind.

The temperature of the ice was also taken, and for this, he placed thermometers at a depth of 30, 50, and 100 cm.

It was during this work of assisting Nordenskjöld that a disagreement arose between Sobral and his mentor. It was an intellectual disagreement, regarding Sobral's interpretation of an atmospheric phenomenon as an aurora. The date was April 5th, and Sobral was sure of his theory, but Nordenskjöld did not agree. Perhaps this deepened Sobral's sense of divide. But he read voraciously, quoting other explorers and analyzing their findings. He took some solace in transcribing the descriptions made by Fridtjof Nansen and Matthew Henson of similar phenomena they had observed in the Arctic north, convincing himself that their findings agreed with his. It must be noted that he read these explorers' accounts in their original respective languages.

Sobral also spent productive time photographing the surrounding area, until his camera was blown by the wind and damaged. Fortunately, Ekelöf was able to repair it. As he watched the birds fly northward, he recalled verdant landscapes in his home country.

April 9th gave the men a much-needed respite as the group made a special effort to celebrate Nordenskjöld's birthday.

On April 21, Sobral ventured out for his daily exercise routine on skis, which he had learned to master at Snow Hill. Accompanied by Bodman, he headed for the glacier, and, from an altitude of approximately 40 meters, the two men skied along the soft snow. Despite the fact that the temperature was 20° below zero, the brisk exercise heated them up, so much so that soon they were shedding their outerwear and gloves. The next thing Sobral realized, he had lost feeling in the finger on his left hand. Bodman and he tried to rub snow on the finger, but, soon the other fingers went numb as well. The two men immediately turned back to the station for medical treatment. The frostbite effect was so quick that, by the time they reached the station, the rest of Sobral's hand had begun to freeze as well. Only with extensive massaging to help the circulation was Ekelöf able to restore life to Sobral's hand. It was a lesson Sobral would not forget.

April 27th marked Nordenskjöld's first foray into fossil hunting on Seymour Island during his first winter at Snow Hill. On this day, he left with Ekelöf and Jonassen on an excursion to survey Seymour. Sobral, along with Bodman, accompanied him for the first 5 km, then turned back home. Nordenskjöld returned to the station two days later "not having seen a single bird" (Sobral 1904: 149). He did, however, discover many fossils—"mollusca, ammonites, etc."—and it began to dawn on him that Seymour Island had a "richer" treasure trove of fossils than did Snow Hill and that it would have served him better as his wintering station (Nordenskjöld et al. 1905: 156).

For Sobral, back at the station, another close call occurred on May 3rd, when, during temperatures of 25° below zero and winds of 36 m/s, one of the dogs was buffeted by the wind and fell into the ice-filled sea, almost immediately dying from the cold and ice. Bodman quickly pulled him from the water and took him into the house—a rarity for the dogs. He was able to warm him in time, and thus save the dog. Sobral and the men valued each dog for their motive power, as these dogs were the expedition's sole means of transportation across the ice. But Sobral, as well as Nordenskjöld, also valued these dogs as friends during their time in Antarctica (Sobral 1904; Nordenskjöld et al. 1905). They were valuable companions both on the water and on the ice and held a special place in the men's regard (Tahan 2016a, b).

Holidays were occasions for celebration and festivities, at which times the six men enjoyed food and listening to the gramophone. Sobral was a food aficionado and greatly relished the feasts that would be served for Swedish and Argentine holidays. Initially, Sobral had not expected any type of recognition for his country's day of independence, but on May 25, Nordenskjöld made great effort to make Sobral feel at home. He had the cook include as many items on the day's menu as possible that were close to Argentine food. This included canned peaches to emulate those from Tigre, and *choclo*—a sweet corn. Sobral appreciated the effort and the meal, although he refrained from eating the *choclo*, as it was very acidy and badly preserved—only Bodman ate it.

June was one of the coldest months, and Sobral faithfully recorded the temperatures and made all the magnetic and meteorological measurements, as well as tidal measurements. This was a physical effort, in that he had to go out into the elements on an hourly basis to make the observations. A tide gauge had been set up about 200 m from the station house, and Sobral would brave 30 m/s winds and −30° temperature, in the middle of the night, to take measurements. At times, he would have to crawl on all fours, against the strong wind, to make the measurements. To compound the complexity of the situation, sometimes his lantern would become extinguished from the wind, making him unable to work in the dark, and so he would have to return to the house and back out into the elements again to make the measurements.

July saw the coldest temperatures that first winter, with the thermometer indicating −35°. It was the height of winter.

4.6   A Stellar Sledge Journey

At the beginning of that month of July, plans were made to undertake a sledge excursion to Mount Haddington. The goal was to test the equipment and to experiment with emergency signals that could be seen from the station. It would be Sobral's very first sledge excursion. He left on July 25, along with Nordenskjöld, Jonassen, and Åkerlundh. The three men took five dogs and one sledge. The temperature, at the time of their departure, was 32° below zero. The air around them remained frozen. "Everything was deep silence," said Sobral (1904: 157). The only sound that could be heard was the sledge runners against the vast vista of snow and ice. But, for Sobral, it would prove to be a forging of spirit by fire. The extreme difficulties he encountered presented challenges for both the men and the dogs. Ice waves of Sastrugi and other obstacles paved the way and made the terrain difficult to traverse. Standing for more than a certain amount of time brought on a severe coldness that Sobral worked to outrun; yet marching quickly brought on an uncomfortable perspiration that soaked his clothes and made them even more damp still. He felt thirst he had not experienced before. The dogs, although strong and sturdy, needed help to re-start the sledge each time it was brought to an abrupt stop due to the terrain. The tent they used had no flooring, and so the coldness was even more bitter at night. During a storm, the frightened dogs attempted to climb inside the tent and sleep with the men. And the dampness that soaked through their clothes and their sleeping bags, even through Sobral's guanaco-skin gloves, made everything have the "consistency of a stone. How unpleasant it is to feel those hard and cold clothes on the flesh!" (Sobral 1904: 159). But the sledge party did encounter some spectacular days; they worked together to scope out their surroundings and photographed the giant basalt hills that emerged out of the snow surface, and that held loose basalt stones, upon which grew moss and lichen in a colorful array of reds, browns, white, black, and orange. In the night sky, they observed the brilliant Southern Cross and Argo constellations, as they looked toward the more "obscure zenith" above the terrestrial South Pole (Sobral 1904: 159–160):

> … it seems that the creator, to make it more desolate and cold, had placed in the surroundings around the Polar circle, the most beautiful constellations of the Universe, leaving in the center that gloomy cap, populated only by stars of small magnitude.

Sobral saw no wildlife on this first sledge trip, nor any traces indicating that they had been there. But he returned from the journey with a new awareness, a realization that the toughest experience, even harder than working at the winter station or navigating on a ship, was making a sledge excursion, especially over a long period of time, and in low temperatures. He had completed this physically demanding journey, and it had even further strengthened his resolve.

It was during this month of July, back in the Malvinas/Falklands, where most of the *Antarctic* crew wintered, that expedition member Axel Ohlin became seriously ill and departed for Sweden. He would sadly die a year later.

4.7 The Taking and Giving of Life

August 6th, a windy and bitterly cold day, marked the lowest temperature of the first winter. It was a brisk 41.4 °C below zero.

Ten days later, in this freezing weather, one of the Greenland dogs named Fía gave birth to eight "great" puppies. Sobral commiserated: "Poor little animals, how much cold they have to suffer before summer arrives!" (Sobral 1904: 164).

A break in the weather occurred on August 18th, when the temperature rose to 0°. Some of the men took this opportunity to go hunting for seals, finding four of them, and killing three. Sobral rejoiced in having fresh meet again after seven months of canned food. The dogs, too, were treated to the meat, and ate so much that they could barely move; they settled down for a long nap. The following day, Sobral accompanied Nordenskjöld and Ekelöf on a 5 km sledge trek to butcher and retrieve the seals that had been killed. Here he found that one of the seals—a Weddell—had a fetus inside of her. He proceeded to cut the meat from the three, noting that this was not a very agreeable task, as it was necessary to immerse his hands in "the blood and the viscera of the animal," which caused his hands to freeze (Sobral 1904: 165–166). Perhaps between the lines here is a wistful lament, an uneasiness about the necessary killing and butchering of these passive animals. Nevertheless, Sobral loaded one of the seals and the fetus on the sledge, and pulled the carcasses home.

4.8 Back to Cape Depósito

As Spring approached on the horizon, Sobral ruminated about its potential, about his hopes of developing more exciting projects at the wintering station, and about waiting for the return of the *Antarctic*. What kind of news would the ship bring about the countries and the families of the Snow Hill expeditioners, he wondered?

What happy or sad tidings awaited them? This rumination risked the onslaught of a melancholy disposition, and Sobral was self-conscious of this tendency and of the need to not think too much lest he delve into a downward spiral of emotion. And so he constantly sought work and productive occupation. In his moments of loneliness, he gave incessant thought to his homeland of Argentina. Yet, in spite of this, he yearned to continue exploring this new home of Antarctica.

The expeditioners planned an excursion to Cape Depósito to take place at the end of August. The goals of this sledge journey were to gauge the conditions of the ice beyond Lockyer Island and check on the status of their provisions left at Haddington in March.

On the morning of August 30, Sobral, Nordenskjöld, and Jonassen departed from the station with a sledge pulled by three dogs—two Greenlanders, and one from the Malvinas/Falklands, named Curry, who was the only dog from those islands who participated in sledge excursions. The sledge party carried 150 kilos of supplies pulled by the three dogs with the help of two of the men, while one man walked ahead and checked the ice for durability. During this journey, Sobral made many observations on the landscape, storm events, and meteorological conditions. He also recorded sightings of seals and the presence of birds. On one occasion, he experimented as cook, but, being nervous and too hasty while cooking pemmican, he accidentally broke a thermometer, thus losing the ability to make measurements of temperatures. Another fiasco concerned the dogs—two of the Greenlanders disappeared, returning home to the station on their own, leaving "poor Curri," the dog from the Malvinas/Falklands, as the only dog to pull the sledge on the return journey, for which he received a beating as motivation (Sobral 1904: 172). The men ended up pulling the sledge themselves on the final leg of the journey.

The days became warmer after September 11th. Sobral believed that if this trend continued, the sea ice would break up during the following month.

4.9   The Spring Trip South

Sobral's most arduous sledging excursion was yet to take place. It was scheduled for late September. The destination was King Oscar II Land, which Carl Larsen had discovered during his voyage on the *Jason* in 1892, at which time he had seen the east coast of Graham Land. Larsen was also credited with discovering several volcanic islands in this area, which he named the Seal Islands. It was Nordenskjöld's goal to go survey this geographical area and to scrutinize its geological features. As promised to the Argentine government, he wanted to take Sobral with him.

Preparations for the journey were an event unto themselves. Sobral carefully prepared his clothing, his bedding, and his equipment. He wanted to leave nothing to chance, and to protect himself as best as he could.

Besides Sobral, Nordenskjöld decided to take Jonassen as the third member of the team. They would bring with them two sledges—one pulled by five dogs and driven by Jonassen, and one pulled by Nordenskjöld and Sobral.

By now the weather was changing for the better, and the days were growing longer, which made this a good time for travel.

Simultaneously, preparations were also made for the arrival of the *Antarctic*. The six men were very aware that the ship was scheduled to begin its voyage back to the station within the next one or two months, and that they themselves would soon be returning north to the Malvinas/Falklands, and then onwards to home.

September 30th arrived, and the spring sledging trip began. The month-long trek to the south and back was an ambitious one. The three men and five dogs traveled from their station at Snow Hill to Christensen Island and back, covering a total distance of approximately 340 miles according to Sobral (and 650 km according to Nordenskjöld). On ski, sledge, and foot, they succeeded in reaching the furthest south latitude achieved by the expedition—66° South. And they successfully surveyed the until now uncharted eastern coast of the Antarctic Peninsula, which hitherto had never been trod-upon.

It was a difficult trip. Along the way, they experienced many challenges, for which they had diligently prepared. Through careful planning, the men had brought with them the exact amount of supplies and equipment necessary. In order to comply with the least amount of weight possible, Sobral had been forced to leave his camera at the station, as the addition of its weight would exceed the total amount of weight each individual could carry.

For Sobral, the journey was one of extreme highs and debilitating lows. He recorded in detail the storms that the sledge party endured, the sheer beauty of the atmospheric phenomena he witnessed, the merciless hunting of seals to supplement their food, and the effect that the phenomenon of refraction had on them and on their ability to accurately perceive the Antarctic landscape that lay around them, as well as the very steps they had to take on the ice. Sobral himself, as well as his two mates, suffered from ophthalmia brought on by snow blindness, although Sobral to a lesser degree than Nordenskjöld and Jonassen (Sobral 1904). Even with the below-zero temperatures, the sun made trekking a very warm endeavor, and Nordenskjöld would later write that Sobral had exclaimed it was as hot on the ice as it was in Buenos Aires (Nordenskjöld et al. 1905). The men also suffered from extreme thirst, which caused Sobral to nearly hallucinate, seeing mirages of breezy Buenos Aires streets lined with cool cafés and happy people drinking cold drinks. Later he, too, would describe this trip in great detail in a speech he would give for the Centro Naval in 1903.

Importantly for science, on this trip, Sobral and Nordenskjöld made significant observations about the so-called Seal Islands—Christensen, Robertson, Oceana, Lindenberg, Castor, Hertha, Jason, and Larsen—that had been discovered by Carl Anton Larsen. Through careful observation and thought, Sobral strongly asserted that these islands were not islands at all, but were nunataks—snowless, rocky, volcanic peaks emerging from the icy glacier field. In another contradiction of Larsen's discoveries, Sobral also asserted that Larsen did not see Graham Land from Christensen; Sobral himself climbed to the top of Christensen—some 300 m —and was able to observe the land of Haddington from there.

In addition to the geological and geographical studies, Sobral conducted atmospheric measurements. These scientific and geographical discoveries were of great importance, and Sobral was certain that, once studied and analyzed by other scientists, they would become an invaluable contribution to human knowledge.

On October 15th, the party made a fateful decision that could have cost them their lives. They left one of their sledges and most of their food in a spot that they did not mark adequately, and embarked on an eight-day mini excursion with few supplies, returning on the 21st and nearly missing the spot where they had left their provisions. Luckily, at the last moment, they found it; had they not, they would have starved.

In addition to this near-miss, Sobral struggled with inadequate clothing, and his boots broke along the way. On top of that, during a stay in their tent, a violent wind ripped a hole in the side of the tent, right next to where Sobral had been resting his head; he and Jonassen teamed up to repair the tent in the middle of the storm.

The sledging party had covered the distance from the wintering station to Christensen in seven days; on their return, they covered the same distance in only three days. It was a herculean effort. Upon their arrival to the station on November 4th, all three men felt "strange" being indoors again, and Nordenskjöld had to fight off an "attack," but it was Sobral who outright collapsed in exhaustion (Nordenskjöld et al. 1905: 239):

> Sobral had sunk down in a chair near the door, but suddenly said that he felt ill, and the next moment fell fainting to the floor. We rushed to his aid; he was undressed and carried to bed, where he soon recovered.

He had lost seventeen pounds during the arduous excursion. Sobral immediately sought rest and nutrition. All three men drank and ate hungrily to recovery. But the scars of their travel remained—Nordenskjöld experienced several physical difficulties, Sobral's eyes gave him trouble for many days, and the men's beards had turned prematurely white. Even Kurre (Curri), the sledge dog from the Malvinas/Falklands, and Nordenskjöld's favorite dog, literally paled from the trip, turning from a "bright yellow" color to "a pale yellowish grey" (Nordenskjöld et al. 1905: 241).

The discoveries they had made were worthy of their herculean efforts. In Nordenskjöld's words (Nordenskjöld et al. 1905: 239–240):

> We had discovered an extensive stretch of coast and thereby proved the connection between Louis Philippe Land and the tracts seen by Larsen, the charts of these regions becoming completely changed in consequence of our expedition. We had made meteorological and biological observations, and had, above all, made collections, which were of great value for a knowledge of the geology of the tracts in question, whilst our proving the existence of the great ice-terrace could be considered as in itself worth a great part of the labour expended on the journey.

Although he and his expedition had been the first to see it, Nordenskjöld would go on to name the ice-terrace the Larsen Ice Shelf.

The sledging party quickly recounted the important events of their journey to their mates awaiting them at Snow Hill, who was quite surprised to see the state in which their companions had returned. During the party's absence, Bodman had continued recording all the meteorological measurements himself, and Ekelöf had pursued his bacteriological studies, while Åkerlundh had kept house and performed the cooking chores.

During the return leg of their trip, just before arriving at the wintering station, Nordenskjöld had asked Sobral if he thought that they would see the ship awaiting them upon their arrival at the station, to which Sobral had replied in the negative (Sobral 1904: 212):

> … but even though we both believed this, that did not stop our hearts from pounding at the thought of the possibility of such an event.
>
> To think that we would have news of the homeland and the family, that we would all embark and that, concluding the oceanographic work, we would head north!

But, of course, the ship was not there when they returned. Now all six men raised their eyes to the water in anticipation of the *Antarctic*. They made bets on when the ship would arrive. And they looked to the north.

On the following day, November 5, 1902, far to the north of Snow Hill, the *Antarctic* departed from the Malvinas/Falkland Islands, heading south. The ship stopped in Ushuaia, Tierra del Fuego, on November 17th, to pick up Johan Gunar Andersson, who had made an excursion there two months prior, along with the young sailor Ole Wennersgaard, to study the area and its lakes, specifically Lago Fagnano. By November 22, the *Antarctic* and its full complement and crew were making their way to South Georgia, and then on to Cape Murray, heading toward Snow Hill.

4.10   Fossils and Philosophers

The stormy weather at Snow Hill continued, but it was now not as extreme. More birds began to return, less snow covered the area. And the day was filled with additional light.

Sobral climbed the basalt cliff to survey the water, but could only see a sender belt that was ice-free.

It was now November 21st, and Bodman, Ekelöf, and Jonassen prepared to set off for their own sledge excursion—to Cape Seymour and Cockburn Island, where they would see if the penguins had returned and gather penguin eggs. As the ship had not yet returned, the expedition thought it best to augment their food supply, and so this trip was planned. Everything was ready, and all that now needed to be done was to harness the five dogs. But two of them had gone off on a walk. Two hours later, they reappeared—Sugen and Amager—covered in blood. Sobral surmised that they had been out hunting some little seal who "had surrendered its life to the voracity of these wolves" (Sobral 1904: 214), and he promptly harnessed them to the sledge, and off they all went. In accompanying the sledge party part of the way, up to Seymour Island, Nordenskjöld found two dead seals, covered in blood. Sobral had surmised correctly.

Nordenskjöld's solo jaunt at this time resulted in one of his significant findings. In traversing the southern part of Seymour Island, he discovered "A beautiful collection of fossils, including some uncommonly well-preserved ammonites" which he brought back to the station with him (Nordenskjöld et al. 1905: 243).

Back at the station, six of Fía's eight puppies, born three months prior, died of a disease believed to be caused by the food they ate. The remaining two survivors, two "cute puppies," were utterly "spoiled" (Sobral 1904: 214). Sobral observed that:

> … little by little, and without any disgust, they have been eating their dead brothers and sisters. The poor puppies are real philosophers, true Antarctic dogs. Their time is spent in playing, eating, and sleeping, but one never observes a quiet moment with them.

The sledge party returned from Cape Seymour on November 25, bringing with them fresh provisions back to the station, including penguins, penguin eggs, and cormorants. They informed Sobral and the remaining men that the majority of life had not yet returned to the penguin rookery on Seymour Island. Sobral added to the station's growing larder of fresh meat by killing a cormorant of a different kind from that which was brought back from Cockburn, as well as a skua. He prided himself on these contributions, and perhaps this was his way of assuring himself that he was continuing to aid in the group's survival.

Shortly after, Sobral visited Cockburn Island to photograph penguins, seals, and cormorants. Enthralled by the birds and the wildlife, and curious about the ice and the terrain, he observed how the ice had changed with the warming of the weather and navigated the ice mounds and crevasses that filled the area. This walkabout on Cockburn, shortly on the heels of his bird hunting, was more of a bird watching endeavor. He visited the nest of a megalestris, and observed the behavior of parents with their eggs. (He did take one egg on this visit, as a specimen, but it broke in his pocket when he bent to pick up a stone.) He photographed penguins tending to—and protecting—their own eggs. And he photographed seals in various different positions. It was Sobral's attempt to appreciate the life that abounded in Antarctica, against all odds, and to spend what he thought to be his last month in the frozen region, in harmony with that nature.

The collecting of bird and penguin eggs was a dietary necessity for the men at the time, but a fact that brought some element of unease. Nordenskjöld observed that, after the men repeatedly robbed the adult penguins of their eggs that they had been so carefully nursing, some parents would return to the sea presumably bereft, some would sleep indifferently in their nests, and others would carry broken or rotting eggs back to their nests in a desperate attempt to nurture these empty shells into their expected young chicks (Nordenskjöld et al. 1905).

On December 2, Nordenskjöld made another trip to Seymour Island, ostensibly to collect more bird eggs, but also to revisit the location for fossils. This time, Sobral did not accompany his mentor on a field trip, but stayed at the station and hunted seals. Nordenskjöld left with Jonassen and Åkerlundh. When Nordenskjöld and his companions returned, Sobral reported that they brought with them penguin eggs and penguins to eat. But the real find was the Giant Penguin fossilized bones that Nordenskjöld had discovered in the north-eastern part of the Island on December 3rd. These were the first fossils of vertebrates found in Antarctica and proved the existence of a Giant Penguin that had roamed the area long ago. "A more important discovery of its kind could scarcely be made by such an expedition as ours" Nordenskjöld would later write (Nordenskjöld et al. 1905: 246). In addition, he found further fossilized plants—"large and quite distinct leaves ... belonging to a variety of different forms of exogenous trees, firs and ferns" (Nordenskjöld et al. 1905: 248) that were similar to those found in South America, proving that Antarctica had had a warmer climate, that it had been "a land with luxuriant vegetation and extensive coasts ... where ... many types of animals and plants were first developed" (Nordenskjöld et al. 1905: 252), and that at one time it had been joined with its adjacent continents, including South America, Africa, and Australia. These fossils were of the late Cretaceous period and the early Tertiary, with the tree fossils definitely being Tertiary vegetable fossils. Sobral must have been very happy for his mentor in finding these fossils, and perhaps a bit regretful that he had not been accompanying the great scientist on this great find.

A year had now passed since the men had departed from the port in Buenos Aires. On December 21st, the six men at Snow Hill decided to celebrate this one-year anniversary. While the men partook of the festivities, to Sobral, it was a bittersweet celebration. He reflected on what he viewed as a sad departure for him, compared with other expeditionary trips. Perhaps it was the tug of his home and family that brought on this note of melancholy. Besides the contrast of the environment and culture, Sobral was very much feeling the contrast in temperatures between Buenos Aires and Snow Hill. But again, he immersed himself in his work, making incessant azimuth observations, describing the quality and quantity of fog, and keeping tabs on the ice-locked sea. He was hoping to see spaces of ice-free water begin to appear, but the ice did not seem to be budging. These hoped-for open spaces of water represented the way home.

Christmas came, along with a celebration that included a variety of food in which Sobral gustily indulged. It was a hopeful time. For him, at this Antarctic home, it was a time of "joy and movement" (Sobral 1904: 226). Sobral took this festive opportunity to share with his mates at Snow Hill the two boxes of "exquisite chocolates" from New York that the painter F.W. Stokes had given to him prior to leaving on the ship. It was a gift of friendship from the one friend that Sobral had made on the *Antarctic*. Now he was sharing it with his colleagues, whom he had hoped would also be his friends, here at the wintering station, as the expedition year was ending. Sobral felt an almost childlike satisfaction from the enthusiastic response he received at the unveiling of these gifts, and at the addition of one of the box covers to their pin-up collection. It was a form of acceptance by his companions (Sobral 1904: 227):

When I presented them in front of my friends, they were greeted with applause. One of the boxes had a figure that represented the gold statue of the *North American Girl*, presented at the exhibitions of Buffalo, and this served to increase the collection of beauties, that from the wall were mute witnesses of our joy.

Sobral and his five companions hoped with joy that they would end the year with the ship at their door. But the year ended without the arrival of the ship. For, at this very time, Captain Larsen and the *Antarctic*, blocked by ice, and unable to find a passage to Snow Hill, were initiating Plan B. On December 29, the ship deposited second expedition leader Johan Gunnar Andersson, cartographer Samuel A. Duse, and sailor Toralf Grunden at Hope Bay (Bahía Esperanza) in the hopes that those three men would be able to cross the ice on foot to reach the expeditioners at Snow Hill. Meanwhile, the *Antarctic* would attempt to find a different way through the pack ice from the Erebus and Terror Gulf. "Terror" became the operative word, as neither party would be able to reach Snow Hill, for now.

4.11 New Beginnings

A new year arrived at Snow Hill, and Sobral—an individual accustomed to all the customs and traditions of joyous holiday celebrations with family and friends—bemoaned the lack of fanfare and festivities to commemorate New Year's Day. Here, in Antarctica, it was just one more day in the calendar year.

The disquietude continued as the group experienced a growing unrest over the lack of appearance of the *Antarctic*. Their religious climbs up the basalt cliff to scope out the sea and the ice were rewarding them with more and more views of open ice-free water. But still the ship did not appear. The waiting and the doubting were insidious, and a true torture for Sobral.

In spite of these feelings, he and his mates did enjoy the arrival of warmer days and continued with all their observations. Sobral took special note of the beauty of the day of Epiphany on January 6, 1903. And he took the time to remember his homeland, his family, and his friends.

It was now austral summer, and their dearly missed ship was expected at any time. But as December turned into January, the beloved *Antarctic* did not appear. And as they waited for the arrival, anxiety mounted.

Sobral found himself in low spirits. He abstained from writing in his diary, from recording his experiences. As late January arrived, his worst suspicions took root. He turned his voyage diary over, and began writing from the back toward the front, now documenting his second, unexpected year in Antarctica.

Figures 4.1, 4.2, 4.3, 4.4, 4.5, 4.6, 4.7, 4.8, 4.9, and 4.10.

Fig. 4.1 House on Snow Hill Island, seen here from the front, at the foot of the glacier. (Photo from *Dos Años Entre los Hielos, 1901-1903*, by José María Sobral (1904), page 118)

Fig. 4.2 The six expeditioners at Snow Hill, at the beginning of the first overwintering in 1902. Front, *left* to *right*: Gösta Bodman, Otto Nordenskjöld, and José María Sobral. Back, *left* to *right*: Ole Jonassen, Erik Ekelöf, and Gustaf Åkerlundh. Photographed by Gösta Bodman. (Departamento de Estudios Históricos Navales, Archivo Histórico, Archivo Fotográfico, Ex-0033 d, Expedición Nordenskjöld. 1902. En Snow Hill—Dr Bodman (fotografo)—Nordenskjöld—Sobral. Atrás Jonassen—Ekelöff—Åkerlundh)

Fig. 4.3 Little house on Snow Hill, and its snowy surrounding area, as seen from the north. (Photo from *Dos Años Entre los Hielos, 1901-1903*, by José María Sobral (1904), page 125)

Fig. 4.4 Ole Jonassen at work, accompanied by some of the sled dogs. Photographic image by José María Sobral. (Departamento de Estudios Históricos Navales, Archivo Histórico, Archivo Fotográfico, Ex-0039 v, Expedición Sueca—Dibujos de Sobral)

Fig. 4.5 Going on a sledging expedition, courtesy of the sled dogs. Photographic image by José María Sobral. (Departamento de Estudios Históricos Navales, Archivo Histórico, Archivo Fotográfico, Ex-0040 m, Expedición Sueca—Dibujos de Sobral)

Fig. 4.6 Skiing on the Antarctic ice. Photographic image by José María Sobral. (Departamento de Estudios Históricos Navales, Archivo Histórico, Archivo Fotográfico, Ex-0039 n, Expedición Sueca—Dibujos de Sobral)

Fig. 4.7 A seal hole in the ice, leading to the water below. Photographic image by José María Sobral. (Departamento de Estudios Históricos Navales, Archivo Histórico, Archivo Fotográfico, Ex-0039 f, Expedición Sueca—Dibujos de Sobral)

Fig. 4.8 A sleeping Weddell seal on the Antarctic ice. Photographic image by José María Sobral. (Departamento de Estudios Históricos Navales, Archivo Histórico, Archivo Fotográfico, Ex-0039 r, Expedición Sueca—Dibujos de Sobral)

Fig. 4.9 Portrait of a penguin. Photographic image by José María Sobral. (Departamento de Estudios Históricos Navales, Archivo Histórico, Archivo Fotográfico, Ex-0040 s, Expedición Sueca—Dibujos de Sobral)

Fig. 4.10 A Megalestris (Antarctic skua) resting at the winter station. Photographic image by José María Sobral. (Departamento de Estudios Históricos Navales, Archivo Histórico, Archivo Fotográfico, Ex-0040 t, Expedición Sueca—Dibujos de Sobral)

References

Nordenskjöld O, Andersson JG, Skottsberg C, Larsen CA (1905) Antarctica: or two years amongst the ice of the South Pole. The Macmillan Co, New York

Sobral JM (1903) Conferencia del alférez de navío José M. Sobral [Conference by under-lieutenant José M. Sobral, sponsored by the Naval Center, and read in the Politeama Argentino on December 19, 1903]. In *Boletin del Centro Naval, Tomo XXI*, ed. Capitán de fragata Juan I. Peffabet, 485–512. Buenos Aires: Taller Tipográfico de la Escuela Naval Militar

Sobral JM (1904) *Dos años entre los hielos, 1901-1903 (Two years amidst the ice, 1901-1903).* Buenos Aires: Imprenta de J. Tragant y Cia., Bolivar 319

Stokes FW (1903). An artist in the Antarctic. The century magazine august 521–528. (http://www.
    antarctic-circle.org/stokes.pdf). Accessed 18 Apr 2015
Tahan MR (2016a) Paw prints in human Antarctic history. Scientific Committee on Antarctic
    Research (SCAR). The 34th SCAR biennial meetings and 2016 open science conference in
    Kuala Lumpur, Malaysia. Abstract and poster. Abstract Book, 20–30 August 2016, p 1017.
    ISBN 978-0-948277-32-0. http://www.scar.org/scar_media/documents/outreach/communications/
    SCAR_OSC_2016_Abstracts.pdf
Tahan MR (2016b) Paw prints in human Antarctic history: the use of sled dogs in Antarctic
    exploration. Latin American Antarctic Historians. The XVI Encuentro de Historiadores
    Antarticos Latinoamericanos in Buenos Aires, Argentina. 28–29 October 2016. Paper and
    presentation. http://dna.gob.ar/xvi-encuentro-de-historiadores-antarticos-latinoamericanos

Chapter 5
Expedition Diary: The Second Winter in Antarctica

Abstract The second winter at Snow Hill for the members of the Swedish-Argentine Antarctic Expedition, January 25, 1903, through October 24, 1903, was documented in José María Sobral's Diary. The entries were made during the time of realization that the ship *Antarctic* may not return to retrieve the expedition, the desperate attempt to stockpile enough food to survive the winter, and the numerous sledging journeys to collect fossils and conduct scientific studies.

The following daily journal entries are from the expedition diary of José María Sobral of the Swedish-Argentine Antarctic Expedition, written during his second winter at Snow Hill in the Antarctic Peninsula, from January 25, 1903, through October 24, 1903. These diary entries were made during the time of realization that the ship Antarctic may not return to retrieve the expedition members, the desperate attempt to stockpile enough food to survive the winter, and the numerous sledging journeys to collect fossils and conduct scientific studies.

Note: Sobral turned his voyage diary over and began writing the second year's entries from the back of the journal forward (Sobral 1901–1903, diary pages 376–289). In this way, he documented his second, unexpected winter in Antarctica.

5.1 This Uncertainty about Whether or Not One Will Have to Stay Here for One More Year is Unpleasant

5.1.1 *January 25, 1903*

Since the 12th of this month, I have not written in my diary, such is the state of spirit that I am in. This uncertainty about whether or not one will have to stay here for one more year is unpleasant, very unpleasant. Here, regarding rationing, we have started to pay attention and to diminish the quantities of some things. We have very little sugar, and in order for it to be enough, we have to consume no more than 7 "squares" per day. Each small square weighs 5–1/4–grams. Clearly, this isn't a

M.R. Tahan, *The Life of José María Sobral*, Springer Biographies, https://doi.org/10.1007/978-3-319-67268-7_5

privation, since one can gradually become accustomed to having coffee and all other things without it. Right now, I don't take more than 21 grams of sugar per day —5–1/4 in the morning coffee, 5–1/4 in the afternoon coffee, and 10–1/2 in the cocoa at night. And in a few days, I expect to be able to have black coffee without displeasure.

Another thing that we are scarce on is fuel, the coal that we have won't last more than 4 months. Some say the kerosene that we have, together with the coal, is enough for cooking, for illuminating the house, and for heating it up a little. All the seals that we find are killed in order to store their fat that we will use for burning. All the birds that are found also are hunted. And we're planning on killing over 500 penguins for the winter. And it's also thought we should wait for the ice to break in order to make a trip by boat, because they [the seals and penguins] can't be brought by sledge, since it's not possible to reach the rocky parts of Cape Seymour from anywhere ________ [unreadable] a pretty long stretch through land, since the rocky part is to the E, where the open free water is. But I think the best thing to do is to go now, without waiting any longer, and kill whatever is needed, and then go get it on a sledge when the ice allows us to. During all these past days, which have been the warmest (from the 20th to the 25th, the average was +1 °C, every day the temperature being above 0), the ice seems to have weakened quite a lot, breaking in some parts and augmenting the old clearings. Nevertheless, the water being free of ice isn't significant. If this heat continued for another 15 days, and some storm blew, at a speed of over 20 m, surely the ice wouldn't resist, but this heat doesn't look like it's going to last, and the wind blowing is quite weak, not exceeding 15 m per second. For one more year, I don't have any shoes, because I don't have but a pair of worn-out boots and a pair of short boots that are breaking. Today, I pulled the fat off a seal skin and I've staked it by hammering it to the house that used to be an instruments shed and is now used for storing what we hunted and the meats. I'm planning to make ankle boots with it. But before winter comes, I'm going to prepare 2 or 3 more skins in order to have enough quantity.

Unbeknownst to Sobral, but most likely suspected by him, is the fact that the ship Antarctic *is at this time trapped in the ice off Paulet Island. Encountering heavier pack ice than is normal for this time of year, which had impeded the ship's progress, and having had difficulty returning to Snow Hill, Captain Carl Anton Larsen had dropped off three expeditioners—Johan Gunnar Andersson, Samuel A. Duse, and Toralf Grunden—at Hope Bay the previous month, on December 29. Unable to return to pick them up, the ship had become frozen into the sea at the beginning of January. Its side had cracked on January 10, and water had begun to enter the ship on January 21. The* Antarctic *crew at this very moment is still diligently striving to save the ship. With no sight of the* Antarctic, *the party awaiting the ship's return at Snow Hill is becoming anxious. Sobral is coming to terms with the reality that he and his companions may have to spend another winter at Snow Hill. The litany of scarcities is alarming—there is not enough food to sustain them, and so the men devise a plan to kill the surrounding wildlife in order to survive the winter.*

5.1.2   January 26, 1903

The sky has been completely overcast with stratus clouds, around 80–100 m high. These clouds came to break on the mountain side, taking the most precious shapes and giving a beautiful aspect to the land. Wind WSW 4–5 m power. And at the last minute, it moved to the NE and it became variable and weaker. At 8 pm, some snow fell. Åkerlundh and Jonassen went to get a seal that was seen from here, killing it and bringing it on the sledge. It was a pretty big Weddell seal. The small boat was put into the water at high tide. Bodman and Ekelöf made an excursion through land into the strait that separates this land from Seymour Island, picking up a great amount of seaweed. They also found a lot of dead crabs (red crustaceans, very similar to the crab) on the land close to the water. No doubt they were left there by the tide, and hundreds of Sterna [terns] were eating them. These crustaceans seem to be the food for seals, penguins, and other birds. Some of the penguins I found had their stomachs filled with them. These crabs swim between the surface and some centimeters close to the surface; next to them I saw another type of fish, approximately 5 cm long, and very similar to the bream; one of them fell from the beak of a cormorant that we killed. No doubt, the poor animal was taking the fish to feed its fledglings. I think these fish must serve as food for the Weddell seal, while the aforementioned red crustaceans must be food eaten by the Lobodon [crabeater seal]. I believe these crustaceans are the Eufasia [Euphausia]. In the past days, we've killed some seals. During the month of January, since the new winter started, 11 seals—2 crabeaters and 9 Weddell—have died. The latter seem to be the more common in this region. The average temperature of the 26th is +0.2. These last days, after the ones in September, are the hottest that we've had, but in all the year, there has not been one day in which the temperature hasn't gone below zero. The barometer from January 19 to January 23 has remained between 750 and 744, but since the 24th it has started to decrease, as of 12 am; on the 25th at 1 pm, it was 728; the 26th at 7 am bar = 732.6, at 2 pm = 734, and at 9 pm 735.

Sobral is learning the biology and zoology of the wildlife as he kills it in order to survive. But, also telling about his character, he cannot help but empathize with the bird who had been taking fish to its young. Meanwhile, he is meticulous in his meteorological observations and record-keeping.

5.2   We Waste a Lot of Penguin Meat

5.2.1   January 27, 1903

Average temperature today is −1.2. It hasn't been above zero. The barometer started to decrease. At 7 am bar = 734.5, at 2 pm bar = 732, at 9 pm = 729.7. It started to snow at 2 pm. It stopped at 8 to start again at 9:30. The wind was SE and

ESE from 1 am until 10 am, E from 10 until 7 pm, 8 and 9 variable, 9:30 pm SSW and snow.

Today, I've worked a little bit on pulling some fat from the seal skin. It still has a little and it's very difficult to take it off with the bad knife I have. Yesterday, I pulled the skin from the penguin, and today, in the morning, I found one of the dogs eating it. We waste a lot of penguin meat; we take only the breast, and the rest is thrown away. And not only with the penguins, but the same is done with the rest of the birds.

The other day, a very light Megalestris [skua] was killed. The head and the stomach were of a whitish color. Each wing was 64 cm long and then 58 cm from the tip of the tail up to the beak. The Megalestris, as well as the other birds in this region, vary in color and size. I don't think that it's right to consider another type in the ________ [unreadable] because we have killed both small and large ones of light color.

Sobral was adamant about killing only what was needed to eat and criticized the wasting of more animals' lives by not eating every part of the animals that the men did kill.

5.2.2   *January 28, 1903*

All day the wind has been SW, and it has been snowing at intervals; at 9 pm it changed to WSW and it started to increase, reaching 12 m. Today's average temperature has been −1.25°. The barometer is increasing very slowly. At 7 am 729, at 2 pm 29.6, and 9 pm 30.0. The boat that had made it to the top of the basalt with a high tide yesterday was pulled today to free open water in the strait that separates this island from Seymour. Five dogs pulled it easily in spite of how bad the snow was. I think it would have been better or faster with sledges. After putting it into the water, it was taken up to a little cove in the ENE part of the island and was beached. The space [of water] with no ice has increased more and more and is inserting itself into this other strait like a tongue. If it continued increasing in this same progression for several more days, and a channel was formed that goes through the strait, the ice would crack with the first strong wind.

5.2.3   *January 29, 1903*

WSW wind blowing all day with a strength of 12 m per second, middle term, going up to a maximum of 13 m at 11 am and the minimum of 10.5 at 5 pm. Snowing. Average temperature = −1.6°, remaining below −1° all day. Bar 730.4 at 7 am, bar 2 pm = 731.5, bar 9 pm = 733.3. At night, it stopped snowing, and it cleared up in such a way that you could see Cockburn perfectly. On these stormy days, everyone is inside. One goes outside every once in a while to look at the thermometer, and after jogging for a while in order to do some exercise, inside again. The poor dogs, curled up leeward of the house, sleep all day, getting up every once in a while to shake off the snow that covers them. Someone shouts "Megalestris!" and at this shout, even if it's actually another bird, someone grabs the gun and, after a while, comes back with an increment of our daily portion. The increasing distance of the sun is becoming noticeable. At 10 pm, you can't read inside, and the lamp has to be lit.

A year ago on this same day, we also had bad weather, but it was from NE and the temperature at 2 pm was exactly the same as today, $-1°$.

I forgot to say that in these past days, 3 big flies have been found—obviously this is a big event. I found the first one; I was reading, and it landed on my book. The second appeared in the kitchen. And the third one was seen by Bodman as it was flying above our heads while we were having lunch. These little animals remind us of a milder climate, and they are around the whole world; how happy we are to see them here, and how disgusted when we see them there! Now in Buenos Aires, for example, they must be full of them, they are also [surrounded] with *coup de chaleur* [heatstroke], while here we have a nice temperature, it's true. But I would like to go back to that uncomfortable heat again, because there I have the world, I have life. I think I see Avenida de Mayo with its cafes, full of little tables, and people sitting in the heat to drink their nice ice-cold drinks. But the world is like this. We always want what we don't have. I'm sure that those on Avenida de Mayo would wish for a little bit of air at $0°$ now, in the month of January. But this would be the pleasure of only a few minutes. They would get tired of that, and then, they would want warmer air. A year ago, I was on board the "Antarctic," in the middle of the Antarctic ice. Two years ago, I was in the cape of San Antonio on a meteorological and hydrographic commission. Three years ago, I was in Greece and 4 years ago on the "Sarmiento" sailing between Buenos Aires and Golfo Nuevo [New Gulf].

The colors of the three crabeater seals that we have killed are like [Emil] Racovitza describes, that is, a yellowish white on the surface of the stomach and on the chest, and, on the back, a girdle of a greenish brown color forms, and between these two colors, like a dividing line starting from the base of the head, and continuing to the sides and up to the back fins, there is a girdle formed of great spots somewhat darker than the general color of the back.

The wind started to ease at 9; at 10 pm, it was only 8 m per second and at 11 pm 6.5 m per sec. One can still see outside at 12 pm.

January is austral summer in Argentina, and very hot in Buenos Aires. Avenida de Mayo is the major avenue filled with restaurants and stores and fashionable sites.

This flight of fancy by Sobral is telling. He is longing for the warmth and sense of belonging of Buenos Aires. But he is wise enough to realize that "the grass is always greener on the other side."

5.2.4   *January 30, [1903]*

All day blowing WSW, snowing at intervals. The sun has shone quite a lot. The wind has kept a speed between 6 and 8 m per second. Some hiked the hills, bringing news that in many parts there were areas with free open water, and this was to the north, near Sidney Herbert Bay; to the south and toward the clearing behind Cockburn, it is much bigger. This is encouraging, and it increases our hopes. This morning, a seal appeared about 200 m from the house close to where there is located the hole for the observations of the tide. It was killed immediately. Today, magnetic observations are made. Bodman, Nordenskjöld, and I, 8 h each.

Today, the average temperature has been = $-1.0°$. Bar 7 am = 735.7, at 2 pm bar = 737.3, at 9 pm bar = 738.9. Today, colored clouds have been seen (Ci-Cu) [cirrus–cumulus].

5.2.5   *January 31, [1903]*

At 11 am, after *frukost*, I took my photographic apparatus, my knife, and the gun, and I went toward the free water that was in the strait that separates this island from Seymour. On my way there, I took a picture of the beautiful gullies on this island near the strait. I had planned to go to the iceberg afterward, but after walking for a little while, and after having an "interview" with some penguins who had just come out of the water, I saw 12 ossifragas [Giant Petrels] that were sleeping in the warmth of the sun. As I came closer, some of them were ready to leave, but, since it seemed that they had just had a feast over the corpse of a dead seal, they all started vomiting, undoubtedly in order to lighten their weight, and then, they began to run around 100 m, some on the ice and some on the water, attempting finally to resume their flight with their heavy weight. I was about 30 or 40 steps away, I fired twice at the most laggard one, and I wounded her twice. As I got closer, she dragged herself painfully toward the edge of the ice, and since two of the dogs ran over to her, she jumped onto a piece of ice that was floating nearby, and when I went to hit her with a stick, she jumped into the water. Having lost my prey, I was about to go back, when I saw some seals at a distance of 100 m. They were five crabeater seals and one Weddell seal. They were all together. And a little further, there was one, which I think was a Leopard. After seeing these seals and having nothing else to do, since I had run out of photographic plates, I went back with the intention of returning after lunch to kill the seals. So I did. At 3 pm, after eating lunch, I set off. As I got near the water, I found four penguins. And since they wouldn't let me get close enough to kill them by hitting them with a stick, and they had the intention of jumping into the water, I shot at them with my carbine, I killed three and the other one jumped into the water. Then, I went back to where the seals had been, not finding more than four: two crabeater seals and two Weddell seals. I killed the four of them. I cut off the head of one crabeater and of the Weddell, with the idea of keeping them. The crabeater seal whose head I had cut off was white and did not have any other patches of lighter color on its back. The color was between yellow and white; that is, it was of a creamy white color. The other was whitish, too, in the stomach, and the girdle on its back was olive color. These crabeater seals are close to the Weddell, but that doesn't mean that they eat the same things. The difference is the food—crabeaters eat dead crabs, while the Weddell eats fish.

Average temperature of the day = $-0.1°$, bar 7 am = 740.5, bar 2 pm = 741.4, 9 pm = 742.2.

Some also say that the crabeater is sometimes mistaken for the Weddell, but I think that's impossible. The crabeater seal cannot be confused with any other kind. First, its teeth are different from all the other mammals. Second, its head is much longer than the Weddell's and its frontal bone being much lower. Even without seeing a seal, and having no idea what they look like, it would still be impossible to confuse the crabeater with the Weddell as being the same type of seal. As for their cry [the sound they make] that Mr. Racovitza finds special in the ones from this Ross Island, I think that all seals have it [make this sound]. I have heard the crabeater's and the Weddell's roar, so to speak (it is very similar to the roar of a cow when it makes it a short roar) in a way that is analogous to that of the Ross seals described by Mr. Racovitza. When I was about to cut the head of the Weddell, a Megalestris came 3 m away from me, and I was going to shoot at it, but failed (that is, the gun failed to fire). I placed the two heads in a sack, as well as the Megalestris, and I took the road again to go back to the house. The weight was pretty heavy, and I had hurt my toes with the stitching of the boots. So these all made the return pretty painful. At 9 pm, after more than 5 h, I was finally back in the house. I have the intention of taking the skin of the white seal, but I don't know where I'm going to keep it until the ship comes, and where to dry it. If I nail it to the wall, the dogs are going to find it and destroy it. I thought of putting it at the top of the basalt and nailing it somewhere where the dogs can't find it. But if these don't come, the birds are going to find it, and they will destroy it. I think the bullet has spoiled the head of the crabeater a little bit. When I finished with the seals, I started to observe the movement of the tide. The tide was rising, and the direction of the current was, more or less, W. It must have a speed of no less than 4 miles per hour. It was an interesting spectacle to see the large and small floes being dragged by the current, some with penguins on them and others with seals that were sleeping calmly. Undoubtedly, these currents must be a major factor in the disintegration of the ice in the summer. Near the edge of the ice field, you can see large spaces of a shady color, but if one tries to go through them, the ice may resist, as happened in my case, or it may not, but if you apply a little force with a stick, the ice gives way, and one can measure it to the water. I cannot completely explain the movement and the transformation of the ice in this state, but I think it's due to the action of the water from above and the absorption [through its pores] of the sea water from below through capillary action; maybe an excess of salinity in the ice, with the mixture of ice particles, is what facilitates the absorption and at the same time keeps the ice in its state of hardness for a certain amount of time.

No, the ice forms because of the action of the sea water in a state of thickness in the adjacent layers, similar to the state of the early ice in Autumn, but this thick mass has the characteristics of sea ice, but in a state of more rigidity, due to a series of upper layers of freshwater and freshwater ice resulting from the melting of the snow. Therefore, a man who tries to walk on it doesn't experience the same sensation as walking on Autumn ice. On it [Autumn ice], one suffers the sensation of walking on something very flexible. I don't know what to compare it to; with every step you take, the ice moves, forming a kind of hole that disappears once you lift your foot, while on the Summer ice, one's steps sink into layers of snow ice and water down to the thick ice underneath, which is the same ice as the one in Autumn, but apparently doesn't seem to have the elasticity due to the upper layers of snow, water, and ice.

The wide range of topics undertaken by Sobral in this entry is impressive. He is photographing, hunting, observing wildlife, making meteorological observations, correcting noted zoologists about their study of seals, studying and analyzing the different types of ice, and strategizing how to outsmart the dogs over a piece of seal skin. He also reveals that he had cobbled together new shoes, recalling the concern he had expressed earlier about having only one pair of boots that would not last the winter, but these new shoes now hurt (recall his foot X-Ray in Buenos Aires).

5.2.6 *February 1, [1903]*

In the morning, wind blew from the NE quite intensely. At midday, it began to snow, and the wind moved to the E. SSW started in the afternoon after some time of calm, and it came with snow. Around midday, the temperature was above 0, but it went down with SSW. I'm afraid that my white seal may have been taken by the devil with this wind.

To diminish the boredom, cards are now being played. Patience games and _______ [unreadable] are the most used. The average temperature of the day is −0.9°, bar = 739.5, 2 pm = 737.4, 9 pm = 735.4.

5.2.7 *February 2, [1903]*

WSW blowing strongly, generally between 8 and 11 m per second, but it drops down to 5 m at 6 pm, and then it increases again. Overcast sky all day, with some snowfall. This afternoon, the three Greenlanders and Cain caught Curri, and the poor thing was bitten in several areas [of his body]. Temp = −1.4, bar 7 = 35.3, 2 pm = 36.6, 9 pm = 38.2.

5.2.8   *February 3, [1903]*

The sky was quite clear in the morning, but at 2 pm it became cloudy again; soft hail fell (just a little). Wind from SW, with an approximate strength of 6 m. The wind began to ease at 6 pm; at this time, it was 6 m strong, and it was only 5 m at midnight. The average temperature of the day was $-2.6°$; bar 7 am = 739.1, 2 pm = 739.3, 9 pm = 741.1. It can be said that the temperature curve of January is typical of summer, contrasting the regularity of its variations with that of winter. Thus, in this month, we can see how the thermometer punctually marks its maximum after midday, around 2 pm, and its minimum after midnight. Variations are generally quite small, 3 or 4 degrees, except for two days, 21st and 22nd, when they were 8 and 9 degrees, respectively. The minimum of the month was $-5.5°$ on the 15th at 10 pm, and the maximum was on the 22nd $+5.5°$. The temperature has not reached above zero on the following days: 29, 27, 18, 17, 16, 15, 11, 10, 9, 8, 6, 3, 1. On the 29th, WSW 11 and 12 m. On the 27th, ESE at 3 m.

On this same day, the ship Antarctic *is beginning its slow death. As the ship attempts to break free from the ice, turning over on its side, it also begins to take on water. The crew begins to lower the boats onto the surrounding ice, while still trying to take the ship to land. Meanwhile, the three expeditioners at Hope Bay are attempting to make a journey to Snow Hill over the ice, and spend the first two weeks of January sledging, but are stopped by open water they cannot cross. (Nordenskjöld et al. 1905).*

5.2.9   *February 4, [1903]*

The SSW that was blowing last night moved to the south at 1 am, remaining in this direction until 4 am; at 5 am, it was blowing very softly from the E; it was calm at 6; NE at 7; and WWE at midday. Overcast sky in the morning, clearing up at 5 pm to become completely overcast with Ci and Ci-S that came from the NW at 7 pm. At 9, the sky became overcast with A-S from the ENE. Wind speed has been between 4 and 5 according to the anemometer. Average temperature was = -0.8, barometer at 7 am = 742.5, 2 pm = 742.4, 9 pm = 742.4.

Today, Bodman went to the "water" of the little strait and killed four seals, some penguins, and one ossifraga. Of the penguins (3) that I killed 4 days ago and left on the ice, I can't find anything but their legs: The ossifragas had eaten them. I found a group of around 20 of them on the seals that I had killed. At 3, Bodman went back with seven [dogs], taking the sledge and bringing back four seal skins. Some crabeater seals had a darker girdle [of color] on their backs, from the head down to the tail. Others also had a girdle of large spots on each side of their body that ended at their back fins. The hide of the one that I killed has neither one thing nor the other. You could see only some reflections that no doubt are produced by the hair which in some places went another direction. It reminded me of a piece of velvet.

I think that the crabeater seal is an animal that, in its proportions, is an intermediate between the Weddell and the _______ [unreadable] and, not as the Antarctic manual says, between the Weddell and the Ross.

5.2.10 February 5, [1903]

All day, the wind has blown between N and NNE. At 6, it moved to the N for some moments, then to the NW, and then calm, and at 6:15, it suddenly blew from the WSW, and the temperature dropped rapidly. From the last SW to this one, the wind has made a full turn clockwise, establishing itself only in the NE and SW, and stabilizing for some time from the ESE and SE. Overcast sky all day, fog coming with fog SW at 6 pm. The NE reached its maximum of 10 m at 8 am; it remained between 9 and 10 m until 4 pm. SW blew up till an approximate speed of 6 m. At 12, when the wind was blowing NE, Bodman measured 13 m. The average temperature of the day is +0.35; it has been a very warm day. Temperature has been above 0 from 7 am to 6 pm. Bar 7 am = 740.6, 2 pm = 740.4, 9 pm = 741.1.

Nordenskjöld is about to make a trip to Cape Seymour with J and E [Jonassen and Ekelöf]. So today they have put everything in the boat that is at the great gorge, but they didn't set off until late because of the wind. Today, I have opened the skulls of the three seals, one crabeater, and two Weddells.

5.2.11 February 6, [1903]

After the *frukost*, N, J, and E [Nordenskjöld, Jonassen, and Ekelöf] set off to Seymour Island to get penguins for the new winter. B and Å [Bodman and Åkerlundh] went to accompany them up to where the boat was docked. While the boat was being put into the water, the dogs disappeared, returning afterward, the Greenlanders full of mud, and stains of blood; because Curry was missing, it was assumed that they had killed him.

After having lunch, B and Å [Bodman and Åkerlundh] went to the open [ice-free] water of the strait to butcher the seals that I had killed last time, but they only found one, and they could only pull out two pieces of hide—because it had somewhat sunk into the ice yesterday, freezing afterward. B [Bodman] hunted a white ossifraga that had only some black and brown spots on its body. But seeing the birds fly, one got the impression that it was completely white. He also brought six Megalestris and six penguins. It's nighttime now, and poor Curri hasn't come back yet; most likely the other dogs have killed him. It has been planned to go search for him in the morning, in order to skin him. Because of the negligence with which things are treated, three of the seals that I had killed have been lost. If, on the same day that I had killed them, their skins would have been pulled off, and they would have been picked up with a sledge, then this wouldn't have happened. In the morning, weak variable wind and calm. It started to blow NNE, moving to the ESW at 5 and varying between this direction and NE. At 9 pm, it again started blowing from the NE with growing strength. At 12 pm, the speed from WE was 9 m and temperature was +1°. Overcast sky all day. Average temperature = −0.9°, bar at 7 am = 743.3, 2 pm = 743.3, and 9 pm 738.5. The barometer drops abruptly. The temperature is high; everything indicates that a southeast blast is coming.

5.2.12   February 7, [1903]

Average temperature of the day = −2.0°, bar 7 am 729.2, 2 pm = 730.3, and 9 pm 735.5. At 2 am, NE blowing at a speed of 11 m, but since then it began to decline to finally calm at 5. At 5:30, SW started to blow strongly. The thermometer, which was marking a temperature above zero, dropped abruptly. On the paper of the data recorder, an almost vertical line could be seen, which indicated the path of the SW. At the same time, the barometer started to go up, its minimum being at around 729 m. The wind was of 7 m at 7 am, 14 m at 8 am, and continuing like that until 12 at night, when it started to escalate. At 8 pm, it started to snow.

In the morning, when I was going to make the observations, I saw a black spot on the ice at around two miles of distance. No doubt it was a seal. After *frukost*, I took a sledge and four dogs, and I went with Åkerlundh to get the seal. It seemed like it was a Ross seal. Short, wide, with almost no neck. Small fins, but its teeth were the same as the ones of the Weddell type. Just in case, I decided to keep the head. With one shot to the dorsal pine, I finished it off. We laid it on the sledge and we took it to the station. The stomach was opened. No fish were found, but these white and soft animals of this shape [draws long cone shape on the page], and others of a reddish yellowish color with a worm shape [draws worm shape]. The seal was female; I have to look at its skin—I haven't examined it yet. In the afternoon, Bodman saw another seal and we went to get it. It was an enormous Weddell of 2 m and 80 cm. After expending great effort to kill it, we laid it on the sledge. But after walking a little ways, we decided to put her on the ice and pull her. It had 8 cm [layer] of fat, and it was a female. A lot of scars could be seen on its body, some were up to 23 cm long.

5.2.13   February 8, [1903]

The average temperature of the day is the same = −1.7, bar = 746.7–744.1–748.9. WSW blowing all day. Wind at 1 am was 19.5. At this hour, the anemometer broke, but Bodman measured 22 m 8 at 3 am, diminishing from this hour; it was 18.5 at 8 am and it kept going down until 4 pm = 15 m, 5 pm = 12 m, 9 pm = 7 m, 11 pm = 2.5 m, and it was completely calm at midnight. The sky was overcast until 7 pm, and it completely cleared up at midnight, becoming a sky full of stars. The first stars I had seen were on February 5. Today, Bodman was on the basalt top, and he saw a lot of water. He says that to the east, the area of ice-free water has a width of around 20 km.

One of the basalt hills they had spied from the ship now serves as their observatory, being the highest point of observation at Snow Hill, just behind their winter station house. This is where Sobral and his companions religiously go to scope out the water, determine the situation of the sea ice, and fervently hope to sight the ship coming back to pick them up.

5.2.14   February 9, [1903]

Average temperature of the day = −0.8, 748.3, 743.4, 739.1. Calm lasted until around 3 or 4 am, when NE wind started to blow. At 5 am, it began to blow at a speed of 3 m, and it remained between 10 and 4 m during the whole day. It's been raining since 6 pm. NE wind eased at 12:45 at night. Today, the same amount of [ice-free] water as yesterday has been seen from the basalt top.

5.2.15   February 10, [1903]

Average temperature = −2.2°, 733.6–731.7–732.3. At 5 am, after some minutes of calm, WSW suddenly started to blow at a speed of 7.5 m. I was outside at 4:50 am and a soft breeze from NE and ________ [unreadable] the noise of the WSW. The intensity of the wind continued increasing until 1 am, when it reached a maximum of 13 m.

Today, there has been very variable cloudiness and, something very rare, I got the chance to observe the direction of upper and lower clouds.

In the high regions of the atmosphere, the sky was covered with beautiful cirrus that had a N direction. Then, I could see clouds that I would call *fracto-cirrus-*________ [unreadable]. They are much lower than the cirrus and had a S direction. They were not A-S, since they had a very filamentous shape; they didn't form banks but passed in large ________ [unreadable]. The lowest F-S had the direction of the wind, which was blowing in SW direction. Today, Bodman was up on the basalt and noticed more [ice-free] water, and so he enthusiastically expects that the ones who set off on the boat to Seymour Island will come back with news about the ship any time now.

5.2.16   *February 11, [1903]*

Average temperature = −2.7, 732.0–731.4 − ________ [blank]. The wind that was WSW last night has been decreasing its intensity and has moved to the S. At 5 am, it was SSW with 4 m; at 5, it moved to SWE and started to blow very weakly from all sides. At 7 am and 8 am, it was calm, snowing. At 9, SSW started to blow again, and it continued until 3 pm, when it began to increase in intensity. At that hour, it started to blow from the WSW. At 9 pm, it had an intensity of 9 m and at 10 pm 10 m. It has snowed all day, but intermittently. The full moon appears every now and then behind a flake of Ni [Nimbus cloud], illuminating with yellowish light and ________ [unreadable]. Today, Bodman climbed to the basalt again, and the ice is in the same condition. Here in the strait, it's a compact field of ice and it doesn't look like it is going to break. Bodman killed a seal approximately one mile away from the station. Cain has disappeared; we haven't seen him since yesterday. I think the Greenlanders have killed him.

As Bodman searches for the ship from the top of the basalt overlook, back in Hope Bay, J.G. Andersson and his companions return from their sledging trip to their drop-off point, and, seeing no sign of the ship (despite the fact that Larsen had assured them he would return), begin to build their stone hut in which they would reside for the winter.

5.2.17   *February 12, [1903]*

WSW blowing all day at a speed between 9.5 and 14.0 m. Overcast day, almost all day; however, something shone, leaving an impression on the data recorder. The barometer is slowly going down, and an increase of wind speed is expected. Bar = 741.1–740.7–729.5. Average temperature = −4.3. Today, I have been looking for Cain, but with no good results. No doubt the poor thing has been killed by the Greenlanders. But I think the disappearance of this dog is curious.

On this fateful day, the day of Cain's disappearance, the ship Antarctic, *frozen into the ice 25 miles from Paulet Island, sinks. Despite the crew's efforts to beach the polar vessel, the ice is too much for it, and crushes it, leaving a gaping hole that foretells its doom. All the crew and scientists—20 men including Carl Skottsberg— as well as the ship's trusty cats, who, Skottsberg writes, "are carried down in a state of terror"—leap to safety onto the drifting ice below, taking as much of their equipment and belongings with them as possible (Nordenskjöld et al. 1905: 532). Most of their scientific findings go down with the ship, a fact that Skottsberg deeply laments. Sobral would later be greatly affected by Carl Skottsberg's retelling of the sinking of the* Antarctic—*including it as a final chapter in Sobral's own book* Dos Años Entre los Hielos, 1901–1903 (Sobral 1904).

5.2.18 February 13, [1903]

Average temperature = 4.2. Bar = 727–725.4–724.9 with a tendency to decrease. Today, the wind has blown fiercely, reaching a speed of 21 m per second, which was the maximum of the day, but in general, it was between 15 and 19 m all day. We haven't had such a strong storm since November.

Temperature all day has been −4 and −5. On February 9 of last year, we had a storm from SW in front of Cockburn that got to 21 m. It looks like these strong winds start in February and end in November. Today, the tide was very high, even reaching up to my trunk and covering the coal and the seal hides. A year ago, I wrote the last letters to my people, because the ship was about to leave; that night I did not sleep.

5.3 I Think That We Are Not Going to Escape Another Winter

5.3.1 February 14, [1903]

Average temperature = 3.1°. Bar 7 am = 724; 725.4; 728. Today, the barometer has gone up a bit. Between ____ [blank] and 7 am, the wind has been between 17 and 19 m WSW per second. At 8 am, it only had 14 m of speed and it continued diminishing slowly until 7 pm, when it was only 10.5 m, but then it started to escalate again. Today, it has been a year that we have been on land in this place. A year ago, we started working on our house with the hope that, at this time, we would be leaving it. But that evil ice, or, better said, the cold, doesn't want it to be that way. Bodman went up the basalt again and says that, toward Paulet [Island], there's a great canal, where the ice cannot be seen, and then, to the south and the east, the water is free of ice and is now wider than in previous times. But the clearings of Sydney Herbert Bay have closed. He [Bodman] is really hopeful that the ship will come, but I think that we are not going to escape another winter. Today, the food was very rich in variety, and abundant, to celebrate the year; the gramophone also worked at night, allowing us to hear the best of its repertoire.

Despite his friend Bodman's optimism, Sobral assesses the situation correctly. He reads the ice well. His words are prescient—he and his colleagues will be forced to spend another winter in Antarctica.

5.4 If They Would Have Taken All the Meat that the Penguins Have, They Would Have Avoided Killing So Many

5.4.1 February 15, [1903]

Average temperature = −2.2. Bar = 730.5–732.2–734.7. Today, at 3 pm, those who had gone on the trip to Cape Seymour came back in the boat. They have killed 400 penguins, opening them and not bringing back more than the breasts (wasting the legs, the back, etc.). If they would have taken all the meat that the penguins have, they would have avoided killing so many; half of that [the number killed] would have been enough. They have brought some zoological samples that I think are very interesting: sea spiders and some type of crab of a brownish color.

The wind has been between SSW and WSW, with overcast sky. It snowed a little.

As for the state of the ice, those from Cape Seymour bring news that could be said is no news, because it's more or less what Bodman saw two days ago, which is that a great canal now extends from the east side of Snow Hill to as far as the eye can see toward Paulet.

Sobral greatly respects Nordenskjöld but does not stop from criticizing him when he feels that criticism is due. He condemns the carelessness of wasting so many penguin lives; yet he also marvels at the interesting zoological samples that his mentor has just brought back with him.

For his part, Nordenskjöld laments the need to kill penguins, describing in his book their innocent lack of suspicion, and, once the men's intent was known, their bravery—how some of the adults and sentinels valiantly attempted to defend their flock or stand their ground; he writes (Nordenskjöld et al. 1905: 264):

Poor things! They little knew our evil intentions.

... It would be difficult to imagine a more disgusting task ....

It was only bitter need which could compel us to this horrible slaughter; nothing else could have prevailed upon me to take part in it .... Much as I had longed for the Antarctic during the course of this summer, I never did so more than during these few days ...

5.4.2 February 16, [1903]

Average temperature = −3.4. Bar 7 am = 747.5; 748.4–750.3. I don't know whether the wind has been variable or from the SE, and due to the mountains, it seems to come from all directions of the _______ [unreadable], continuing like that until 8 pm, when it establishes from the SSW. Until 7 pm, the maximum speed was 3 m; at 8, it was 5 m; and at 12, it was 8 m. Overcast sky all day. There was soft hail, and at 11, snow, and it continued snowing for the rest of the day. Two trips were made on the sledge to the ice-free water that there is in the little strait, bringing 300 penguins that had been left there yesterday. Some [of the penguins] were put on salt in barrels, others are in the dispensary. Cain was found dead on the other side of the magnetic instruments house.

5.4.3   February 17, [1903]

Wind between the SSW and W all day. Between 1 am and 9 am, it blew at a speed of 9 and 10 m. It maintained between 9 and 11 m until 8 pm, when it definitely started to increase, blowing at a speed of 13 m at 12. Overcast sky all day and snowing (it didn't snow between 8 am and 3 pm). Average temperature of the day = −4.7. Bar 7 am = 739.1–738.7–740.1. Today, Bodman went up to the top of the basalt again and says he saw a lot of ice; that is, that the water that had been ice-free there is now covered with floes, but toward Paulet, he could not see well, because there was a lot of fog in that direction. According to Nordenskjöld, the penguins at Cape Seymour (the chicks) already have almost as much fur as the adults, and the latter have already abandoned their chicks that are able to go into the water this season. They have seen some sick Megalestris. Probably, they are hurt by ________ [unreadable] here in the station.

The temperature started to go down at midday, having escalated to −2°. At 6 pm, it was below −5°.

5.4.4   February 18, [1903]

Average temperature of the day is −9.4°. Bar 7 am = 741.5–743.5–746. Wind from SSW has blown all day at a speed between 14 and 15 per second, bringing a lot of snow. We haven't had a storm with such cold since November.

At this same time last year, the thermometer was at around the same height, also with a storm from the same direction, which is the one that had torn down the magnetic instruments house.

In the morning, inside, the thermometer drops down to +3.0°, and this is not felt in the body but feet get cold quickly when one sits down for a moment.

5.5 What a Splendid Night

5.5.1 February 19, [1903]

At 2 am, the wind from WSW moved to the SW, blowing in that direction until 9, when it moved to the SSW and blew in that direction for the remainder of the day. Overcast sky until 11 am; it started to clear up at 1 pm, and then Cockburn, the land of Mt. Haddington, and Seymour Island could be seen. At 6 pm, only the north half of the sky was covered with an extended bank of clouds which displayed all the transitions between the S-Cu and the A-Cu. At 10 pm, only a small portion of the sky to the NE is overcast with a bank of Ci-S; the stars of first and second magnitude shine splendiferously, being the masters of this region of the cold for more than 4 hours. The Cross [Southern Cross, or Crux] toward ESE and Orion to the NW shine and stand out because of their beauty. Sirius. Maybe at this same time, there, far away to the north, my poor parents look at the same stars thinking of me. They are sitting in the heat, seated on sofas in the yard, taking in the air, while the children play around them. They are in silence, looking up at the sky and asking God to bring me back [safely]. Nordenskjöld went to the top of the basalt this afternoon and says that the state of the ice is more or less the same as what Bodman observed two days ago. Average temperature of the day = −10.4, bar 7 am = 748.0, 750–751.5. The speed of the wind is diminishing, and it was 16.5 m at 7 am; 12 m at 2 pm; and 7 m at 9 pm. The maximum wind speed today was at 1 am = 19.5. The minimum temperature was at 7 am = −12°. The temperature slowly increased all day, but it sped up to −7°; it dropped one degree, and at 10 pm, it was −7° again, and went down again at midnight. It is now −10° with a downward tendency. At this same time, the wind has moved to the NE and blows very weakly. At that same hour, the thermometer was at −7°, [the barometer] dropped to 30, only to go up again at 10 pm; that is, following the same movement as the thermometer, but in the opposite direction. What a splendid night! At 12:15, it is completely calm, and there is not one cloud in the sky.

5.5.2 February 20, [1903]

Average temperature = −4.85; bar 7 am = 751.4; 747.2; 743. Since 10 pm last night, the temperature has started to drop, reaching −14.5° at 6 am. At that hour, the barometer started to go down after having been at its maximum of 752 m for several hours. The temperature also began to rise, and the NE wind established itself beginning 5 am.

It has been a beautiful day today. Not one cloud has obscured the pureness of the blue sky. The only thing uncomfortable has been the NE, which has increased progressively and with speed. At 9 pm, it had 80 m of strength. It is the first clear day in all of February, so we've taken advantage of it and made observations of the sun in order to fix the chronometers. B, N, E, and J, [Bodman, Nordenskjöld, Ekelöf, and Jonassen] went in groups of two to get seals, killing two and bringing their skins back to the station. At 12, I went to the top of the basalt to see the state of the ice, and it's devastating. There is less free water than in December and January. Part of the free water was covered with new ice, and it [the free water] is only 1 or 2 miles wide by 3 or 4 miles long along the east coast of "Snow Hill." The free water between Paulet and Seymour Island is covered with compact pack ice. I think the new winter is a fact.

I started to skin Cain in the afternoon, but I had to abandon the task, because it [the body] was very frozen, and my knife is not sharp enough. With two dog skins, I plan to make a pair of pants.

I think a new southeast strong wind is approaching fast, because the barometer is decreasing rapidly.

Today, the sun has melted some of the snow, with water running down to the sides of the mountains.

Peridota has started [to be in heat].

5.5.3   *February 21, [1903]*

Average temperature of the day = −2.1°. Bar 739.4, 739.0, 740.0. Weak breeze from the NE, and calm between 1 am and 3 pm, with completely clear sky; the barometer going down, and the thermometer that had gone down to −5.7 went back up to +1.6. Between 3 and 6, weak gusts from N and WSW, the thermometer going down and the barometer going up. A bank of clouds that seem to be S-Cu appeared from the NNE but were covering a very small part of the sky. At 6, a breeze of 4 m per second from the WSW started to blow, moving to the SW shortly after, and before 7 pm, it was blowing from SSW and increasing its intensity. At 10 pm, the SW was blowing at a speed of 9 m per second, and the temperature was −4° and continues decreasing.

Today, it has been a year since we last saw the ship. Bodman was at the top of the basalt, and the state of the ice is the same as yesterday. So another year will go by without our seeing it [the ship]. Today, a seal was hunted near the little strait. The skin was pulled from Cain and nailed onto the magnetic apparatuses house. I took a picture of the termination of the glacier.

5.5.4   *February 22, [1903]*

Average temperature of the day = −6.35°. Bar = 743.2; 745.2; 746.1. SSW blowing till midday. This storm has been with a clear sky. Only at 11, some Cis started to pass by, coming from the S 40 S. The maximum of this storm was at 3 am, 10 m, being between this speed and 7 m until 12, when it moved to the SW and at 4 to the WSW, so rapidly diminishing in strength that at 4 pm it was almost calm. Between 4 and 7, it blew from the S 60 W, so weakly that it wouldn't move the _______ [unreadable]. At 7, it was completely calm. The sky was clear almost all day. The time of maximum cloudiness was at 1 pm, with a fourth part of the sky covered (3 by cirrus from S 40 E and 1 by a bank apparently of S-Cu that has been in the NE since yesterday). Ekelöf and Nordenskjöld went to look for seals to the south of the black glacier, killing two or three. Bodman, who was on top of the basalt, says that the state of the ice is the same.

5.5.5   *February 23, [1903]*

Average temperature = −5.8; bar 746.3–746.9—746.1. Weak gusts from W, SW, NW, NE, but we can say the day was calm. In the morning, there was fog, and *"Brimfrost,"* clearing before midday to leave its beautiful cirrus, which comes from SSE. At 8 pm at night, the sky was completely clear. At 10 pm, gusts from different directions began to come, and at 10:05 pm, the SSW started to blow; between 10:15 and 10:30, it blew at a speed of 7 m. The temperature—which is odd—escalated until 11, dropping from 11 pm to 12 at night. At 10 pm, it was −7°; at 11 pm −4°; and at 12 −7°.

With the objective of killing some seals, at 10:30 I took the ice skis and went to the strait that separates Seymour Island from Snow Hill. Just as I began to walk, I realized it would be better to leave the skis, because, with the snow being new and like sand, and in general hard, it was better to walk. So I did. So I circled the free water that there was in the strait, going near the shore of Seymour Island without killing any seals or any penguins; I saw only five Megalestris [skuas] and one ossifraga [petrel]. Have the penguins begun to migrate? From there, I went toward the *"iceberg,"* around a mile away, and toward the NNW there is another iceberg, much smaller, and next to this one I saw something that looks like a seal. After walking for a moment, I convinced myself that it was. I killed it with a shot from my gun. And I pulled off its skin. While I was in the middle of this operation, I heard some snorts some steps away from me, and I saw that another seal was about to come up through the "hole," but the demon dogs started barking, scaring it away. When I finished skinning out [the seal I had killed], I opened the stomach and found it full of food. This was composed of, first, a great quantity of some kind of worms [draws a worm shape]; then, some kind of spider or star of eight arms and white color, more or less of this shape (!) [draws star], the arms from the round part were 5 cm long. Some of these animals were intact, others weren't. Then, I found two white sticks that seemed to be cartilages, they were very flexible, and they ended in this shape [draws a stick with a ragged end]. I had the intention of keeping them, but, besides the very bad odor that they emanated, one of the dogs dared to eat it, leaving only one piece behind. The seal had the same skin as false leopards, but I don't think it was a Ross. When I arrived back at the station, I found an ossifraga that the dogs had trampled, but it couldn't fly, and I killed it with a stick, regretting this afterward, because I wasn't going to make any use of it all, and because I didn't know whether its inability to fly was because it was sick or because it had eaten too much, which is something common with these birds.

5.5.6   February 24, [1903]

NNE blowing all day. Speed between 5 and 6 m. Overcast sky. Average temperature = −7.2°. Bar = 744.0, 742, 739. No news.

5.5.7   February 25, [1903]

Average temperature of the day = −5.7°. Bar 736, 736, 734.3. From 5 am to 8 pm, blowing SSW. At 9 am, it had a speed of 12 m; at 4 pm 5 m; and at 9 pm calm. Clear sky since midday.

The situation pertaining to provisions looks more serious than I had thought. The coffee is not going to last. Neither will the butter. What will happen if, in addition to this year, we also have to go through yet another?

5.5.8   February 26, [1903]

Average temperature of the day = −1.6. Bar 730.0; 726.5; 726.9. Wind from NNE until 5 pm, when it changed to ENE and continued until 9 pm; variable at 10 pm, and at 11 pm, it began to blow from the S. Between 9 am and 4 pm, the sky was sometimes clear, sometimes not. And it became cloudy at 4, and it continued like this until midnight, when it cleared up again. Between 1 and 7 am, NNE blew at a speed of 4 m. At 8, it calmed completely, only to start again at 9; very weak until 1 pm; then 4 m per second.

The day has been, one could say, nice, especially the moments of calm and of weak wind that we had all morning, which, with the sun shining, invited us to go out for a walk. Bodman and Nordenskjöld went to get seals, but with no results.

5.5.9   February 27, [1903]

The temp = −4.0. Bar = 734.1; 734.4; 734.5.

The wind at 1 am moved to the SSW, blowing in this direction until 6:30 am, with strength between 5 and 7 m, raising the barometer; at 7 am, NE started again, blowing in this direction until 6 pm, with an average strength of 4 m, but sometimes reaching 7 m. At 7, it moved to the E, and until 12 at night, it would vary between S and ENE with a speed of 4 m. (I think the wind is from the SE, and due to the land that we have to one side, it swirls and thus seems as if variable.) The sun has shone, and there has been a great movement of clouds in the sky. The sky has been overcast at times and clear at others. The clouds (S-Cu-Ci) came from the SW.

5.5.10   February 28, [1903]

Cloudy day. SW blowing until 10 am (since 2 am) and then NE with a strength of around 5 m. These days, the smokers have noticed that the tobacco will not last either, and since some smoke more than others, they have divided it. Now, they also take their tea and, after drinking it, dry it, smoking it [the tea leaves] afterward.

Today, two seals have been killed in the "small strait," bringing two skins and one body to the station.

The average temperature is −4.0. Bar = 741, 742, 739.5.

On this day, the expedition members from the ship Antarctic *finally reach Paulet Island. After 16 days of trekking along the drifting ice with their equipment and 11 boats, leaping over crevasses as the ice broke around them, and avoiding icebergs that were smashing into the drifting ice on which they stood, they take refuge on the penguin-populated island, along with at least one cat who has survived the ordeal. Over one thousand penguins would give their lives to now save the lives of these expeditioners (Nordenskjöld et al. 1905).*

5.5.11   March 1, [1903]

Average temp = −1.1. Bar = 732.8; 729; 727. Blowing between NE and ENE from yesterday until 6:30 pm. It calmed for some minutes, some gusts from NW blew, and SW came blowing between 5 and 8 m. At 2:30, it started to rain (!!) quite heavily, continuing like this until 5:30 pm. Wind from the SW moved to the SSW.

5.5.12   March 2, [1903]

Average temperature = −4.1. Bar = 731;734.0;732. SSW blowing until 6:10 pm, when it calmed; the NE beginning very weakly at 11 pm. All day, around 6 to 7 m. After midday, the sky partly cleared at times.

5.5.13   March 3, [1903]

Average temperature = −1.3. Bar = 737.7; 738; 742.4. NE until 9 am; calm between
this time and 12, and gusts from the NE, at 1 from the S, at 2 calm, and at 5 SSW
started. Between 5 pm and 12 at night, the speed from SSW was between 6 and 9 m.
Overcast sky until 12 m, clearing up in the afternoon and completely clear after 11.

5.5.14   March 4, [1903]

Average temperature of the day has been +1.5°, escalating up to +5.5°.
Bar = 742.5; 739.7; 741.5. From 6 am to 3 pm, wind from NS with partially clear
sky. Wind speed between 1 and 4. Calm at 4, and then until 8, gusts from all
directions. A cool breeze from the SSW started to blow at 8:10 and stopped at 12 at
night. Today, Bodman killed a Weddell seal near the little strait, finding in it a [fully
formed] fetus of around 2 decimeters long.

5.5.15   March 5, [1903]

Wind from NE and ENE from 9 am, being ESE during the first hours of the
morning. At 6:30 pm, after some minutes of calm, SW came with speeds between 7
and 8 m and a sudden drop in temperature. Overcast sky all day and raining heavily
from 9:30 am to 6 pm. At times, rain eases and mixes with snow. With SW, snow
started to fall copiously. At 12 at night, SW had a speed of 12 m.
    Average temperature of the day = −1.1. Bar 743.5; 741.2; 739.5.

5.5.16   March 6, [1903]

Average temperature of the day = −9.1. Bar 736.5; 737.4; 738.8.
    All day blowing S 55 W with snow, with strength of 21 m, but sometimes
reaching 23 m.

5.5.17   March 7, [1903]

Average temperature of the day = −11.4. Bar = 738.0; 737.3; 735.6.
    All day blowing from WSW with a strength between 20 and 22 m. The axis of
the anemometer broke, and Bodman has spent all day working until he made
another one.

5.5.18   *March 8, [1903]*

Average temperature = −11.3. Bar = 733.3, 732.5; 742.6.

Wind from WSW with snow all day. The speed went down to 15 or 16 m only to escalate again up to 22.

5.5.19   *March 9, [1903]*

Average temperature = −11.7. Bar = 731.4; 731.3, 735.0.

The barometer, which until today had small ups and downs but with an overall downward tendency, started to escalate at 2 pm, apparently in a permanent way.

The SW continued, a little weaker than yesterday, between 15 and 20 m and with a lot of snow. We could see some stars shinning last night, and it made us think that today would be a nice day, but it doesn't look like it [the snow storm] is going to end; today is the fourth day.

5.5.20   *March 10, [1903]*

Average temperature of the day = −11.9. Bar = 735.0; 738.0; 741.0. Wind varying from SW and WSW, with a strength between 18 and 20 m per second, bringing a lot of snow. In the afternoon, the sky started to clear up a bit, and the intensity of the wind increased, making the rising of the barometer occur more quickly. The thermometer started to go down at 10 am, the temperature at that time being −10°. At 2 pm, it was −11.1° and at 11 pm −14.5°. The wind will very likely ease tomorrow, because that is the pattern that the meteorological instruments indicate: The SW comes, the thermometer goes down to a certain limit, which in this case was −10°, and it remains at this "stormy" temperature, so to call it, until the bad weather is about to end; then, the wind tends to move S, barometer goes up, and thermometer goes down from the "stormy" temperature.

Today, the Eskimo [Greenland] dogs have killed poor Elenita.

5.6   The Storms Come Earlier by Several Days

5.6.1   *March 11, [1903]*

Average temperature = −14.25. Bar = 745.22; 745.8; 747.4.

Wind from WSW and sometimes SW until 9 am. Between 10 and 12, it was from the SW, and at 1 pm, it moved to SSW, continually diminishing its intensity.

At 7 am, it was of 124 points (14 m or 15.5?); at 2 pm, it was 122 points; and at 7 pm, it was of 8 m. At 8, it calmed at times, and strong gusts from the W blew, and then, it calmed again, and they came from the SSW after a few seconds. At this hour, there were 30 points. At 9, it became established from the SSW at 6 m and with a tendency to increase. The sky has been clear all day. It is remarkable how the amount of snow on top of the glacier has increased; it now has half the height of the glacier; it will very likely get to the same height by the end of the year. It seems that the storms come earlier by several days (7 days) compared to last year's. Thus, the big storm in March started on the 12th at night, while this year it started on the 6th.

At 10:45, the thermometer was at $-17°$; clear sky, wind from the SSW, 5 m speed. On March 11 of last year, the minimum was $-3.5°$ and the maximum was $+7.8°$. On the 12th, the maximum was $= +6.4$ and the minimum $= -13.2$ (at night at the start of the storm); on the 13th maximum $= -11.4$ and minimum $= -16°$.

As Sobral watches the snow on the glacier increase, and the winter storms begin to arrive at Snow Hill, the three expeditioners at Hope Bay finally take shelter in their stone hut which they have built around their tent. They, too, see winter arrive in Antarctica and now settle into their new shelter for a soot-covered but warm abode. Penguins and seals (700 and 21, respectively) would give their lives there as well, providing fuel for heat and meat for food. Less than two dozen fish would be caught as well—ice fishing not being an easy task (Nordenskjöld et al. 1905).

5.6.2 March 12, [1903]

Average temperature $= -16.6$. Bar $= 747.13$ (2 pm); 747.33 (7 am); 746.13 (9 pm). Variable wind between SSW and WSW. At 7 am 8.5 m, 2 pm 6 m, and at 9 pm _______ [unreadable]. The wind from the SW, calm at 8 pm, weak gusts from SSE. WNE and calm until 10. At 12, weak gusts from N, NE with intervals of calm. Until 2 pm, weak gusts from N, NE with intervals of calm. Until 2 in the afternoon, the sky was clear at times, but then it became completely cloudy. Today, "Fía" started to give birth. This morning, the thermometer dropped down to $-18.2°$.

5.6.3 March 13, [1903]

Variable winds and calm. At 2 pm, EE blew up to 5 m. Overcast sky all day and snowing at times. Bar $= 742.5$, 738.73, 736.8. Average temperature $= -10.4$. Barometer going down since the day before yesterday at 2, but descent has accelerated today.

5.6.4   *March 14, [1903]*

Average temperature = −14.4. Bar = 734.0, 734.0, 734.6. At 1 am, SSW started to blow with a strength of 7 m; at 7 am, 12 m, overcast sky, being now 10, there is cloudiness; at 2 pm, speed 15 m and overcast sky. SSW blew all day. At night, speed diminished a little, to 10 m.

5.6.5   *March 15, [1903]*

Average temperature = −15.3. Bar = 733.7; 732.2; 729.9. Since 10 am, the wind has moved to the SW, quickly increasing its speed. At 11, it broke the axis of the anemometer. At 11 am 7 m, at 2 pm 18 m, and at 9 pm between 24 and 25 m. The wind comes with a huge amount of snow. At 12:30 at night, the wind had a speed of 24.5 m. The direction is between SW and S 55 W. I make magnetic observations until 2 am.

5.6.6   *March 16, [1903]*

Average temperature = −14.8. Bar = 730.8; 733.5; 735.7. SW and WSW all day. At 3 am 25 m, at 6 am 24, at 2 pm 20, and at 9 pm 18. And at 12 at night 13. The sky has been quite clear between 10 am and 6 pm today. The sun has hit the heliograph between 10:30 am and 4:30 pm. The anemometer was fixed. "Amager" gave birth today.

5.6.7   *March 17, [1903]*

Average temperature of the day = −14.2. Bar = 734.6; 732.7; 732.1. The sky started to clear up at midday. At 4 pm = 7, at 6 pm = 7, but it again gets cloudy at 12.

  SW and WSW wind at 3 am 15 m, at 9 am 15 m, at 2 pm 14 m, and at 9 pm 11 m.

5.6.8   *March 18, [1903]*

Average temperature = −12.9. Bar = 729.5; 729.2; 728.6. Wind from the SW all day, with a lot of snow. Until midday, the wind was between 10 and 11 m per second, but then it started to increase quickly; at 9 pm, it was at 17 m, and at 12 at night, it was between 19 and 20 m.

5.6.9 March 19, [1903]

Average temperature = −13.7. Bar = 728.6; 731.3; 734.4. Wind from SW with a lot of snow all day. At 1 am, the speed of the wind was 20 m; at 3 am 22 m; at 6 am 23.2 (maximum); at 7 am 22; it began to diminish at this hour, at 2 pm 19 m; and at 9 pm 18.3 m.

5.6.10 March 20, [1903]

Average temperature = −13.1. Bar = [+ 4.3, 734.4–744.0], 743.5, 742.4.

SW and WSW at 7 am 18 m, 2 pm 17 m, and 9 pm 21 m. Lots of snow all day. Today, in the morning, I was sitting next to the table reading, and I was stomping because my feet were cold, when the stupid doctor [Ekelöf] started mocking me by stomping harder, and he told me, opening those eyes of his and spitting foam out of his mouth: "You're always bothering people by stomping like that!" For a couple of seconds, I was astonished and didn't answer because I couldn't understand his sudden spurt of rage, and I told him: "You are a fool and stupid, why don't you tell me not to do it in another way?" to which he did not reply. He was smoking that nasty tobacco, and I told him: "You also bother me with your smoke!" These people spend their time playing until 12 at night, not letting me sleep, and during the day, the moments that I don't sleep weigh heavily on me.

Sobral has reached the boiling point with Dr. Ekelöf, who had already formed an adversarial relationship with the young lieutenant back in Port Stanley (on December 31, 1901), when he first insulted Sobral and all the "mixed race" Argentine people. This time, however, Sobral fights back and, in a more mature manner than Ekelöf, points out the good doctor's own shortcomings. The incident provides insight as to Sobral's sense of fair treatment, and how he calmly handles an incendiary situation.

There is a sentence begun and then thoroughly crossed (and smudged) out at the end of this entry. Perhaps Sobral had a few more choice words to say on the subject, that he then thought better of saying. There are also drops of what looks to be oil or some other material dropped and imbedded onto the page. This entry marks the last time, for some time (till the end of this month), that Sobral muses on more than just the weather in his diary. Perhaps the outburst relayed here caused him to withdraw for a while. More likely, as he was reading and continuing to learn the Swedish language, he was focusing on his books and took solace in his studies.

5.6.11 March 21, [1903]

Average temperature = −14°. Bar = [+ 1.5, 732.7–742.5], 744.1, 746.2.

SW and WSW 7 am 20 m, 2 pm 18 m, 9 pm 17 m. Wind tends to diminish.

5.6.12   *March 22, [1903]*

Average temperature = −10.6. Bar = 738.2, 739.96, 740.3.

Wind was from the SW until 2 pm; at that moment, it moved to SSW, at 7 am 11 m, at 2 pm 8 m, at 6 pm 5 m, and at 9 pm 15 m. Overcast sky all day, with A-S. At 9 pm, the wind started to calm and moved to the S. At 11 pm, a very weak breeze blew from ENE, being completely calm at 12 at night.

5.6.13   *March 23, [1903]*

Average temperature = −10.55°. Bar = 739.4, 735.86, 729.48.

At 4 am, it started to blow from the E until 9 am, when it moved to the NE, variable between 2 and 5, and at 5:15, it started from the SSW; at 10:45, it started to snow; and at 12 am and 5 in the afternoon, the land was completely covered in snow, which fell between 10:45 and 5:45. Wind speed at _______ [blank] was 10 m, at 12 17 m.

5.6.14   *March 24, [1903]*

Average temperature = −15.98. Bar = 721.8, 721.6, 726.6, wind from WSW and S 55 W with a lot of snow. Speed of the wind at 7 am was 21 m; at 2 pm 25 m, which was the maximum; and at 9 pm 21 m.

5.6.15   *March 25, [1903]*

The average temperature = −14.4. Bar = 730.0, 732.2, 734.5. Wind from SW with snow. At 7 am 11 m, 18 m, and 15 m. Between 9:30 am and 4:30 pm, the sun has left an impression on the paper of the data recorder. After midday, the sky has almost cleared up. At 7 am, the amount of clouds was 6; at 2 [pm] 8; and at 9 pm 1.

5.6.16   *March 26, [1903]*

The average temperature = −12.6°. Bar = 736.43, 738.74, 740.06. SSW and SW blowing all day. The sky has been very clear. At 7 am 6 (clouds), at 2 pm 3, and at 9 pm 8. The speed of the wind at 7 am was 18.5 m; at 2 pm 18 m; and at 9 pm 14 m. During the day, there has been little *snöfyk* [snowfall] and Cockburn could be seen every so often. The *snöfyk* was on the glacier and on the sea ice.

5.6.17   March 27, [1903]

The average temperature = −11.1°. Bar = 740.35, 741.1, 741.05. Wind from the SW with little *snöfyk* (in the afternoon). The sky in the morning was quite clear, but it clouded over at midday. At 7 am 8, 2 pm = 10, 9 pm <3, but at 12 at night it became overcast again. The speed of the wind at 7 am was 13 m; at 2 pm 12 m; and at 9 pm 12 m.

5.6.18   March 28, [1903]

The average temperature = −13.1°. Bar = 740.76, 738.95, 738.00.
   Wind from the SW (sometimes it would move toward WSW and SSW).
   A little snöfyk on the glacier and on the strait. Sky overcast with Cu-Ni and A-Cu. Wind speed around 13.5 m all day.

5.6.19   March 29, [1903]

The average temperature = −14.6. Bar = 735.67, 734.0, 731.94. Wind from SW and WSW. Sky overcast with A-Cu and Cu-Ni (but not completely). At 7 am neb [nebulosity] 8, at 2 neb 4, and at 9 pm neb 2.

$$7\,\text{am}\begin{bmatrix}-15.3\\-15.4\end{bmatrix}\quad 2\,\text{pm}\begin{bmatrix}-13.9\\-14.0\end{bmatrix}\quad 9\,\text{pm}\begin{bmatrix}-14.7\\-14.8\end{bmatrix}$$

5.6.20   March 30, [1903]

$$7\,\text{am}\begin{bmatrix}-15.45\\-15.45\\732.6\\15\,\text{m}\end{bmatrix}\quad 2\,\text{pm}\begin{bmatrix}14.7\,\text{m}\\-15.4\\-15.4\\732.2\end{bmatrix}\quad 9\,\text{pm}\begin{bmatrix}22\,\text{m}\\-17.4\\-17.4\\732.5\end{bmatrix}$$

   Wind from the SW, WSW, with much snöfyk. The maximum strength of the wind has been 22 m at 10 pm and the minimum 15 m at 5 am, but in general it was between 19 and 20 m.

5.6.21   *March 31, [1903]*

$$7\,\text{am}\begin{bmatrix}22.0\\-17.8\\-17.9\\735.4\end{bmatrix}\quad 2\,\text{pm}\begin{bmatrix}21.1\ \text{m}\\-17.16\\-17.7\\737.9\end{bmatrix}\quad 9\,\text{pm}\begin{bmatrix}21\ \text{m}\\16.0\\18.2\\740.75\end{bmatrix}$$

Wind from WSW with snöfyk. The anemometer broke at 7 am. The maximum strength of the wind has been 22 m and the minimum 19 m.

5.6.22   *April 1, [1903]*

$$7\,\text{am}\begin{bmatrix}[-]16.2\\ [-]16.3, 9\,\text{neb}\\45.4\\16.5\ \text{m}\end{bmatrix}\quad 2\,\text{pm}\begin{bmatrix}[-]14.4\\ [-]14.6\\49.47, 2\,\text{neb}\\12.0\ \text{m}\end{bmatrix}\quad 9\,\text{pm}\begin{bmatrix}-15.5\\-15.5, 0\,\text{neb.}\\51.2\\ \text{calm}\end{bmatrix}$$

Until 10, sky completely overcast and a little snöfyk on the glacier and the strait. At 10, it started to clear up toward the south, and the Ni-Cu and H-Cu that covered the sky began to disappear toward the NE. At the same time, the wind would diminish, moving toward the SSW at 6 pm. At 7:30 pm, it was calm and the sky was completely clear. At 10:30 pm, it started to blow from the ENE, and the barometer has been dropping since 9 pm. At 12 at night, the sky became completely overcast.

The maximum temperature during the month of March was +5.4 on March 4 at 6 pm, and the minimum was −18.2 on March 12 at 7:10 am.

5.6.23   *April 2, [1903]*

$$7\,\text{am}\begin{bmatrix}-6.8\\-6.9\\742.86\\ \text{neb}\,10\\7.5\ \text{m}\end{bmatrix}\quad 2\,\text{pm}\begin{bmatrix}+0.4\\0.0\\ \text{neb}\,10\\734.04\\6\ \text{m}\end{bmatrix}\quad 9\,\text{pm}\begin{bmatrix}-0.2\\-0.9\\ \text{neb}\,3\\752.74\\1\ \text{m}\end{bmatrix}$$

The ENE kept blowing until 11 am. At 12 NNE; at 1 N; and at 2:45 pm, the temperature quickly dropped 5°. The wind moved to the NNW, W, and SW. It snowed from 6 am to 2 pm. At 4, 5, and 6 pm, calm; then gusts from E and ENE that started at 10 pm.

Today, two Megalestris were seen, one was killed. Life in the station is monotonous and disgusting. And to think I'll have to be here for another 10 months at least. N, E, and B [Nordenskjöld, Ekelöf, and Bodman] play cards until 11 or 12 at night. And since they're laughing out loud and talking, I have to be on guard duty until that hour. They sleep either all morning or all afternoon. And at night they bother everyone else. I'm making these little boots of tarpaulin, with soles made of seal skin, for [going outside to make] the magnetic observations, because the ones I have had nails.

The barometric minimum for the month of March was 721.3 on the 24th at 8 am.

The barometric maximum for the month of March was 747.7 on the 11th at 11 pm.

5.6.24 April 3, [1903]

	7 am		2pm		9 pm
	-1.75		-8.5		-11.2
	-2.75		-9.6		-11.4
	$+9.0$		75.8		713.0
	neb 2		733.5 min $= -8.7$		735.3
	730.5 max $= +1.2$		var		var
	NE		0 m		0 m
	5 m		Neb 1		neb $= 0$

Very nice day. Clear sky and little wind. Today, Nordenskjöld went to take the temperature of the sea at different depths, and they killed a seal. Bodman went toward the little strait and killed four seals, all of them Weddell, female, and pregnant, big and very fat. I made observations corresponding to the sun and one observation of declination and azimuth of the basalt top from the station point.

5.6.25 April 4, [1903]

	7 am		2 pm		9 pm
	-0.35		-0.2		-5.7
	-1.0		-2.9		-6.15
	$+7.7$		$+13.7$, min $= -15.1$		714.2
	729.7 max $= +0.1$		729.35		732.5
	N		calm		SSE
	8 m		neb 7		5 m
	Neb 5				neb 2

The day has been pretty hot. All morning and part of the afternoon, the temperature was around _______ [blank]. In the morning and until 2 pm, wind blowing from NNW, N, and NNE. At 10 pm, the sky became overcast. Today, I took a walk that lasted 5.30 h _______ [unreadable—hunting or searching for] seals. I found three open holes, a sure sign that the seals are not far away, but I couldn't kill more than one of the Weddell kind, a female. In its stomach, I found a large quantity of fish similar to the sand smelt. In the different parts of the _______ [unreadable], they must have been 10–15 cm long. Probably, those fish bones that I found in the stomach of a seal were parts of the spine of one of these fish. A lot of _______ [unreadable] are seen. The amount of water that is free of ice is still considerable. In the little strait, I saw hundreds of cormorants today, some standing on the floes and others swimming or flying in flocks of 20 or 30. There were also several Megalestris [skuas] and seagulls. Pagadromas [snow petrels] and sternas [terns] are also seen.

5.6.26   April 5, [1903]

	7 am		2 pm		9 pm
	−18.7		−13.9		−15.9
	−18.8		−14.0		−15.9
	+7.0 max = +0.2		+17.3 min = 18.5		+10.6
	731.55		734.25		734.75
	SW14m		SW12 m		SW18.5 m
	Neb = 10		neb 10		neb 7

All day wind from the SW, but not with a lot of snow. SW started at 3:30 am, after it had been blowing intermittently from the SE and ENE.

5.6.27   April 6, [1903]

	7 am		2 pm		9 pm
	−17.1		−17.0		−17.1
	−17.15		−17.1		−17.2
	+3.1 max = −13.1		+18.6 min = −17.0		+11.0
	738.2		742.65		743.95
	SW16 m		SW		SW10.5 m
	neb 5		neb = 9		neb = 8

SW and WSW blowing all day with snow. At night it cleared.

5.6.28   *April 7, [1903]*

$$
7\,\text{am}
\begin{bmatrix}
-16.0 \\
-16.2 \\
+3.2 \\
743.2\,\text{max} = -15.0 \\
\text{SSW} \\
\text{neb}\,10 \\
9.2\,\text{m}
\end{bmatrix}
\quad
2\,\text{pm}
\begin{bmatrix}
-16.4 \\
-16.5 \\
+13.8 \\
744.7\,\text{min} = -18.0 \\
\text{SSW} \\
\text{neb}\,0 \\
\text{snowing} \\
4.5\,\text{m}
\end{bmatrix}
\quad
9\,\text{pm}
\begin{bmatrix}
-17.0 \\
-17.0 \\
+13.7 \\
744.0 \\
\text{SSW} \\
\text{neb}\,2 \\
4.5\,\text{m}
\end{bmatrix}
$$

Wind from the SSW all day, and it snowed a little in the afternoon; at 9 pm, it cleared up almost completely.

5.6.29   *April 8, [1903]*

$$
7\,\text{am}
\begin{bmatrix}
-16.75 \\
-16.9 \\
+5.3 \\
741.9\,\text{max} = -15.2 \\
\text{SSW} \\
\text{neb}\,6 \\
35\,\text{m}
\end{bmatrix}
\quad
2\,\text{pm}
\begin{bmatrix}
-17.4 \\
-17.4 \\
+13.7 \\
741.4\,\text{min} = -18.8 \\
\text{SW} \\
\text{neb}\,10 \\
5\,\text{m}
\end{bmatrix}
\quad
9\,\text{pm}
\begin{bmatrix}
-17.3 \\
-17.3 \\
+12.2 \\
739.75 \\
\text{SSW} \\
\text{neb}\,10 \\
\text{snowing} \\
7.5\,\text{m}
\end{bmatrix}
$$

Before midday, the sky was clear, but a bank of S-Cu started to move forward from the W, covering the entire sky by 2 pm. At 5 am, it was calm, and in the next hours, it blew from the SSW; at 9:30 am, it moved to the NE for some seconds, going back to the SW and blowing from the SSW since 1 pm. At 6 pm, it started to snow. Magnetic observations are made today, and great disturbances are noted; in the three apparatuses, quick movements that come suddenly are recorded. Observations during the whole hour are taken by Bodman and me, one at a time. In the morning, N [Nordenskjöld] went with J [Jonassen] to get the seal skin that was still left from the ones Bodman had killed the other day. I told N [Nordenskjöld] where the seal was that I had killed on the 4th so that he would show J [Jonassen] the place. But he thought there would be a lot of snow, and so he didn't look for it. So I went and I found the skin as well as the body, with no snow on it; also, there was another seal on the ice; I killed it and I pulled off its hide. This operation I did with gloves, because skinning at 17° below zero was not very pleasant. It is said that the cat with gloves traps no mouse. That doesn't apply to humans. In these regions, the saying fails, because here one even hunts seals with gloves on.

It's curious how little instinct these animals have; they climb up onto the ice, they see another one [of them] dead, and, with no distrust, they sleep calmly in the sun.

We are able to get close to them, even while making noise, and they wake up only when they're about to drop dead from a bullet or a spear. In this season, it looks like all the females are pregnant, at least the ones that have been killed; except for one, they all were. I think the term of the pregnancy is between 10 months and a year. They begin to give birth in October, but they spend a lot of time with the baby, more than a month, and not like Racovitza says, 2 or 3 days. Today, "Abel" and "Matilda" have been shot dead, so the only dog left from the Malvinas [Falklands] is "Skottsberg."

Sobral's excitement at the "great disturbances" detected via the magnetic observations can be felt, and his persistence at pursuing seals for food is notable, even as he studies their anatomy and behavior. The unfortunate dogs from the Falklands, as he notes, were weakening in the harsh Antarctic environment and were also no match for the warring Greenlanders. Skottsberg, the last remaining dog from that southern island, was named in honor of fellow expeditioner and botanist Carl Skottsberg.

5.6.30   April 9, [1903]

7 am		2 pm		9 pm	
	[−]17.6		−14.0		−14.5
	[−]17.6		−14.5		−14.5
	+8.0		+10.1		+11.2
	741.55 max = −14.6		745.1		746.8
	SSW		SW		25 m
	8 m		8 m		E
	Neb = 3		neb 3		neb = 1
			Min = −18.0		

The sky had cleared up in intervals, but at 6 pm, it cleared completely. The SSW blew until 6 pm. At 7 am, it was ENE, and it kept blowing from the ENE, E, and ESS.

5.6.31   April 10, [1903]

7 am		2 pm		9 pm	
	−10.95		−12.5		−7.0
	−11.0		−12.6		−7.0
	+3.0		+16.1		+13.4
	742.55		743.0		738.8
	Max = 9.0		min = −16.3		NNE
	ENE		calm		neb 10
	2.5 m		neb 10		6 m
	Neb 10		it snows a little		
	snowing				

Until 8 am, wind from the ENE, E, and ESE, and snowing. Since then, and until 7 pm, it was almost calm, with gusts being felt from different directions. At 7 pm, NE started to blow at a speed of 5 m, and at 9:10, after some seconds of calm, the SW came in a gust of 7 m per second; the temperature, which was −7°, quickly dropped down to −19.0°, and a lot of snow came.

5.6.32   April 11, [1903]

$$
7\,\text{am}
\begin{bmatrix}
-15.75 \\
-16.1 \\
+4.2 \\
745.05 \\
\text{Max} = -6.8 \\
\text{SSW(very weak)} \\
\text{Neb} = 7
\end{bmatrix}
\quad
2\,\text{pm}
\begin{bmatrix}
-12.9 \\
-13.0 \\
+14.1 \\
745.6 \\
\text{min} = -19.4 \\
\text{ESE} \\
5.5\,\text{m} \\
\text{Neb} = 10
\end{bmatrix}
\quad
9\,\text{pm}
\begin{bmatrix}
-15.5 \\
-15.45 \\
+12.7 \\
743.5 \\
\text{SW} \\
7\,\text{m} \\
\text{neb} = 10 \\
\text{it is snowing}
\end{bmatrix}
$$

The day dawned with an overcast sky and a weak wind from the SW. At 5 and 6 pm, some patches of blue sky could be seen and also the Ci and Ci-Cu marching at a great speed from W to E. At the same time, a bank of S appeared from the NE. At 7 am, the SW calmed completely, and weak gusts blew successively from the WNW, NW, N, NE, and ENE. At 11 am, isolated beautiful snow crystals began to fall, perfect, entire pieces, and a thick snow fog came from the ENE. At 1 pm, the air started to clear up, and the sky became covered with a bank of A-S. Until 7 pm, the wind was between ENE and ESE. At 4:30, it started to snow, and from this time until 8:30, there was 1 mm 6 of precipitation in the form of snow, fully covering the land in a beautiful white, which is very rare here. At 8:40, SSW began to blow with 7 m of strength; before that, the wind from the ESE calmed completely at 7 pm; at 8, it would blow very weakly from the W for some seconds, and then it calmed. At night, whisky was drunk to celebrate Easter.

5.6.33   April 12, [1903]

$$
7\,\text{am}
\begin{bmatrix}
-17.8 \\
-17.8 \\
+4.2 \\
741.2 \\
\text{Max} = -11.6 \\
\text{SW} \\
-15.5\,\text{m} \\
\text{snöfyk} \\
\text{Neb} = 10
\end{bmatrix}
\quad
2\,\text{pm}
\begin{bmatrix}
-18.2 \\
-18.2 \\
+16.6 \\
741.65 \\
\text{min} = -19.0 \\
\text{SW} \\
12.5\,\text{m} \\
\text{snöfyk} \\
\text{neb} = 10
\end{bmatrix}
\quad
9\,\text{pm}
\begin{bmatrix}
-15.7 \\
-15.7 \\
+10.2 \\
737.05 \\
\text{SW} \\
18.5\,\text{m} \\
\text{snöfyk} \\
\text{neb} = 10
\end{bmatrix}
$$

All day SSW and SW blowing with a lot of snow and growing intensity. At 12 at night, the wind was blowing at 24.5 m (206 points), the maximum of the day. At midday, we ate three dishes [courses], part of which was Argentine corn and dessert. Anyway, it was an authentic feast to celebrate Easter. These nights are very clear because we have a full moon. On April 9 at 10 pm, with the sky being clear, I was able to read perfectly in the moonlight. Among the seals that have been killed today, some had 8 cm of fat. Today, we have eaten smoked penguin (raw); it is excellent.

5.6.34   April 13, [1903]

$$
7\,\text{am}
\begin{bmatrix}
-14.0 \\
-14.0 \\
+4.1 \\
735.25 \\
\text{Max} = -14.0 \\
\text{SW} \\
23\,\text{m} \\
\text{Neb} = 10 \\
\text{snöfyk}
\end{bmatrix}
\quad
2\,\text{pm}
\begin{bmatrix}
-14.0 \\
14.05 \\
+16.9 \\
739.5 \\
\text{min} = -18.25 \\
\text{S 55 W} \\
9 = \text{neb} \\
13.5\,\text{m} \\
\text{snöfyk}
\end{bmatrix}
\quad
9\,\text{pm}
\begin{bmatrix}
-13.8 \\
-14.0 \\
+13.3 \\
739.6 \\
\text{S 11 W} \\
11\,\text{m} \\
\text{neb} = 10 \\
\text{snöfyk}
\end{bmatrix}
$$

5.6.35   April 14, [1903]

$$
7\,\text{am}
\begin{bmatrix}
-14.2 \\
-14.45 \\
+5.1 \\
741.75 \\
\text{Max} = -12.1 \\
\text{SW} \\
7.5\,\text{m} \\
\text{A-Cu}
\end{bmatrix}
\quad
2\,\text{pm}
\begin{bmatrix}
-11.0 \\
-11.7 \\
+18.2 \\
744.2 \\
\text{min} = -14.7 \\
\text{calm} \\
\text{neb} = 4
\end{bmatrix}
\quad
9\,\text{pm}
\begin{bmatrix}
-12.1 \\
-12.0 \\
+14.37 \\
741.7 \\
\text{calm} \\
\text{neb} = 10
\end{bmatrix}
$$

Today is Sobral's birthday, but he does not mention it, either in his diary, or to his Antarctic companions. Perhaps it is his modesty that restrains him, as well as his discomfort with his present situation as "persona non grata."

5.6.36   *April 15, [1903]*

$$
7\,\text{am}
\begin{bmatrix}
-6.5 \\
-8.0 \\
+8.4 \\
737.5 \\
\text{max} = -3.7 \\
\text{calm} \\
\text{neb} = 9 \\
\text{Ci; Ci}-\text{s}
\end{bmatrix}
\quad
2\,\text{pm}
\begin{bmatrix}
+3.4 \\
+1.5 \\
+16.0 \\
735.65 \\
\text{min} = -13.5 \\
\text{calm} \\
\text{neb} = 10 \\
\
\end{bmatrix}
\quad
9\,\text{pm}
\begin{bmatrix}
+1.5 \\
+1.0 \\
+14.5 \\
735.05 \\
\text{SSW}, 7\,\text{m} \\
\text{neb} = 10 \\
\ \\
\
\end{bmatrix}
$$

At 7 pm, it started to rain, stopping at 9 to continue with some snow. Last year, we did not have this rain. The amount was around 1 m 5. The temperature went down, and quite a thick layer of ice formed (on the land), which made walking quite difficult. At night, the temperature dropped quite a lot, to $-14.0°$. The SW reached its maximum speed of 18 m at 12 at night.

5.6.37   *April 16, [1903]*

$$
7\,\text{am}
\begin{bmatrix}
-9.5 \\
-10.6 \\
+7.9 \\
745.2 \\
\text{Max} = +5.0 \\
5.8\,\text{m} \\
\text{Neb} = 1
\end{bmatrix}
\quad
2\,\text{pm}
\begin{bmatrix}
-8.9 \\
-9.2 \\
+18.6 \\
746.1 \\
\text{NE7 m} \\
\text{neb} = 9 \\
\text{min} = -14.1
\end{bmatrix}
\quad
9\,\text{pm}
\begin{bmatrix}
-5.9 \\
-6.5 \\
+14.5 \\
741.7 \\
\text{calm} \\
\text{neb} = 10 \\
\
\end{bmatrix}
$$

Today, the weather is pretty nice. In the morning, the SW diminished a lot, blowing with a clear sky. At 8 am, it moved to the SE and then to the NE. Between 4 pm and 10 at night, complete calm reigned. Tonight, I have made two meridian observations, one of [alpha] α Eridani and one of α Argus, resulting in $4 = 64°\ 22'\ 3.$"

5.6.38   *April 17, [1903]*

$$
7\,\text{am}
\begin{bmatrix}
-4.0 \\
-5.7 \\
+9.1 \\
744.45 \\
\text{Calm} \\
\text{Neb2}
\end{bmatrix}
\quad 2\,\text{pm}
\begin{bmatrix}
-0.9 \\
-3.0 \\
+17.7 \\
746.85 \\
\min = -9.5 \\
\text{NE very weak} \\
\text{Neb} = 3
\end{bmatrix}
\quad 9\,\text{pm}
\begin{bmatrix}
0.0 \\
-2.5 \\
+15.6 \\
743.75 \\
\text{calm} \\
\text{neb} = 9.5
\end{bmatrix}
$$

The morning was precious. Very weak variable winds and calm all day.

5.6.39   *April 18, [1903]*

$$
7\,\text{am}
\begin{bmatrix}
+4.0 \\
-0.2 \\
+11.2 \\
740.3 \\
\max = +6.1 \\
\text{calm} \\
\text{neb} = 10
\end{bmatrix}
\quad 2\,\text{pm}
\begin{bmatrix}
+3.6 \\
+0.7 \\
+17.4 \\
743.6 \\
\min = -3.0 \\
5.9\,\text{m} \\
\text{neb} = 3
\end{bmatrix}
\quad 9\,\text{pm}
\begin{bmatrix}
-1.2 \\
-3.5 \\
+16.1 \\
747.4 \\
\text{ENE1 m} \\
\text{neb} = 2
\end{bmatrix}
$$

Until 10 am, very weak variable wind. At 10, it blew out suddenly from the S with a high temperature. At 3 pm, it had 10 m of speed, and it started to ease and to be calm at 10 pm.

Today, there has been a conference between N, E, and B [Nordenskjöld, Ekelöf, and Bodman] behind the magnetic instruments house.

At night, I was invited to play cards by Bodman and Ekelöf, an invitation that I courteously declined. I observed the meridian of Spica and another one of Mars. Birds are still seen around. On the afternoon of the 16th, I saw a snow petrel which, after flying around the station, headed toward the E. On the night of the same day, as I was observing, I sensed a bird passing by; I think it was a cormorant.

By this time, Nordenskjöld et al. were sensing Sobral's discomfort and lack of complete integration into the circle of colleagues, and efforts were made to bring him into the fold and to make him feel more comfortable. Nordenskjöld was a mentor to Sobral, and Bodman was as close a companion as Sobral would have on this excursion. Ekelöf was most likely cooperating at the urging of Nordenskjöld. Yet Sobral still maintained somewhat of a quiet and polite—perhaps wary—distance, occupying himself instead with astronomical observations and bird watching, and increasing the sophistication of his record-keeping.

5.6.40   April 19, [1903]

$$
7\,\text{am}
\begin{bmatrix}
-5.3 \\
-6.3 \\
+10.7 \\
744.4 \\
754.2 \\
\text{max} = +7.8 \\
\text{calm} \\
\text{neb} = 10 \\
\text{Ci-S, Ci, Cu-Ni} \\
\text{Direction WSW} - \text{ENE}
\end{bmatrix}
\quad
2\,\text{pm}
\begin{bmatrix}
+4.1 \\
+0.95 \\
+17.7 \\
739.6 \\
\text{min} = -8.5 \\
\text{var} \\
1\,\text{m} \\
\text{neb} = 9 \\
\text{Ci-S, Ci, Cu-Ni} \\
\text{Direction W}
\end{bmatrix}
\quad
9\,\text{pm}
\begin{bmatrix}
-10.0 \\
-10.1 \\
+13.2 \\
747.5 \\
\text{SW} \\
\text{neb} = 10 \\
10.5\,\text{m} \\
\text{snowing a little}
\end{bmatrix}
$$

Variable wind all day, and 8–10 parts of the sky covered with Ci, Ci-S, and Cu-Ni. Ci and Ci-S would come from the WSW with radiation for many hours, WSW–ENE. Cu-Nis always accumulates between the NE and a 45-degree angle from the zenith, while the other clouds are in the rest of the sky: Ci, Ci-S, etc. Temperature and humidity undergo sudden changes of many degrees, while the barometer drops rapidly. At 4 pm, the barometer started to go up, and a bank of S-Cu covered the sky. At 4:35, after some seconds of calm, the WSW came in gusts of around 17 m per second; at the same time, the thermometer would go down rapidly when the wind ________ [unreadable] was at +4.8, 25 min later; that is, at 5 pm, it was −3.0°; at 6 pm −7°; and at 9 pm—10.0°. Ten minutes after the wind had started, snow began to fall, diminishing the wind for a while before it began to go up again later. The barometer would go up rapidly: At 4, it was at 736 mm and at 9 pm 746 mm; that is, 10 mm in 5 h. At 6 pm, the wind was at 18 m, and right away it started to diminish rapidly, so much so that at 9 it was only at 13 m. The snow that has fallen is very little. It seems that the direction of the wind coincides with the orientation of the belts of cirrus; that is, the bearing of the radiation points. Today, I have seen a Scoresby's gull and a snow petrel going toward the ENE. Nordenskjöld asked me if I had any complaints or if there was something that he could do so that I could have a better time, to which I replied that I had no complaints. At 11 pm, the sky cleared up completely, and the wind was from SSW with [a strength of] 5.5 m.

5.6.41   *April 20, [1903]*

$$
7\,\text{am}
\begin{bmatrix}
-16.8 \\
-17.0 \\
+7.1 \\
751.25 \\
\text{max} = +4.8 \\
\text{var}\ 2.5\,\text{m} \\
10 = \text{neb} \\
\text{Ni}
\end{bmatrix}
\quad
2\,\text{pm}
\begin{bmatrix}
-15.7 \\
-15.7 \\
+16.6 \\
750.15 \\
\text{min} = -17.4 \\
\text{E4.5 m} \\
\text{neb} = 10 \\
\text{In 2 h 30 m\_fell}
\end{bmatrix}
\quad
9\,\text{pm}
\begin{bmatrix}
-16.8 \\
-16.8 \\
+10.8 \\
745.6 \\
\text{var} \\
\text{neb} = 10 \\
\text{snowing} \\
3\,\text{m}
\end{bmatrix}
$$

Until 1 am, wind from the S blew, and the sky was clear. At 3, the wind was NE 1 m and clear sky. During the rest of the day, the wind has been between the ENE and the SE, snow falling since 11. This last storm has been remarkably short, and it didn't last more than 1 h.

5.6.42   *April 21, [1903]*

$$
7\,\text{am}
\begin{bmatrix}
[-]18.3 \\
[-]18.3 \\
+5.8 \\
738.85 \\
\text{max} = -14.6 \\
\text{SW}, 5.5\,\text{m} \\
\text{snowing} \\
\text{neb} = 10
\end{bmatrix}
\quad
2\,\text{pm}
\begin{bmatrix}
-20.0 \\
-20.0 \\
+17.7 \\
736.8 \\
\text{min} = -20.0(\_) \\
\text{SW}, 16\,\text{m} \\
\text{neb} = 10 \\
\text{snöfyk}
\end{bmatrix}
\quad
9\,\text{pm}
\begin{bmatrix}
-23.2 \\
-23.3 \\
+10.0 \\
733.6 \\
\text{SW}, 20\,\text{m} \\
\text{neb} = 10 \\
\text{snöfyk}
\end{bmatrix}
$$

Until 5 am, the wind blew from the S, SE, but at this hour it started from the SW with a lot of snow and rapidly increased in intensity, augmenting over a meter per hour. The maximum of the day occurred at 9:10 at night: 20 m. This is the first time this year that the thermometer drops to −20°.

5.6.43   *April 22, [1903]*

$$
7\,\text{am}
\begin{bmatrix}
[-]24.95 \\
[-]25.05 \\
+1.1 \\
738.0 \\
\text{max} = -18.8 \\
\text{SSW}\,12.0\,\text{m} \\
\text{neb} = 5
\end{bmatrix}
\quad
2\,\text{pm}
\begin{bmatrix}
[-]24.4 \\
[-]24.55 \\
+13.1 \\
739.8 \\
\text{min} = -24.8 \\
\text{SW}\,15.5\,\text{m} \\
\text{neb} = 10 \\
\text{snöfyk}
\end{bmatrix}
\quad
9\,\text{pm}
\begin{bmatrix}
[-]24.5 \\
[-]24.5 \\
+14.9 \\
736.5 \\
\text{SW}\,20.5\,\text{m} \\
\text{neb} = 10 \\
\text{snöfyk}
\end{bmatrix}
$$

Today, in the afternoon, Peridota gave birth to nine robust descendants. Peridota is a female dog that is [only] 8 months and 7 days old! How fertile! The wind in the morning [blew] up to 11.5 m, and it started to increase at midday; the maximum was 21 m at 9 pm.

5.6.44 April 23, [1903]

The maximum of the day has been 255 m (average of an hour). Almost all day, the wind has been from the WSW with a lot of snow. The temperature has remained at around −23°. With this temperature and this wind, our noses are beginning to be in danger. The fact of not having a W.C. is also starting to be very unpleasant, not because of the cold, but because of the wind; that is, if we had these temperatures and calm [without any wind], we wouldn't suffer at all.

Between 7 am and 2 pm, the wind was between 25.5 m and 23.0 m. The minimum of the day was 20 m at 1 am. The sky cleared up completely after 6 in the afternoon. Six of Peridota's babies have died because of the cold; she has 3 left. Magnetic observations were made today. The temperature in the little house is between 20 and 23 degrees; after having the lights on for 5 min, it rises by 0.5°. It is a very quiet day, there are no disturbances. There seems to be a coincidence between the increases of wind speed and the drop in temperature.

	7 am		2 pm		9 pm
	−23.2		−22.95		−24.0
	−23.3		−23.0		−24.1
	−0.6		+8.9		+9.2
	732.35		734.9		734.75
	max = −23		S55W, 23 m		SW, 23 m
	SW 23.5 m		neb = 5?		neb = 2?
	neb = 10				

5.6.45 April 24, [1903]

	7 am		2 pm		9 pm
	[−]24.5		[−]24.0		[−]23.9
	[−]24.7		[−]24.05		[−]24.0
	+2.4		+9.6		+8.0
	734.7		737.85		739.55
	Max = [−]22.2?		min = [−]24.6		SW 15.5 m
	SW 22 m		SW 17 m		neb 5
	neb 8		neb 10		snöfyk
	snöfyk		snöfyk		

During the first hours of the day, the sky was clear, but at 5, it began to get cloudy, remaining so between 6 and 3, until 12 at midday, when it reached 10. At 8 pm, it started to clear up again, and at 10 pm, the sky was completely clear. The wind was diminishing quickly; the maximum of the day was at 6 am, 24 m; at 10 pm, it was 14 m.

5.6.46   April 25, [1903]

$$
7\,\text{am} \begin{bmatrix} [-]21.3 \\ [-]21.3 \\ 0.0 \\ 733.1 \\ \text{Max} = [-]21.3 \\ \text{E}\,4\,\text{m} \\ \text{Neb} = 10 \end{bmatrix} \quad
2\,\text{pm} \begin{bmatrix} [-]18.3 \\ [-]18.25 \\ +12.9 \\ 725.7 \\ \text{min} = 25.7 \\ \text{SW}\,2\,\text{m} \\ \text{neb} = 10 \end{bmatrix} \quad
9\,\text{pm} \begin{bmatrix} -13.5 \\ -13.95 \\ +11.6 \\ 723.3 \\ \text{SSW}\,12\,\text{m} \\ \text{neb} = 8 \end{bmatrix}
$$

While cleaning the hygrometer, we broke four of its hairs. It was changed for another one. At 8:30 at night, a rushing wind blew from the SW, clearing up the sky almost completely.

5.6.47   April 26, [1903]

$$
7\,\text{am} \begin{bmatrix} -17 \\ -17.1 \\ +4.5 \\ 723.6 \\ \text{Max} = 9.0 \\ \text{N} \\ \text{neb} = 4 \end{bmatrix} \quad
2\,\text{pm} \begin{bmatrix} [-]16.7 \\ [-]16.85 \\ +13.6 \\ 725.75 \\ \text{min} = 19.4 \\ \text{ESE}\,3\,\text{m} \\ \text{neb} = 2 \end{bmatrix} \quad
9\,\text{pm} \begin{bmatrix} [-]22.7 \\ [-]23.0 \\ 11.8 \\ 730.6 \\ \text{SSW} \\ \text{neb} = 1 \end{bmatrix}
$$

Variable wind all day, and quite clear sky. At 12 at night, I observed the circum-meridian altitudes of α Eridani in order to calculate the latitude. The temperature, which was around −25° during the observation, was very unpleasant. What one breathes condenses onto the mirrors, and glass of the _______ [unreadable], forming a layer of ice and making observations very difficult.

### 5.6.48	April 27, [1903]

$$
7\,\text{am}
\begin{bmatrix}
-21.95 \\
-22.0 \\
+2.9 \\
730.95 \\
\text{max} = -12.9 \\
\text{calm} \\
\text{neb} = 8
\end{bmatrix}
\qquad
2\,\text{pm}
\begin{bmatrix}
-23.25 \\
-23.4 \\
+12.9 \\
735.1 \\
\text{min} = -25.5 \\
\text{NNE 1 m} \\
\text{neb} = 1
\end{bmatrix}
\qquad
9\,\text{pm}
\begin{bmatrix}
-28.0 \\
-28.0 \\
+11.8 \\
738.65 \\
\text{calm} \\
\text{neb} = 0
\end{bmatrix}
$$

At 4 am, I observed a meridian of α Argus and I corrected the circle. During the night, from the 26th to the 27th, ice crystals have been [falling] continuously. Today, I have observed one of the nicest luminous phenomena: Ever since the morning, the fog had begun to increase, and by 12, it was a thick fog. Only toward the zenith, a patch of blue sky appeared. At 1 pm, when I went outside to make the time observations, I found this most beautiful phenomenon. [Draws observation pattern.]

### 5.6.49	April 28, [1903]

$$
7\,\text{am}
\begin{bmatrix}
-22.5 \\
-22.5 \\
+2.1 \\
737.8 \\
\text{Max} = -19.6 \\
\text{E 2 m} \\
\text{Neb} = 10
\end{bmatrix}
\qquad
2\,\text{pm}
\begin{bmatrix}
-22.8 \\
-22.9 \\
+11.4 \\
740.1 \\
\text{min} = 27.9 \\
\text{SW 5 m} \\
\text{Neb} = 10
\end{bmatrix}
\qquad
9\,\text{pm}
\begin{bmatrix}
[-]22.8 \\
[-]23.05 \\
+11.4 \\
43.9 \\
\text{SSW} = 10\,\text{m} \\
\text{Neb} = 8
\end{bmatrix}
$$

Variable wind until 1:30, when the SW began. Cloudy between 3 am and 8 pm.

### 5.6.50	April 29, [1903]

$$
7\,\text{am}
\begin{bmatrix}
[-]22.0 \\
[-]22.2 \\
+3.2 \\
746.1 \\
\text{SSW 9.5 m} \\
\text{Neb} = 4 \\
\text{max} = [\text{blank}]
\end{bmatrix}
\qquad
2\,\text{pm}
\begin{bmatrix}
[-]19.75 \\
[-]20.0 \\
+14.0 \\
748.25 \\
\text{min} = -26.6 \\
\text{SSW 5 m} \\
\text{neb} = 3
\end{bmatrix}
\qquad
9\,\text{pm}
\begin{bmatrix}
[-]21.1 \\
[-]21.1 \\
11.5 \\
745.45 \\
\text{SSW 14 m} \\
\text{neb} = 5
\end{bmatrix}
$$

Wind from the SSW all day, with quite clear sky.

5.6.51   *April 30, [1903]*

$$
7\,\text{am}
\begin{bmatrix}
[-]18.0 \\
[-]18.05 \\
13.2 \\
742.2 \\
\text{max} = [-]16.7 \\
\text{SSW } 17\,\text{m} \\
\text{neb} = 3
\end{bmatrix}
\quad
2\,\text{pm}
\begin{bmatrix}
[-]19.1 \\
[-]19.2 \\
+11.8 \\
743.1 \\
\text{min} = -21.6 \\
\text{SSW } 11\,\text{m} \\
\text{neb} = 3
\end{bmatrix}
\quad
9\,\text{pm}
\begin{bmatrix}
[-]18.6 \\
[-]18.6 \\
11.8 \\
743.4 \\
\text{S } 3\,\text{m} \\
\text{neb} = 10
\end{bmatrix}
$$

The maximum of the wind was at 1 am, 19 m, and from that time, it began to diminish and to move to the S. Since 7 pm, the fog could be seen moving forward from the NE. At 9, the sky was completely overcast, and everything was covered with frost. At 8:30, the wind moved to the SE and blew very weakly. Taking advantage of a moment when the wind was calm, and before the sky became covered with fog, I was able to observe an altitude of α Argus to calculate the time.

5.6.52   *May 1, [1903]*

The fourth month of the year is now finished. And although it's been somewhat stormy, it's the nicest [month] that we've had in 1903. Today is a holiday for the Swedish people, because it's the beginning of Spring in Sweden. Magnetic observations are taken, and it's quite a calm day. At night, I tried to make altitude observations in order to calculate the time, but in vain; the condensation of breath and moisture over mirrors, glasses, ________ [unreadable], and magnifying glasses makes it impossible to make good observations. To correct this, they're thinking of installing the transit theodolite; this will make it much easier and faster. According to what they say, there are some clearings in the water ENE. Last night, there was a great freeze. The amount of precipitation that it has brought is incredible; the land is completely white. The average temperature for the month of April was −14.05. Last year it was −13.37

$$
7\,\text{am}
\begin{bmatrix}
-20.0 \\
-20.0 \\
+3.2 \\
739.6 \\
\text{max} = -17.2 \\
\text{NE } 2\,\text{m} \\
\text{Fog } 8\_ \\
\text{Neb} = 10
\end{bmatrix}
\quad
2\,\text{pm}
\begin{bmatrix}
-20.9 \\
-20.85 \\
+13.3 \\
740.8 \\
\text{min} = -22.3 \\
\text{ENE } 1\,\text{m} \\
\text{Neb} = 10 \\
\text{Ci, Ci-S, S}
\end{bmatrix}
\quad
9\,\text{pm}
\begin{bmatrix}
-23.5 \\
-23.5 \\
+10.0 \\
740.2 \\
\text{SW } 2\,\text{m} \\
\text{Neb } 10
\end{bmatrix}
$$

5.6.53   *May 2, [1903]*

$$
7\,\text{am}\;\begin{bmatrix}[-]25.1\\ -25.1\\ +4.4\\ 737.75\\ \text{Max}=-19.4\\ \text{Calm}\\ \text{neb}=8\\ \text{Ci\_ENE, WSW}\end{bmatrix}\quad
2\,\text{pm}\;\begin{bmatrix}[-]23.1\\ [-]23.25\\ +9.0\\ 737.05\\ \min=-25.0\\ \text{W 5 m}\\ \text{neb}=7\end{bmatrix}\quad
9\,\text{pm}\;\begin{bmatrix}[-]23.9\\ [-]23.9\\ +11.6\\ 737.4\\ \text{NE 0.2 m}\\ \text{neb}=8\end{bmatrix}
$$

From 9 am to 1 pm, snow fell, although it was very little. At 5 pm, it started again and continued, stopping at times. At 11, Nordenskjöld and Jonassen set off toward Seymour Island with a sledge pulled by four dogs. The objective is to bring back more of the provisions stored in that location. They have been trying to set off for a week now, but are waiting for "fint väder" [nice weather] that never comes. [The departure] was resolved just today.

5.6.54   *May 3, [1903]*

A pretty awful day, with changing winds and some snowfall. At 1:30 pm, wind from NE became established, and the temperature rises and more snow starts to fall. At night, it looks like it has the intention to clear up. I saw something around the moon that drew my attention. I don't know whether it's a coincidence in the positioning of two cirrus and the moon, or a luminosity phenomenon. [Makes a drawing.] Today at 3 pm, the ones from the Seymour Island trip came back. You can still see snow petrels around here. These birds probably spend the entire year here because they find food in any seal hole.

$$
7\,\text{am}\;\begin{bmatrix}[-]20.5\\ [-]20.5\\ +4.4\\ 736.65\\ \max=[-]19.8\\ \text{neb}=10\end{bmatrix}\quad
2\,\text{pm}\;\begin{bmatrix}[-]17.0\\ [-]16.95\\ +13.2\\ 736.7\\ \min=-25.0\\ \text{ENE 3 m}\\ \text{haze of snow}\end{bmatrix}\quad
9\,\text{pm}\;\begin{bmatrix}-14.6\\ [-]14.6\\ +11.6\\ 734.4\\ \text{NE 6 m}\\ \text{neb}=9.5\end{bmatrix}
$$

5.6.55   *May 4, [1903]*

$$
7\,\text{am}\;\begin{bmatrix}[-]24.4\\ [-]24.4\\ +3.0\\ 736.6\\ \max=-15.4?\\ \text{SW}\,18\,\text{m}\\ \text{neb}=10\\ \text{snöfyk}\end{bmatrix}
\qquad
2\,\text{pm}\;\begin{bmatrix}[-]23.75\\ [-]23.85\\ +13.2\\ 740.7\\ \min=-26.1\\ \text{SW}\,15.5\,\text{m}\end{bmatrix}
\qquad
9\,\text{pm}\;\begin{bmatrix}[-]25.0\\ 85.0\%\\ +8.3\\ 741.2\\ \text{SSW}\,11.5\,\text{m}\\ \text{neb}=8\end{bmatrix}
$$

At 2:30 am, the SW started to blow with snow. The thermometer dropped, and the barometer began to rise at the same time. It continued blowing with strength all day. The maximum was 20 m at 10 am, and since that time, it began to diminish its intensity; at 12 at night, it was of 12.5 m.

5.6.56   *May 5, [1903]*

$$
7\,\text{am}\;\begin{bmatrix}-27.0\\ -27.0\\ +0.6\\ 742.65\\ \text{Max}=[-]22.8\\ \text{SSW}\,13\,\text{m S}\\ \text{Neb}=10\end{bmatrix}
\qquad
2\,\text{pm}\;\begin{bmatrix}-28.1\\ -28.2\\ +9.2\\ 764.4\\ \min=-28.0?\,(\text{the wind})\\ \text{SSW}\,11\,\text{m}\\ \text{Neb}=4\end{bmatrix}
\qquad
9\,\text{pm}\;\begin{bmatrix}-27.1\\ -27.3\\ +8.8\\ 746.5\\ \text{SSW}\,13.5\\[4pt] \text{Neb}=9.5\end{bmatrix}
$$

5.6.57   *May 6, [1903]*

$$
7\,\text{am}\;\begin{bmatrix}[-]27.6\\ [-]27.6\\ +1.3\\ 748.5\\ \text{Max}=-26.0\\ \text{SSW}\,6\,\text{m}\\ \text{Neb}=1\end{bmatrix}
\qquad
2\,\text{pm}\;\begin{bmatrix}[-]28.1\\ [-]28.5\\ 750.9\\ +14.6\\ \min=-28.1\\ \text{NE}\,1\,\text{m}\\ \text{Neb}=2\end{bmatrix}
\qquad
9\,\text{pm}\;\begin{bmatrix}[-]27.3\\ [-]27.3\\ +9.0\\ 749.55\\ \text{E}\,2.5\,\text{m}\\ \text{Neb}=1\end{bmatrix}
$$

5.6.58 *May 7, [1903]*

$$
7\,\text{am}
\begin{bmatrix}
[-]27.0 \\
[-]27.0 \\
+0.8 \\
745.7 \\
\text{Max} = 25.2 \\
\text{ENE } 1.5\,\text{m} \\
\text{Neb} = 10
\end{bmatrix}
\quad
2\,\text{pm}
\begin{bmatrix}
-24.9 \\
-24.9 \\
+9.2 \\
745.5 \\
\text{min} = -29.3 \\
\text{variable } 0\,\text{m} \\
\text{Neb} = 8
\end{bmatrix}
\quad
9\,\text{pm}
\begin{bmatrix}
-24.5 \\
-24.5 \\
+10.1 \\
744.1 \\
\text{calm } 0.5\,\text{m} \\
\text{Neb} = 9.5
\end{bmatrix}
$$

5.6.59 *May 8, [1903]*

$$
7\,\text{am}
\begin{bmatrix}
-8.35 \\
-8.4 \\
+2.5 \\
739.5 \\
\text{Max} = -7.1 \\
\text{Calm } 0.5\,\text{m} \\
\text{Neb} = 8
\end{bmatrix}
\quad
2\,\text{pm}
\begin{bmatrix}
-2.4 \\
-3.5 \\
+14.2 \\
737.0 \\
\text{min} = -28.5 \\
\text{NE } 7.5\,\text{m} \\
\text{Neb} = 9.5
\end{bmatrix}
\quad
9\,\text{pm}
\begin{bmatrix}
-21.65 \\
-21.7 \\
+12.0 \\
735.6 \\
\text{SW } 15.5\text{m} \\
\text{Neb} = 9.5 \\
\text{snow}
\end{bmatrix}
$$

At 6:50 pm, the NE calmed, and the temp = −1°, and after the usual turn, the SW began. At 7:50, the temperature was −21.0°, that is, 20 degrees lower in 1 hour.

5.6.60 *May 9, [1903]*

$$
7\,\text{am}
\begin{bmatrix}
-24.95 \\
-25.0 \\
+5.2 \\
739.3 \\
\text{Max} = 0.8 \\
\text{SW } 15.5\,\text{m} \\
\text{Neb} = 10
\end{bmatrix}
\quad
2\,\text{pm}
\begin{bmatrix}
[-]24.9 \\
[-]25.0 \\
+8.3 \\
743.05 \\
\text{min} = -25.0 \\
\text{WSW } 14.5\,\text{m} \\
\text{Neb} = 8
\end{bmatrix}
\quad
9\,\text{pm}
\begin{bmatrix}
[-]24.25 \\
[-]24.4 \\
+7.5 \\
745.65 \\
\text{SW } 11\,\text{m} \\
\text{Neb} = 10
\end{bmatrix}
$$

The maximum of the wind was 19 m at 1 am, and the minimum was at 12 at night, and at 9 pm, it was 11 m. After 2, the sky cleared up quite a bit, but, since 5, it began to become overcast with A-Cu (there are moon coronas with these clouds). All day, the wind has been SW and WSW. This morning, I saw a snow petrel flying around the station.

5.6.61   May 10, [1903]

$$
7\,\text{am}
\begin{bmatrix}
-22.8 \\
-23.0 \\
+3.0 \\
748.4 \\
\text{WSW 13 m} \\
\text{Neb} = 10
\end{bmatrix}
\quad
2\,\text{pm}
\begin{bmatrix}
-23.05 \\
-23.3 \\
+7.6 \\
750.2 \\
\text{min} = -25.8 \\
\text{SW 14.5 m} \\
\text{Neb} = 1
\end{bmatrix}
\quad
9\,\text{pm}
\begin{bmatrix}
[-]24.1 \\
[-]24.3 \\
+6.9 \\
751.85 \\
\text{SSW 11 m} \\
\text{Neb} = 7
\end{bmatrix}
$$

The maximum of the wind was at 2 pm and the minimum at 4 pm, 8.5 m. During the day, the wind has been from the SSW and SW.

5.6.62   May 11, [1903]

$$
7\,\text{am}
\begin{bmatrix}
-24.9 \\
-25.0 \\
-0.8 \\
752.8 \\
\text{Max} = -22.7 \\
\text{SW 13 m} \\
\text{Neb} = 10 \\
\text{haze with snow}
\end{bmatrix}
\quad
2\,\text{pm}
\begin{bmatrix}
-23.9 \\
-23.9 \\
+10.6 \\
757.05 \\
\text{min} = -25.7 \\
\text{WSW 18.5 m} \\
\text{neb} = 10 \\
\text{snöfyk}
\end{bmatrix}
\quad
9\,\text{pm}
\begin{bmatrix}
-22.0 \\
-22.05 \\
+9.7 \\
759.75 \\
\text{SSW 13 m} \\
\text{neb} = 10 \\
\text{S}
\end{bmatrix}
$$

During the first hours of the morning until 10 am, [the wind] blew from the SW, but from that time it was from the SSW.

5.6.63   May 12, [1903]

$$
7\,\text{am}
\begin{bmatrix}
[-]25.9 \\
[-]25.9 \\
+1.0 \\
762.0 \\
\text{max} = -19.9 \\
\text{calm} \\
\text{clear sky}
\end{bmatrix}
\quad
2\,\text{pm}
\begin{bmatrix}
+13.6 \\
761.9 \\
-21.3 \\
-21.5 \\
\text{min} = -25.95 \\
\text{calm} \\
\text{neb 1 Ci}
\end{bmatrix}
\quad
9\,\text{pm}
\begin{bmatrix}
+12.0 \\
761.6 \\
-20.7 \\
-20.6 \\
\text{calm} \\
\text{Ci neb 1}
\end{bmatrix}
$$

A very nice day. Clear and calm but cold. The sun has shone for 6 h from 9 am to 3 pm, but it doesn't warm up any more. Today, I made some observations of stellar altitudes to calculate the time.

5.6.64   *May 13, [1903]*

$$
7\,\text{am}\left[\begin{array}{l}
-16.55 \\
-16.5 \\
+6.0^{\circ} \\
759.45 \\
\text{Max} = -16.2 \\
\text{Calm} \\
\text{Frost} \\
\text{Very weak variable wind}
\end{array}\right.
\qquad
2\,\text{pm}\left[\begin{array}{l}
89 = \underline{} \\
7 = 21.8 \\
+15.0 \\
759.8 \\
\min = -25.3 \\
\text{calm} \\
\text{neb} = 0 \\
\text{precipitation in ice} \\
\text{crystals}
\end{array}\right.
\qquad
9\,\text{pm}\left[\begin{array}{l}
-18.95 \\
-18.9 \\
91 \\
+10.3 \\
755.25 \\
\text{calm} \\
\text{neb } 9.5 \\
\text{Ci P.R N-S} \\
\text{corona}
\end{array}\right.
$$

Calm or very weak and variable winds all day.

5.6.65   *May 14, [1903]*

$$
7\,\text{am}\left[\begin{array}{l}
-3.75 \\
-3.85 \\
+4.5 \\
744.6 \\
\max = -3.15 \\
\text{ENE 3 m} \\
\text{Neb} = 10 \\
\text{Wind}
\end{array}\right.
\qquad
2\,\text{pm}\left[\begin{array}{l}
-14.1 \\
-14.4 \\
+13.8 \\
744.3 \\
\min = -18.2 \\
\text{SSW 18 m} \\
\text{Neb} = 2
\end{array}\right.
\qquad
9\,\text{pm}\left[\begin{array}{l}
-12.05 \\
-12.90 \\
+12.4 \\
748.2 \\
\text{NE 5 m} \\
\text{Neb} = 0
\end{array}\right.
$$

At 7 am, NE began to blow with completely overcast sky. At 10 am, the intensity started to diminish, and it began to grow calm. From 10:35 to 11:10, it was calm. At this last hour, the WSW started, moving to the SW immediately and then to the SSW at a speed of around 10 m which, after an hour, was 13.5 m. But with the same speed that it had come, it started to diminish: at 3 pm 2.7 points, at 4 pm 100, at 5 pm 89, and at 7 pm 55. At 9 pm, after some time of calm, it began to blow from the NE.

5.6.66   *May 15, [1903]*

$$
7\,\text{am}\left[\begin{array}{l}
-15.5 \\
-15.6 \\
+5.7 \\
749.4 \\
\max = +2.0 \\
\text{calm} \\
\text{neb} = 1
\end{array}\right.
\qquad
2\,\text{pm}\left[\begin{array}{l}
-13.0 \\
-13.0 \\
84.5 \\
+17.0 \\
751.0 \\
\text{ESE 1 m} \\
\text{neb} = 2 \\
\min = -20.5
\end{array}\right.
\qquad
9\,\text{pm}\left[\begin{array}{l}
95.5 \\
-16.4 \\
+10.3 \\
750.35 \\
\text{calm} \\
\text{neb} = 2
\end{array}\right.
$$

Calm and very weak variable wind all day. The maximum speed of the wind was 4.5 m at 5 pm. Today, I have made some observations with the transit instrument. It has been a very nice day.

5.6.67   May 16, [1903]

$$
7\,\text{am}
\begin{bmatrix}
-11.45 \\
-11.95 \\
+6.6 \\
747.5 \\
\text{Max} = -5.9 \\
\text{NE}\,9\,\text{m} \\
\text{Neb} = 2
\end{bmatrix}
\quad
2\,\text{pm}
\begin{bmatrix}
-8.3 \\
-9.0 \\
+14.8 \\
745.3 \\
\text{min} = -19.5 \\
\text{NE}\,11\,\text{m} \\
\text{Neb} = 9
\end{bmatrix}
\quad
9\,\text{pm}
\begin{bmatrix}
-3.9 \\
-4.9 \\
+11.2 \\
739.8 \\
\text{NE}\,11\,\text{m} \\
\text{Neb} = 2
\end{bmatrix}
$$

The NE began at 2 am and blew all day with very constant strength, between 9 and 11 m per second. Temperature and humidity undergo very sudden changes.

Today, there has been a big racket. When we were finishing breakfast, J [Jonassen], who was in his room, started to cough and spit. Then, E [Ekelöf] closed the door of his [Jonassen's] room. J [Jonassen] opened it again, saying something that I didn't understand. After a little while, he closed it again, saying that if he wanted to have it closed, *he* would close it (directing this to E [Ekelöf]). After that, E [Ekelöf] went to speak badly of J [Jonassen] with Å [Åkerlundh]. And he [Jonassen] started to answer from his room. Then, N [Nordenskjöld] came out and started fighting with E [Ekelöf]. The result was that, at lunch time, N and E [Nordenskjöld and Ekelöf] changed places [so as to have Nordenskjöld, rather than Ekelöf, sit near Jonassen], but J [Jonassen] did not show up at the table either at coffee time or at lunch time. J [this should be E, Ekelöf] and N [Nordenskjöld] went back to occupying their former places, and J [Jonassen] did not show up again at the table and ate his meals in his room.

It seems that Sobral's view of Dr. Ekelöf as one who stirred up trouble was shared by some of the others, at least by dog-driver Jonassen, and that his mentor Nordenskjöld served as the parental figure in this chilly tableau of domestic strife.

5.6.68   May 17, [1903]

Poor Skottsberg, the last dog from the Malvinas [Falklands] that was left, was found dead this morning, killed by the other dogs [the Greenlanders]. Now, counting the puppies, there are 8 males and eight females left.

Since 4 in the afternoon, the SSW has been blowing; first, it was an outgoing wind, but then the temperature began to drop from $0°$ to $-21°$ at 9 pm.

$$
7\,\text{am}
\begin{bmatrix}
-4.45 \\
-5.90 \\
+7.8 \\
736.7 \\
\text{Max} = -15 \\
\text{SSW and SSE } 0.5\,\text{m} \\
\text{Neb} = 3
\end{bmatrix}
\qquad
2\,\text{pm}
\begin{bmatrix}
-0.75 \\
-3.1 \\
+17.1 \\
737.95 \\
\text{min} = -8.9 \\
\text{SSW } 8\,\text{m}
\end{bmatrix}
\qquad
9\,\text{pm}
\begin{bmatrix}
-21.0 \\
-21.8 \\
+9.0 \\
740.95 \\
\text{SSW } 19.5\,\text{m} \\
\text{Neb} = 0
\end{bmatrix}
$$

The wind from the SSW reached its maximum at 9 pm, and from then, it began to diminish in intensity; at 12 at night, it was blowing at a speed of 16.5 m.

It seems that the month of May of this year is going to stand out because of the shortness of the storms and the clearness of the weather; so far, that is all we've had.

Keeping track of the dogs, Sobral notes that, with the demise of the last Falkland dog Skottsberg, there are now 16 dogs, including puppies born during the voyage and during the time at Snow Hill. The eight from Port Stanley have all died, either from weakness or from attacks waged by the Greenlanders. In his book written after the expedition, Nordenskjöld laments the loss of the dog Skottsberg, though he does not refer to him by name. He writes (Nordenskjöld et al. 1905: 278):

> I looked for him a long time in vain, but at last found him lying dead among some blocks of
> ice on the shore. I felt his loss very much; he had followed me faithfully on all my
> wanderings, and was the last of the dogs who was a companion and a friend.

Nordenskjöld's words echo the viewpoint and sentiments of many of the explorers of the early twentieth century who relied on their dogs not just for mobility but also for companionship in the cold isolation of Antarctica. While attitudes toward the use of dogs for exploration in Antarctica differed from country to country, many personal diaries and recollections of early and mid-century explorers emphasize the importance of the presence of dogs on these expeditions, not just for the physical safety, but also for the mental well-being of the humans (Tahan a, b).

5.6.69   May 18, [1903]

$$
7\,\text{am}
\begin{bmatrix}
-18.9 \\
-19.5 \\
+2.75 \\
744.35 \\
\text{Max} = -0.1 \\
\text{S } 5\,\text{m} \\
\text{Neb} = 3
\end{bmatrix}
\qquad
2\,\text{pm}
\begin{bmatrix}
-17.75 \\
[-]17.85 \\
+9.7 \\
737.25 \\
\text{min} = -21.5 \\
\text{E } 5\,\text{m} \\
\text{Neb} = 10 \\
\text{snowing}
\end{bmatrix}
\qquad
9\,\text{pm}
\begin{bmatrix}
-21.9 \\
-22.0 \\
+10.4 \\
739.05 \\
\text{SSW } 11\,\text{m} \\
\text{Neb} = 3
\end{bmatrix}
$$

The sky became overcast at 8 am, and snow began to fall at midday. The wind blew from the NE beginning at 9 am, and the barometer dropped quickly. At 3 pm, after a ¼ hour of calm, it started to blow from the SSW with strength. The maximum of the wind was 17.5 m at 6 pm, then it diminished rapidly, and at 12 at night, it was just 5.5 m.

5.6.70   *May 19, [1903]*

Today, the day dawned with clouds and with fog. A very awful day. In the morning, it blew from the NE and SE, but the barometer proceeded to drop quickly, announcing a SW. At 1:45 pm, the SW started to blow violently, calming down completely a quarter of an hour later, without any progressions of any kind. At 5:10, the SW started to blow again, only to calm 2 h 30 m later, and for a quarter of an hour, weak gusts blew from the SE and NE; 2 h 45 m later, the SW returned, each time with a lot of snow.

I'm still at least 8 months away from the day of my freedom. These people constantly argue about food. And it is relatively abundant. What will happen if one day there is no food to eat? [Blocked out something he had written so as to completely obscure it.] Now, they're talking about dividing the food that can be divided; what feelings of a person, such little generosity.

$$
2\,\text{pm}
\begin{bmatrix}
-24.9 \\
-25.0 \\
+13.6 \\
729.8 \\
-24.7 \\
\text{SSW 10 m} \\
\text{Neb} = 10
\end{bmatrix}
\quad
7\,\text{pm}
\begin{bmatrix}
-19.8 \\
-19.8 \\
734.4 \\
+3.2 \\
\text{ENE 4 m} \\
\text{Neb} = 10
\end{bmatrix}
\quad
9\,\text{pm}
\begin{bmatrix}
-24.8 \\
-24.8 \\
+11.3 \\
726.9 \\
\text{SSW 8.5 m} \\
\text{Neb} = 10
\end{bmatrix}
$$

Until 12 at midday, variable wind between NE and SE, with a lot of haze and frost. At 12:45, it suddenly started to blow violently from the SSW, with snow or haze, and it was very difficult to distinguish between them. An hour and a half later, it calmed suddenly and started to blow very weakly from the ENE and SE; 1 h 45 m later, the SSW started again. The maximum speed was 14 m at 3 pm. These storms of such short duration are truly curious, since we have never had them before. In this year's compass rose, I think the component of the winds from the SE will be much larger than last year's.

5.6.71   *May 20, [1903]*

$$
7\,\text{am}
\begin{bmatrix}
[-]26.45 \\
[-]26.55 \\
+1.8 \\
726.2 \\
\text{Max} = 14.9 \\
\text{SSW}\,7\,\text{m} \\
\text{Neb} = 10
\end{bmatrix}
\quad
2\,\text{pm}
\begin{bmatrix}
[-]21.3 \\
[-]21.3 \\
+14.7 \\
729.45 \\
\text{min} = -26.4 \\
\text{Neb} = 7 \\
\text{E}\,2\,\text{m}
\end{bmatrix}
\quad
9\,\text{pm}
\begin{bmatrix}
[-]26.25 \\
[-]26.25 \\
+12.9 \\
727.0 \\
\text{SSW}\,8\,\text{m} \\
\text{Neb} = 8
\end{bmatrix}
$$

At 7 am, the SSW that had begun yesterday calmed down, and between 8 and 3 pm, breezes blew, varying between NE and SE. At 4, the SSW started again, and it calmed at 11 pm. The maximum strength of the wind was 11.5 m at 7 pm. These small storms follow the same rules as the big ones in terms of barometer and thermometer; that is, the barometer rises ________ [unreadable—most likely "and the wind"] blows from the SSW, only to drop again, moving to the NE, E, and SE. Today, the bread appeared divided into as many pieces as there were people at the table. Ever since the incident with E [Ekelöf], J's [Jonassen's] room has its door constantly closed. But they've built a little window that looks to the kitchen, where the air comes from, and through which food is passed to him; from what I've heard said, it is special.

Sobral tracks the short storms raging outside across the Antarctic ice and the mini storms raging within the house that cause turbulence among the men. In addition to the cabin fever that is starting to catch fire among the expeditioners, and that is resulting in the fraying of nerves, the shortage of food now necessitates that they ration out portions for each individual—a concept that Sobral initially does not seem to warm up to. His meticulous record-keeping, however, does not falter during these turbulent times.

5.6.72   *May 21, [1903]*

Of all the individuals whom I have heard singing, the one who sings the worst is N [Nordenskjöld]. On days of great events, when those moments favor him, he lets loose into the air his fa-la-la-las, an ensemble of sounds that have no harmony whatsoever, and that make anyone whose ear is even moderately educated, positively see stars. Today, at night, he "sang," and I, amazed, couldn't find the reason for it, because what has been happening during these last few days would cause anything but give him a reason to reward us with his beautiful voice. But when I was already in bed, and while I was reading, I heard a conversation being held in such a loud voice, that one, located at a 5-m distance away, could easily understand everything. According to this conversation, E [Ekelöf] and J [Jonassen] today have become friends, dispelling the Cu- Ni that has been hanging over the head of their indulgent "papa" [Nordenskjöld].

$$7\,\text{am} \begin{bmatrix} -18.4 \\ -18.4 \\ +2.2 \\ 729.65 \\ \text{Var}\,2\,\text{m} \\ \text{Neb} = 4 \end{bmatrix} \quad 2\,\text{pm} \begin{bmatrix} -6.5 \\ -6.8 \\ +12.5 \\ 725.7 \\ \text{calm} \\ \text{Neb} = 10 \end{bmatrix} \quad 9\,\text{pm} \begin{bmatrix} -15.05 \\ -15.35 \\ +12.6 \\ 725.3 \\ \text{calm} \\ \text{Neb} = 10 \end{bmatrix}$$

Until 4 pm, very weak breezes from all directions, but, since that time, complete calm. The sky was overcast until 6 pm, when it started to clear up, and at 10:30 pm, it cleared up completely.

This is a clever little piece of writing by Sobral, facetiously yet endearingly poking fun at the expedition leader Nordenskjöld and using meteorological record-keeping "cloud" terminology to describe the until now stormy relationship between Ekelöf and Jonassen. It gives a nice insight into Sobral's own sense of humor among the dire circumstances and his mastery of interpersonal earth and human relationships.

Besides his usage of meteorological terms, Sobral at this time was also learning Swedish, even using the classic comic book Strix *to expand his Swedish vocabulary (Skottsberg 1961). One can only imagine the conversations he held using that reference source as his matrix.*

5.6.73   *May 22, [1903]*

$$7\,\text{am} \begin{bmatrix} -8.4 \\ -9.55 \\ +5.8 \\ 722.95 \\ \text{max} = -3.8 \\ \text{WSW}\,(\text{very weak?}) \\ \text{Neb} = 1 \end{bmatrix} \quad 2\,\text{am} \begin{bmatrix} [-]22.6 \\ [-]22.85 \\ +12.3 \\ 732.3 \\ \text{min} = -23.4 \\ \text{SSW}\,15\,\text{m} \end{bmatrix} \quad 9\,\text{am} \begin{bmatrix} 26.7 \\ 26.75 \\ +9.5 \\ 738.2 \\ \text{S}\,13\,\text{m} \\ \text{Neb} = 10 \end{bmatrix}$$

The wind from the SSW that had started to blow suddenly increased its intensity at 9 am, breaking the anemometer. The sky was clear almost all day (until 8 pm); at 8 h 30 m, it became completely overcast. The maximum wind speed has been 16 m.

J [Jonassen] now keeps the door to his room open all day and talks openly with the others, but he still doesn't eat at the table.

5.6.74   May 23, [1903]

<table>
<tr>
<td rowspan="8">7 am</td>
<td>[−]26.3</td>
<td rowspan="8">2 pm</td>
<td>−23.3</td>
<td rowspan="8">9 pm</td>
<td>−25.5</td>
</tr>
<tr><td>[−]26.25</td><td>−23 − 25</td><td>−25.55</td></tr>
<tr><td>+ 3.8</td><td>+ 15.8</td><td>+ 11.9</td></tr>
<tr><td>740.6</td><td>740.75</td><td>737.7</td></tr>
<tr><td>max = + 7.7</td><td>min = −28.5</td><td>SSW 10 m</td></tr>
<tr><td>calm</td><td>var0.5m</td><td>neb = 10</td></tr>
<tr><td>fog</td><td>fog10</td><td>snowing</td></tr>
<tr><td>Frost 10</td><td></td><td></td></tr>
</table>

Fog until 6 pm; then it started to snow. At 4 pm, it started to blow from the SW, moving to the SSW shortly afterward. At 9 pm, it began to clear up, nebulosity being = 4 at 11 pm.

5.6.75   May 24, [1903]

<table>
<tr>
<td rowspan="8">7 am</td>
<td>[−]28 − 45</td>
<td rowspan="8">2 pm</td>
<td>[−]26.4</td>
<td rowspan="8">9 pm</td>
<td>[−]25.5</td>
</tr>
<tr><td>[−]28.55</td><td>[−]26.45</td><td>[−]25.55</td></tr>
<tr><td>+ 1.7</td><td>+ 14.0</td><td>+ 9.9</td></tr>
<tr><td>735.45</td><td>731.7</td><td>732.95</td></tr>
<tr><td>Max = [−]20.7</td><td>min = 28.4</td><td>SW 15 m</td></tr>
<tr><td>SSW 10 m</td><td>SW 17 m</td><td>Neb = 8</td></tr>
<tr><td>Neb = 10</td><td>Neb = 10</td><td>snöfyk</td></tr>
<tr><td></td><td>snöfyk</td><td></td></tr>
</table>

SSW with snow blowing all day, and at 9 pm, the sky started to clear up. The maximum of the wind was at 2 pm.

5.6.76   May 25, [1903]

<table>
<tr>
<td rowspan="8">7 am</td>
<td>−24.9</td>
<td rowspan="8">2 pm</td>
<td>−25.8</td>
<td rowspan="8">9 pm</td>
<td>−28.0</td>
</tr>
<tr><td>−24.9</td><td>−26.6</td><td>−28.0 ?</td></tr>
<tr><td>+ 4.0</td><td>+ 12.7</td><td>85.6</td></tr>
<tr><td>736.95</td><td>740.75</td><td>+ 9.6</td></tr>
<tr><td>Max = −23.5</td><td>min = [−26.0 / −25.5]</td><td>743.25</td></tr>
<tr><td>SSW 7 m</td><td></td><td>WSW 2 m</td></tr>
<tr><td>Neb = 8</td><td>calm</td><td>neb = 0</td></tr>
<tr><td></td><td>neb = 1</td><td></td></tr>
</table>

Until midday, it blew from the SSW, and since that time, it was almost calm until 9 pm, when it began to blow from the SW. Since 10 am, the sky was very clear; only to the S, a bank of clouds could be seen, which remained in the same position until nighttime. The maximum strength of the wind was 13 m at 6 am. Today is my people's day. God willing, they will celebrate it in peace and happiness. The sun of May rose to the NE like a shining ball of fire through clouds of snow blown by the wind. I had the intention of greeting it with a 21-shot salvo, but I thought that these bullets could be used in a more practical way, for hunting seals or birds that we need.

The May Revolution of 1810, which would lead to the independence of Argentina in 1816, is a date not taken lightly by Sobral and one that he celebrates in the ice fields of Antarctica, not knowing whether his country, at this very time, is at war. With no other Argentine compatriots to share the holiday, Sobral poetically observes the dawning of the day and wisely decides to conserve the 21-shot salute.

5.6.77 *May 26, [1903]*

$$
7\,\text{am}
\begin{bmatrix}
-29.15 \\
-29.15 \\
+1.8 \\
748.3 \\
\max = -23.0 \\
\text{calm} \\
\text{neb} = 2
\end{bmatrix}
\quad
2\,\text{pm}
\begin{bmatrix}
[-]21.9 \\
[-]21.95 \\
+13.1 \\
748.05 \\
\min = -30.4 \\
\text{ENE}\,5\,\text{m} \\
\text{neb} = 7
\end{bmatrix}
\quad
9\,\text{pm}
\begin{bmatrix}
[-]24.5 \\
[-]24.45 \\
+12.2 \\
744.35 \\
\text{SW}\,5\,\text{m} \\
\text{neb} = 10
\end{bmatrix}
$$

The SSW, which started yesterday night, calmed at 4:30 am, and at 10:30 am, the NE started, which lasted until 5 pm; the SSW started at 6 pm. The sky was quite clear until 3 pm, when it became completely cloudy; at 7 pm, a little snow started to fall. The maximum of the wind was 11 m at 12 at night.

5.6.78 *May 27, [1903]*

$$
7\,\text{am}
\begin{bmatrix}
[-]25.9 \\
[-]25.9 \\
+2.4 \\
739.95 \\
\max = -19.2 \\
\text{SSW}\,9\,\text{m} \\
\text{Neb} = 10 \\
\text{snöfyk}
\end{bmatrix}
\quad
2\,\text{pm}
\begin{bmatrix}
-24.0 \\
-24.1 \\
+13.2 \\
731.45 \\
\min = -28.0 \\
\text{SSW}\,7\,\text{m} \\
\text{Neb} = 10
\end{bmatrix}
\quad
9\,\text{pm}
\begin{bmatrix}
[-]23.25 \\
[-]23.45 \\
+13.3 \\
743.45 \\
\text{SSW}\,4\,\text{m} \\
\text{Neb} = 1
\end{bmatrix}
$$

Wind from the SSW all day, and calm at 10 pm. Overcast sky until 8 pm, when it cleared up. The maximum strength of the wind was 10 m at 1 am.

5.6.79   *May 28, [1903]*

$$
7\,\text{am}
\begin{bmatrix}
-21.5 \\
-21.4 \\
+2.8 \\
738.85 \\
\text{Max} = +21.0 \\
\text{E}\,2\,\text{m} \\
\text{Neb} = 10
\end{bmatrix}
\quad
2\,\text{pm}
\begin{bmatrix}
16.9 \\
17.5 \\
+13.17 \\
737.6 \\
26.5 \\
\text{SSW}\,5\,\text{m} \\
\text{Neb} = 10
\end{bmatrix}
\quad
9\,\text{pm}
\begin{bmatrix}
22.4 \\
22.5 \\
+13.0 \\
746.6 \\
\text{SSW}\,9\,\text{m} \\
\text{Neb} = 1
\end{bmatrix}
$$

Fully overcast sky until 5:30, at which time it cleared up completely. Variable wind during the entire morning. At 2 pm, the SSW began. At 7:30, it calmed completely, and, right away, the NE started to blow very weakly, and some fog came. At 8 pm, the SSW came back, calming at 10 pm. The maximum strength of the wind was at 9 pm.

5.6.80   *May 29, [1903]*

$$
7\,\text{am}
\begin{bmatrix}
-20.4 \\
-20.4 \\
+5.2 \\
747.75 \\
\text{Max} = -16.2 \\
\text{Var} \\
\text{Neb} = 6
\end{bmatrix}
\quad
2\,\text{pm}
\begin{bmatrix}
-5.0 \\
-5.95 \\
+15.2 \\
744.0 \\
\min = 24.55 \\
\text{NE}\,4\,\text{m} \\
\text{Neb} = 10
\end{bmatrix}
\quad
9\,\text{pm}
\begin{bmatrix}
-4.0 \\
-5.45 \\
+18.47 \\
748.5 \\
\text{SSW}\,5\,\text{m} \\
\text{Neb} = 10
\end{bmatrix}
$$

Variable wind from the ENE, NE, E, ESE, and SE until 6:30, when it started to blow from the SW. The temperature, which had risen to +1°, started to drop _______ [unreadable] during the night, reaching −12.5°, and it began to rise once the SSW ceased, at 11 pm. We had fog in the morning, which dissipated at 9 pm. All day the day was overcast, starting to clear up at 11 pm.

5.6.81   *May 30, [1903]*

$$
7\,\text{am}
\begin{bmatrix}
-4.75 \\
-5.45 \\
+7.0 \\
\max = +0.6 \\
\text{NE}\,4\,\text{m} \\
\text{Neb} = 4
\end{bmatrix}
\quad
2\,\text{pm}
\begin{bmatrix}
-1.05 \\
-2.3 \\
+16.5 \\
750.1 \\
\min = -12.1 \\
\text{NE}\,6\,\text{m} \\
\text{Neb} = 8
\end{bmatrix}
\quad
9\,\text{pm}
\begin{bmatrix}
-1.3 \\
-2.1 \\
+14.3 \\
743.5 \\
\text{NE}\,8\,\text{m} \\
\text{Neb} = 9
\end{bmatrix}
$$

Completely clear sky and calm until 6 am. The NE started at 7, and at 9, the sky was overcast; the rest of the day, the NE blew, quite strongly at times. At 10 am, the strength was 9 m, and at 11, it was 10 m, but then it diminished until 5 pm, increasing after that time up to its maximum at 9 pm. In the afternoon, nebulosity did not reach below 7, but it suddenly cleared up at 8 pm, only to become overcast again, with a little snow falling at 9 pm. Despite the weather, we were able to observe the passage of three stars. For days now, we have had the great comfort of a latrine made of snow and planks. It's like a grand hall. As always, the dogs are in charge of the cleaning. When one goes there, one becomes surrounded by a dozen hungry dogs that, with nervous movements of paws and heads, lick their chops while they wait for their coveted bite. As can be seen, nothing is wasted in the station. Today, the temperature rose 3 degrees above 0. It's nice to go out for a walk with these temperatures, even though the strong NE wind is uncomfortable. Clearings of water can still be seen to the SE, as well as a large number of snow petrels.

5.6.82   May 31, [1903]

Another month has ended! Oh, what I would give to have all the ones from this year finished! Life among these people is becoming too difficult for me. These people don't do anything else but criticize me and control me, especially during mealtime. The one who demonstrates his anger toward me the most, through his looks, is Nordenskjöld; our main ration is made up of penguin and seal meat. In general, seal meat is eaten in the form of steak, and penguin is cooked; these people use an enormous amount of pepper in their food; therefore, I couldn't eat the seal steak without complaints. I would spend from 9 pm at night until 3 pm the following day on only one cup of coffee and a little bit of bread. One day, Nordenskjöld asked me if I liked seal, and I said yes, that I liked it, but I couldn't eat it because so much pepper was bad for me. So he told me that it was very simple to make special steaks without pepper for me, which naturally I accepted. But it turns out that when I am hungry and I eat the three steaks that they bring to me, he gives me these looks as if he wants to strike me down. Now, things have changed. At the table, my seat is in the middle, and I constantly have to be passing things to the others. When I don't anticipate passing something beforehand, they don't ask me directly to do it, but rather indirectly, by telling one another to pass the salt, etc., that they need, directing the sentence to someone who can't possibly [be in the position to] do it; and so I patiently play the servant, to whom no one does the honor of asking anything.

Today, I went to the boss of the expedition [Nordenskjöld], and I told him that I would be very thankful if he changed my seat due to the aforementioned reason, and because I had heard talk of making changes. He answered that it was very simple, and that, however, he hoped "the food was enough." A pointless hint, since I'm the one who eats the least. Ever since I heard all their continuous arguments over food, I have rationed myself and I eat the least [amount] possible. In the last 15 days, I have lost 3 lb. One of the discussions that they had happened to be about dividing the things that they could, because "some" were eating the things of the others. So they ended up dividing the bread, which gets to the table divided into four parts. At night, they weren't happy with this anymore, and Bodman said that he thought that the bread that was taken to the table was too much (even though it is completely eaten), and Nordenskjöld answered that it wasn't a lot because "when you're hungry, you eat stale bread." And I suppose this was said as a taunt to me, because when he finished saying this, everyone else started laughing. Since then, I eat a third of the bread that they eat and the least amount of food possible.

$$7\,\mathrm{am}\begin{bmatrix}+4.2\\+0.9\\+10.2\\738.6\\ \mathrm{Max}=\begin{bmatrix}+6.3\\+4\end{bmatrix}\\ \mathrm{SSE\,2\,m}\\ \mathrm{Neb}=3\end{bmatrix}\quad 2\,\mathrm{pm}\begin{bmatrix}+0.9\\-1.5\\+11.7\\738.1\\ \mathrm{min}=-4.6\\ \mathrm{ENE\,1}\\ \mathrm{Neb}=10\end{bmatrix}\quad 9\,\mathrm{pm}\begin{bmatrix}+1.05\\-0.05\\+15.9\\734.4\\ \mathrm{ENE\,2\,m}\\ \mathrm{Neb}=10\end{bmatrix}$$

From 9 am to 11 pm, overcast sky. During the day, the wind has consisted of calm and weak breezes from the ENE, ESE, and SE.

Sobral feels that he is suffering many layers of humiliation: being ignored while being expected to serve the others, who do not give him the courtesy of requesting his assistance; being ridiculed for eating anything, as he attempts to sustain himself with stale bread; and being reprimanded for eating too much, even though he feels he is depriving himself of food. How much of this was imagined by him under stressful conditions is a question to consider; however, there does seem to be the element of a "persona non grata" at work here for the young Argentine among his colleagues, and he had a propensity to feel the stab of his social standing (or lack thereof) quite acutely and intensely. Later, in his notes written in 1903, Sobral would say that the reason the men were able to maintain such good health during their second winter was their "healthy diet" of "seal and penguin meat, its only inconvenience being the small amount" (Sobral 1903b).

5.6.83   *June 1, [1903]*

$$
7\,\text{am}
\begin{bmatrix}
-5.7 \\
-6.55 \\
+10.7 \\
734.4 \\
\text{Max} = +5.5 \\
\text{NE}\,0.6\,\text{m} \\
\text{Neb} = 7
\end{bmatrix}
\quad
2\,\text{pm}
\begin{bmatrix}
-7.8 \\
-8.15 \\
+18.3 \\
735.65 \\
\text{min} = -8.0 \\
\text{var}\,1\,\text{m} \\
\text{Neb} = 9
\end{bmatrix}
\quad
9\,\text{pm}
\begin{bmatrix}
-13.5 \\
-13.55 \\
+13.8 \\
729.3 \\
\text{SSW}\,8\,\text{m} \\
\text{Neb} = 10 \\
\text{snowing}
\end{bmatrix}
$$

The SSW started very weakly at 5 pm and continued until 6, when it began to quickly increase its intensity, and snow began to fall. The sky has been overcast since midday. At 12 at night, the [speed of the] wind was 13 m per second.

5.6.84   *June 2, [1903]*

$$
7\,\text{am}
\begin{bmatrix}
-17.5 \\
-17.5 \\
+7.6 \\
720.76 \\
\text{max} = -3.9 \\
\text{SSW}\,12\,\text{m} \\
\text{Neb} = 10 \\
\text{snöfyk}
\end{bmatrix}
\quad
2\,\text{pm}
\begin{bmatrix}
-16.95 \\
-16.95 \\
+17.0 \\
718.5 \\
\text{min} = -17.5 \\
\text{SW}\,13\,\text{m} \\
\text{Neb} = 10 \\
\text{snöyra}
\end{bmatrix}
\quad
9\,\text{pm}
\begin{bmatrix}
-12.55 \\
-12.5 \\
+11.5 \\
714.2 \\
\text{SSW WSW}\,7\,\text{m} \\
\text{snöyra}
\end{bmatrix}
$$

In general, the wind has been from the SSW. In the afternoon, it started to move to the SW and to diminish its intensity. At the same time, it began to blow in gusts; for some seconds, it was completely calm, and on the hills, only the sound of its gusts could be heard. At 8 pm, the average speed per second was 5 m, and it was variable between the SSW and the WSW. The barometer is dropping quickly: At 9 pm, it was at 713 m, the lowest barometric height observed at the station, so far. The thermometer keeps its tendency to rise, and it remains relatively high, so we can forecast—with a good chance of being right—a very heavy storm. In general, great barometric drops bring strong winds, but the winds also blow very strongly, with little distinctions from the average.

5.7 Here It Is As If I Were Alone

5.7.1 *June 3, [1903]*

The SW blows its frozen gusts, piling mounds of snow behind every object. Sitting on my bed, and tired of reading, I look at the fanciful shapes that the snow makes when it's melting on the glass. Now, it's the profile of an old man with the nose of a Roman consul, who looks attentively toward a corner, but after a while, a bit of the snow falls off, taking with it the nose and the beard and leaving only the circle. Or it's a 40-year-old woman, fat and flirtatious, who is sitting and coquettishly rejecting an astonished face that is making an effort to kiss her. I don't know if it's my imagination that causes me to see representations of animated beings in each [bas] relief of snow on the glass, or if it is a true vision. At times, the 40-year-old lady takes on a thoughtful character, the coquettish smile that had promised a thousand things disappears, and the horns and beard of the astonished face fall, [leaving an image that now] looks as if it were a son speaking to his mother, and from the look of her expression, I think that he is asking for money. I'm tired of reading study books. How I wish I had some novels for a change. And music, with what pleasure I would hear the piano play, with what pleasure [I would go to] the theater. If, at least, one had someone to talk to for a while. But here, it is as if I were alone. When I talk or I laugh, I do it with myself.

$$
7\,\text{am} \begin{bmatrix} -16.25 \\ -16.25 \\ +6.6 \\ 717.35 \\ \text{Max.} = -11 \\ \text{WSW}\,14\,\text{m} \\ 10 \\ \text{snöfyk} \end{bmatrix} \quad 2\,\text{pm} \begin{bmatrix} -14.65 \\ -14.65 \\ +11.2 \\ 723.0 \\ \text{min.} = -16.9 \\ \text{WSW}\,17\,\text{m} \\ \text{neb. [nebulosity]} = 10 \\ \text{snöfyk} \end{bmatrix} \quad 9\,\text{pm} \begin{bmatrix} -15.7 \\ -15.7 \\ +8.5 \\ 731.2 \\ \text{WSW}\,20\,\text{m} \\ \text{neb} = 10 \\ \text{snöfyk} \end{bmatrix}
$$

At 1 am, the wind had a speed of 24 m, and then at 2 am, it was at 32 m per second! It's a shame that, at that time, one of the threads of the electric transmission broke. And it couldn't be fixed until 11 am. So we don't know what the maximum [speed] was. But probably it has exceeded that speed, because my trunk (which no wind has ever moved until now) was moved from its place to a distance of 100 m across the ice, and it stopped by itself, because some little mounds interrupted its slide. The direction taken by the trunk was SW [draws an arrow] NE, and it also blew away the lid of the meridian instrument compartment and some boxes full of rocks! At 6 am, it diminished quite a lot, down to 12 m, but it immediately rose again. The temperature hasn't dropped yet; during the day, it has been around $-16°$, so we will still have a lot of wind. The barometer rises very quickly. The amount of snow that has fallen is huge. We haven't had such a strong wind in 9 or 10 nights. The anemometer resisted admirably.

Sobral is feeling the vacuum of loneliness by this time, and certainly the windy, wintry weather adds to his sense of isolation. He seeks company in his creative imagination and sets the scenarios of that imagination free.

In his 1903 notes for his speeches and reports, Sobral would later say that the factors that influenced the health of the expeditioners the most were "exercise and a good library"—the former being "scarce during the winter due to the strong snowstorms and harsh cold," and the latter being insufficient as "some of us had to read the same book up to three times" (Sobral 1903b). One could argue that the health to which he refers is both physical and mental.

5.7.2   *June 4, [1903]*

$$
7\,\text{am}\begin{bmatrix}[-]18.5\\ [-]18.55\\ +5.4\\ 744.75\\ \text{Max}=[-]\,13.7\\ \text{SW}\,20\,\text{m}\\ \text{Neb}=8\\ \text{snöfyk}\end{bmatrix}\quad
2\,\text{pm}\begin{bmatrix}-18.0\\ -18.25\\ +11.9\\ 751.6\\ \text{Min}=-24.5\\ \text{SW}\,11\,\text{m}\\ \text{neb}=10\\ \text{snöfyk}\end{bmatrix}\quad
9\,\text{pm}\begin{bmatrix}[-]15.8\\ [-]15.9\\ +18.6\\ 750.4\\ \text{NE}\,5\,\text{m}\\ \text{neb}\,0.5\end{bmatrix}
$$

Since 6 am, the wind and the nebulosity have been diminishing. After 4 pm, the sky was completely clear. At 5 pm, the wind from the SSW calmed completely, but at 7 pm, variable wind started, although weak, and at 7 pm, it blew from the NE with not just a little strength, bringing [with it] some snow. The NE calmed at 10 pm. The maximum of the wind was 22 m at 6 am. Observations of lunar culminations were made.

Yesterday's strong wind was from the SW, because my trunk was dragged in that direction, as was a seal skin, which I found 1 km away, stuck to the snow.

5.7.3   *June 5, [1903]*

$$
7\,\text{am}\begin{bmatrix}[-]14.95\\ [-]15.5\\ +3.5\\ 746.15\\ \text{Max}=-13.9\\ \text{calm}\\ \text{neb}=5\end{bmatrix}\quad
2\,\text{pm}\begin{bmatrix}[-]18.5\\ [-]18.2\\ 72.5\%\\ +15.0\\ 747.65\\ \text{min}=-23.2\\ \text{calm}\\ \text{neb}=0\end{bmatrix}\quad
9\,\text{pm}\begin{bmatrix}[-]16.9\\ -\\ +12.6\\ 746.45\\ \text{calm}\\ \text{neb}=0\end{bmatrix}
$$

Weak gusts from the NE every now and then, but, in general, the day has been calm and clear. There were some clouds in the morning; the sky has been clear since 10 am.

5.7.4   *June 6, [1903]*

$$
7\,\text{am}
\begin{bmatrix}
-10.26 \\
-10.6 \\
+4.6 \\
724.25 \\
\text{Max} = -9.4 \\
\text{SE} = 5\,\text{m} \\
\text{Neb} = 4
\end{bmatrix}
\quad
2\,\text{pm}
\begin{bmatrix}
-18.0 \\
-17.9 \\
(84.6\%) \\
\text{min} = -23.0 \\
+15.8 \\
747.2 \\
\text{SSW } 14\,\text{m} \\
\text{Neb} = 10
\end{bmatrix}
\quad
9\,\text{pm}
\begin{bmatrix}
-17.0 \\
-17.1 \\
82\% \\
+11.8 \\
750.85 \\
\text{SSW } 12\,\text{m} \\
\text{neb} = 1
\end{bmatrix}
$$

Very little nebulosity during the first hour of the morning, and variable wind between the NE and SE. At 8 am, the sky became completely overcast, and the wind blew in isolated gusts from all directions. At 9:10, after some minutes of calm, the SSW started to blow, with fog and some snow. The maximum of the wind was at 2 pm, and until 12 at night, it continued to blow, although while moving to the S and diminishing in intensity. Since 4 in the afternoon, the sky has cleared up pretty much, and there were clouds (fog) only at around 30° from the horizon.

Today, between 11 and 12:30, N, E, B, [Nordenskjöld, Ekelöf, Bodman], and I (on my initiative) had a conversation, in which I complained about the way that they were treating me. They tried to apologize about everything, but about some things they were silent. I didn't restrict myself to only protesting, but also accused them of everything that they have done to me.

5.8   Maybe These Gatherings Will Make Our Relationships a Little More Friendly

5.8.1   *June 7, [1903]*

$$
7\,\text{am}
\begin{bmatrix}
-26.1 \\
-26.15 \\
83\% \\
+4.8 \\
753.0 \\
\text{Max} = 10.0 \\
\text{Calm} \\
\text{neb} 0
\end{bmatrix}
\quad
2\,\text{pm}
\begin{bmatrix}
21.7 \\
21.7 \\
84\% \\
+15.0 \\
752.35 \\
\text{min} = -27.0 \\
\text{ENE } 4\,\text{m} \\
\text{neb} = 2
\end{bmatrix}
\quad
9\,\text{pm}
\begin{bmatrix}
-15.55 \\
-15.5 \\
+10.5 \\
749.0 \\
\text{ENE } 1\,\text{m} \\
\text{neb} = 10
\end{bmatrix}
$$

The SSW that was blowing yesterday calmed completely around 2 am, and at 3 am, it was already blowing from the NE. From 1 am to 8 am, speed was 0, but at 8 pm, the sky started to become filled with Stratus, and at 9 am, the nebulosity was 10. At 11, neb = 1, and it continued like this until 2 pm. At 3, neb = 8; the rest of the day _______ [unreadable]-S and fog. Between 8 am and 9 pm, the wind has blown from the ENE; 10 and 11 calm; and at 12 from the E with fog of 500 m. Today, E, N, B [Ekelöf, Nordenskjöld, Bodman], and I have played *Whist*. Maybe these gatherings will make our relationships a little more friendly.

Sobral has made a necessary breakthrough in appealing to the sense of fairness in his comrades. An attempt of inclusiveness is now undertaken by the expedition leader and some of his colleagues toward Sobral, and Sobral seems to be willing to try to give it a go for the sake of harmony.

Meanwhile, on Paulet Island, the ultimate tragedy takes place on this day. Navigator and sailor Ole Christian Wennersgaard succumbs to his circumstances and dies, surrounded by the crew of the Antarctic *who are taking refuge on the cold volcanic rock of Paulet. His is the only death to occur during this expedition, and the cause is attributed to heart disease (Nordenskjöld et al. 1905). Still, the environment and desperation must have contributed to his ailment. He is buried on Paulet.*

5.8.2   *June 8, [1903]*

$$
7\,\text{am}
\begin{bmatrix}
-21.7 \\
-21.7 \\
+3.7 \\
744.75 \\
\text{Max} = -14.1 \\
\text{Calm} \\
\text{Neb} = 3 \\
\text{snöfyk}
\end{bmatrix}
\qquad
2\,\text{pm}
\begin{bmatrix}
-25.2 \\
-25.2 \\
+14.0 \\
744.2 \\
\min = -25.8 \\
82.4\% \\
\text{S}\,5\,\text{m} \\
\text{neb} = 4
\end{bmatrix}
\qquad
9\,\text{pm}
\begin{bmatrix}
[-]26.7 \\
[-]26.4 \\
+12.1 \\
745.0 \\
89\% \\
\text{SSW}\,21\,\text{m} \\
\text{neb}\,10
\end{bmatrix}
$$

In the morning, calm and weak variable winds. The sky was quite clear during those hours, but at 1 pm fog started to approach from the NE. At 2 pm, a weak wind from the S started to blow, and suddenly, at 2 hr 26 min, it came from the SW with a speed of around 11 m/s, with fog and snow.

5.8.3 *June 9, [1903]*

$$
7\,\text{am}
\begin{bmatrix}
-28.0 \\
- \\
+1.7 \\
741.5 \\
\text{Max} = -19.4 \\
\text{SW } 15\,\text{m} \\
\text{Neb} = 10
\end{bmatrix}
\qquad
2\,\text{pm}
\begin{bmatrix}
[-]26.15 \\
[-]26.5 \\
+10.9 \\
739.6 \\
\text{min} = -28.1 \\
\text{SSW } 3\,\text{m} \\
\text{neb} = 1
\end{bmatrix}
\qquad
9\,\text{pm}
\begin{bmatrix}
[-]27.0 \\
[-]27.0 \\
+11.7 \\
735.11 \\
\text{calm} \\
\text{neb} = 0
\end{bmatrix}
$$

The sun rises around 10 am, and sunset is around 2 pm. After sunset, it doesn't take long for the moon to come out, so it could be said that we have no nights. This night couldn't be more splendid. The sky is perfectly clear and with no wind. But the thermometer is between $-31°$ and $-27°$. You don't feel the cold much, but the slightest breeze is enough for your nose and your hands to begin to hurt. Today, we have made observations of lunar culminations and of a great number of stars, in order to determine the distance between the ice and the time. This month is very different from the same month last year. Now, the storms are of short duration, and we have relatively more clear days. If the weather continues in this way, it's not improbable that this winter will be colder than the one before.

5.8.4 *June 10, [1903]*

$$
7\,\text{am}
\begin{bmatrix}
26.25 \\
[-]26.0\,(?) \\
79.3\% \\
+1.6 \\
731.35 \\
\text{Max} = -24.3 \\
\text{Calm} \\
\text{Neb} = 0
\end{bmatrix}
\qquad
2\,\text{pm}
\begin{bmatrix}
-18.5 \\
-18.1 \\
+13.18 \\
731.25 \\
\text{min} = -30.5 \\
70.3\% \\
\text{calm} \\
\text{Neb} = 9 \\
(\text{Ci-S})
\end{bmatrix}
\qquad
9\,\text{pm}
\begin{bmatrix}
[-]18.9 \\
[-]18.9 \\
+13.6 \\
732.6 \\
64.2\% \\
\text{var } 0.5\,\text{m} \\
\text{Neb} = 8 \\
(\text{A-Cu})
\end{bmatrix}
$$

What a beautiful morning! To the SW, at around 10, the full moon above the horizon will stop sending us its light, but to the NE, the sky, tinted in purple, can already be seen, and the smallest stars disappear. The sky is very clear. And a little wind from the ENE is beginning to blow, and with $-27°$, being outside is not pleasant. At 9:45 today, a beautiful sunrise. Beginning from this hour, the sky has started to get cloudy, but the cloudiness was very variable during the rest of the day. Between 4 pm and 12 at night, very weak but variable wind has blown; the maximum speed was 0.5 m.

5.8.5   *June 11, [1903]*

$$
7\,\text{am}\begin{bmatrix} -20.7 \\ -21.5 \\ +5.0 \\ 732.0 \\ \text{Max} = -11.3 \\ \text{SSW } 16.5\,\text{m} \\ \text{Neb} = 0 \end{bmatrix}
\qquad
2\,\text{pm}\begin{bmatrix} [-]22.7 \\ [-]22.95 \\ +12.5 \\ 735.1 \\ \text{min} = -24.5 \\ \text{SSW } 10\,\text{m} \\ \text{Neb} = 10 \end{bmatrix}
\qquad
9\,\text{pm}\begin{bmatrix} [-]21.7 \\ [-]21.7 \\ +12.2 \\ 738.6 \\ \text{SE } 2\,\text{m} \\ \text{neb} = 10 \end{bmatrix}
$$

Dawn broke with a clear sky and weak variable wind. The SSW started to blow with strength at 5:30. At 8 am, the sky became completely overcast, clearing up at 5 pm and becoming overcast again at 8 pm. The SSW began to diminish 3 h after it had started, and it calmed at 5:35 pm. It was calm at 7 and 8 pm; at 9, SE; and calm until 12 at night, alternating with gusts from ENE and NE. Overcast sky. The SSW blew for 12 h.

5.8.6   *June 12, [1903]*

$$
7\,\text{am}\begin{bmatrix} [-]\,25.9 \\ [-]\,25(?) \\ +2.7 \\ 732.85 \\ \text{SSW } 8\,\text{m} \\ \text{Neb} = 10 \end{bmatrix}
\qquad
2\,\text{pm}\begin{bmatrix} -27.55 \\ -27.5\,? \\ +9.2 \\ 735.7 \\ \text{min} = \begin{bmatrix} -27.55 \\ -27.3 \end{bmatrix} \\ \text{Hy} = 82 \\ \text{SSW} = 6\,\text{m} \\ \text{Neb} = 10 \end{bmatrix}
\qquad
9\,\text{pm}\begin{bmatrix} -27.25 \\ -27.25\,(?) \\ \text{Hy} = 81.8 \\ 712.9 \\ 738.9 \\ \text{SSW } 8\,\text{m} \\ \text{Neb} = 9 \end{bmatrix}
$$

All day, SSW blowing with fog. The sky was cloudy until 9 pm, when it cleared up considerably, nebulosity being 3. The maximum speed of the wind was 11 m at 5 am.

5.8.7   *June 13, 1903]*

$$7\,\text{am} \begin{bmatrix} [-]29.05 \\ [-]29.25 \\ +1.2 \\ 741.2 \\ \text{Max} = -24.8 \\ \text{SW } 13\,\text{m} \\ \text{Neb} = 4 \end{bmatrix} \quad 2\,\text{pm} \begin{bmatrix} [-]27-5 \\ [-]27.6 \\ +12.5 \\ 744.45 \\ \text{min} = -29.1 \\ \text{SSW } 10\,\text{m} \\ \text{neb} = 8 \end{bmatrix} \quad 9\,\text{pm} \begin{bmatrix} [-]28.3 \\ [-]28.35 \\ [+]12.2 \\ 746.5 \\ \text{SSW } 8\,\text{m} \\ \text{neb} = 6 \end{bmatrix}$$

SSW and SW blowing all day, but with partially clear sky. At 12 at night, it cleared up completely. The maximum of the wind was at 7 am, and the minimum was at 12 at night (7 m).

5.8.8   *June 14, [1903]*

$$7\,\text{am} \begin{bmatrix} -29-4 \\ -29.6 \\ 76.87 \\ +1.4 \\ 744.7 \\ \text{Max} = -25.2 \\ \text{Calm} \\ \text{Neb} = 0 \end{bmatrix} \quad 2\,\text{pm} \begin{bmatrix} -29.5 \\ -29.5 \\ +12.3 \\ 744.2 \\ \text{min} = -33.0 \\ 70.6\% \\ \text{calm} \\ \text{Neb} = 0.2 \end{bmatrix} \quad 9\,\text{pm} \begin{bmatrix} -27.0 \\ -27.2 \\ +9.3 \\ 742.9 \\ 67.5\% \\ \text{calm} \\ \text{Neb} = 0.2 \end{bmatrix}$$

The difference between storms [occurring] this month and the same month last year is truly remarkable. Now, they are of a very short duration and slow speed. It has been a beautiful day today, calm and clear but cold. At 9:30 am, the temperature dropped to $-33.3°$; all day it has remained lower than $-29°$. With these temperatures, it doesn't feel very cold when it's calm.

5.8.9   *June 15, [1903]*

$$
7\text{ am}
\begin{bmatrix}
[-]16.5 \\
[-]17.6 \\
+2.6 \\
740.85 \\
\text{Max} = -16.5 \\
44\% \\
\text{SSW } 6 \text{ m} \\
\text{Neb} = 0
\end{bmatrix}
\quad
2\text{ pm}
\begin{bmatrix}
[-]18.5 \\
[-]19.5 \\
+12.5 \\
742.75 \\
\text{min} = -33.4 \\
44.7\% \\
\text{SSW } 7 \text{ m} \\
\text{neb} = 2
\end{bmatrix}
\quad
9\text{ pm}
\begin{bmatrix}
+10.7 \\
743.1 \\
-19.8 \\
-20.25 \\
65\% \\
\text{SSW } 11 \text{ m} \\
\text{neb} = 0
\end{bmatrix}
$$

Calm and clear sky until 6 am. At that time, the SSW started to blow, and some clouds appeared. At 8 h 30 m, the wind calmed. At 1, it blew very weakly from the NE, and at 2, the SSW started again. The maximum strength of the wind was at 9 pm. Last night at 11 pm, the temperature dropped to $-33.5°$, the lowest that we've had thus far. Today, at 5:30 am, the temperature rose quickly from $-25°$ to $-18°$ and then slowly to $-15.5°$, only to drop precipitously at 9 am. At 10 am, it was $-28°$; during the rest of the day, it has remained at around $-20°$. The relative humidity dropped to around 30% at 9 am. At 10 pm, the sky started to become cloudy, and at 12, nebulosity was 10. Temperature remains around $-20°$.

5.8.10   *June 16, [1903]*

$$
7\text{ am}
\begin{bmatrix}
[-]22.3 \\
[-]22.45 \\
+8.0 \\
746.6 \\
\text{Max} = -15.5 \\
\text{SW } 8 \text{ m} \\
\text{Neb} = 8
\end{bmatrix}
\quad
2\text{ pm}
\begin{bmatrix}
[-]26.8 \\
[-]26.8 \\
+14.0 \\
750.4 \\
[-]27.5 = \text{min} \\
\text{SSW } 6 \text{ m} \\
\text{Neb} = 8
\end{bmatrix}
\quad
9\text{ pm}
\begin{bmatrix}
[-]25.05 \\
[-]25.05 \\
+13.0 \\
751.85 \\
\text{SSW } 10 \text{ m} \\
\text{Neb} = 1
\end{bmatrix}
$$

Fog and snow most of the day. The wind that predominated between 1 am and 9 am was SW. After 10 am, SSW blew continuously. Until 9 am, the sky was overcast at more than 6, but then it remained between 1, 2, and 3.

5.8.11   June 17, [1903]

	[−]22.5		[−]24.8		[−]22 − 65
	[−]22.45		[−]24.9		[−]22.8
	+ 8.0		+ 14.0		+ 11.1
7 am	746.6	2 pm	752.2	9 pm	751.15
	Max = −21.4		[−]27.1 = min		SSW 15 m
	SW 6 m		SSW 15 m		neb = 2
	Neb = 10		neb = 5		

SSW and SW blowing with snow all day. In the afternoon, the sky cleared up somewhat and then became cloudy again from 10 pm on. The maximum speed of the wind was 18 m at 4 pm, and the minimum was 6 m at 6 am.

5.8.12   June 18, [1903]

	[−]24.9		[−]25.5		[−]26.45
	[−]24.9		[−]25.65		[−]26.6
	+ 2.9		+ 10.7		+ 10.8
7 am	753.35	2 pm	754.6	9 pm	755.9
	Max = −21.3		min = −27.6		72.7%
	79.5%		73.9%		SSW 14 m
	SSW 12 m		SSW 11 m		Neb = 1
	Neb = 10		Neb = 10		

SSW blowing all day, and overcast sky until 5 pm. At 6, nebulosity was = 3, and at 7, it was = 1, and it continued to be clear until 12 at night. The maximum speed of the wind was 14.5 m at 4 pm, and the minimum was 8 m at 4 am. There has been fog during the morning and some snöfyk [snow cover].

5.8.13   *June 19, [1903]*

$$7\,\text{am}\begin{bmatrix}[-]25.5\\ [-]25.7\\ +3.1\\ 753.95\\ \text{Max}=-23.1\\ 69\%\\ \text{SSW 4 m}\\ \text{Neb}=10\end{bmatrix}\quad 2\,\text{pm}\begin{bmatrix}[-]28.95\\ [-]29.00\\ +10.4\\ 753.6\\ \min=-30.5\\ (-28.8)\\ \text{SW 8 m}\\ \text{Neb}=1\end{bmatrix}\quad 9\,\text{pm}\begin{bmatrix}[-]29.8\\ [-]29.8(?)\\ +9.4\\ 753.55\\ \text{SSW 3 m}\\ \text{Neb}=0\end{bmatrix}$$

Until 5 am, the wind was from the SW. At that time, the sky became cloudy. At 6 am, the wind was from the W, and the speed had diminished to 2 m (the minimum of the day). Between 7 am and 11 am, it blew from the SSW with more strength, clearing the sky. At 5 pm, the sky cleared up completely (at 12 pm, 1 pm, 2 pm, and 3 pm, it blew from the SW). The maximum speed of the wind was 10 m at 1 am. At 10 pm, the SSW calmed, and at 10:30, a weak breeze started to blow from the NE. Today, the temperature has dropped to −32.1 twice: at 6 pm and at 11 pm.

5.8.14   *June 20, [1903]*

$$7\,\text{am}\begin{bmatrix}[-]33.7\\ -\\ 752.0\\ \text{Max}=-24.8\\ \text{Calm}\\ \text{neb}=0\end{bmatrix}\quad 2\,\text{pm}\begin{bmatrix}-27.5\\ -27.5\\ 81.3\%\\ \min=-35.1\\ ([-]27.3)\\ +12.5\\ 753.2\\ \text{calm}\\ \text{neb}=0\end{bmatrix}\quad 9\,\text{pm}\begin{bmatrix}-31.2\\ -31.2\\ +11.3\\ 755.7\\ 80\%\\ \text{calm}\\ \text{neb}=0\end{bmatrix}$$

Clear sky and complete calm all day. At 11 pm, a weak E started to blow. In the afternoon, despite the apparent purity of the air, frost fell on all objects. Today, the temperature dropped to −35.5. The fact that it was −27.7 at 2 in the afternoon was merely by chance; it had risen at that moment only to immediately drop to −32. The average of the day was below −30°.

5.8.15   *June 21, [1903]*

$$
7\,\text{am}\;
\begin{bmatrix}
-29.5 \\
-29.45 \\
79.70 \\
+0.8 \\
757.45 \\
\text{Max} = -25.0 \\
\text{NW 1 m} \\
\text{Neb} = 10
\end{bmatrix}
\qquad
2\,\text{pm}\;
\begin{bmatrix}
-25.0 \\
-25.1 \\
+15.0 \\
754.95 \\
\text{min} = -35(?) \\
-25.0 \\
\text{NE 11 m} \\
\text{Neb} = 10
\end{bmatrix}
\qquad
9\,\text{pm}\;
\begin{bmatrix}
+13.8 \\
747.6 \\
-19.45 \\
-19.55 \\
\text{ENE} \\
\text{Neb} = 10
\end{bmatrix}
$$

At 4 am, the sky became completely overcast, and the NE started to blow. At 8 and 9 am, it cleared up somewhat, but before midday, nebulosity was 10 _______ [unreadable].

5.8.16   *June 22, [1903]*

$$
7\,\text{am}\;
\begin{bmatrix}
[-]26.2 \\
[-]26.2 \\
+2.2 \\
745.55 \\
\text{Max} = -18.7 \\
\text{Hy} = 77.5\% \\
\text{SSW 10 m} \\
\text{Neb} = 1
\end{bmatrix}
\qquad
2\,\text{pm}\;
\begin{bmatrix}
[-]27.6 \\
[-]27.5 \\
+14.1 \\
748.35 \\
\text{min} = -27.5 \\
([-]27.5) \\
79.8\% \\
\text{calm} \\
\text{Neb} = 10
\end{bmatrix}
\qquad
9\,\text{pm}\;
\begin{bmatrix}
-26.0 \\
-26.0 \\
+13.5 \\
749.0 \\
83\% \\
\text{calm} \\
\text{Neb} = 3
\end{bmatrix}
$$

At 12 h 5 m at night, the SSW started to blow after around half an hour of calm. The sky started to clear up at 3; at 11 am, it was calm, and the rest [of the day] remained the same. The sky was cloudy between 2 pm and 8 pm, at which time it began to clear up, only to become overcast again at 12 at night. Despite the fact that the sun reached its maximum declination north, it marked the paper of the heliograph between 11 am and 12:30 pm.

5.8.17   *June 23, [1903]*

$$
7\,\text{am}
\begin{bmatrix}
-21.0 \\
-20.95 \\
+3.8 \\
746.35 \\
\text{Calm} \\
\text{Neb} = 10 \\
\text{snowing}
\end{bmatrix}
\quad
2\,\text{pm}
\begin{bmatrix}
[-]20.5 \\
[-]20.45 \\
+13.2 \\
744.6 \\
\min = [-]27.1 \\
\text{ENE} \\
\text{neb} = 10 \\
\text{snowing}
\end{bmatrix}
\quad
9\,\text{pm}
\begin{bmatrix}
[-]21.5 \\
[-]21.45 \\
+14.8 \\
743.6 \\
\text{calm} \\
\text{neb} = 10 \\
\text{snowing}
\end{bmatrix}
$$

At 6:30 am, it began to snow [and that lasted] until 12 am, with 3.1 m of precipitation. It continued to snow all day; from midday to 11 pm, 1.87 m of precipitation fell. At 8 am, the NE was blowing; calm between 9 and 11 am; wind from the E at 11; and at 12 wind from the NE. It kept blowing from the E and the NE until 10 pm. At 11 SW and calm at 12 at night.

Today is a holiday for the Swedish people. *"Midsommaren afton"* [Midsummer's Eve]. At night, we eat with a tablecloth (!) and various extra dishes: cooked rice, cod, with peas. Although this is a "cheap" meal for someone who cooks, it's a banquet for us, because at night we usually only have a cup of tea and a bit of cooked penguin. On the wall next to the table, there is a great quantity of photographs depicting the beautiful sex. Today, Nordenskjöld increased the collection by adding some more [pictures] which he had cut out from an almanac. Even in these solitary regions, to the women belong a great part of our thoughts. Today, the house has been extraordinarily warm, because the kitchen has had a fire burning all day long. Generally, it's only lit until 3 pm (from 7:30 am). And the temperature hasn't dropped below $-18°$. Every day, Bodman and I, who are the ones with the upper beds, receive a continuous rain from the ceiling. Today, it has been a downpour. We uselessly place blankets and towels against the ceiling, but the water still goes through it, and our beds are real lakes.

At night, as usual, *Whist* was played, and to honor the holiday, *"Punch"* was served. This is the first time we've seen so much snow on the ground, not because today it has fallen more, but because usually the wind blows it away. Today, as an exception, the wind has been calm. If the weather continues like this tomorrow, I think I will go out on skis.

The barometer remains at around the same height. At 11 am, the thermometer rose up to $-16°$ but then dropped quickly, and for the rest of the day, it remained between $-20°$ and $-22°$.

5.8.18   *June 24, [1903]*

2 pm	$[-]21.75$ $[-]21.80$ 85.3% $+13.9$ 738.55 Min $= -22.5$ WNW Neb $= 10$ snowing				7 am	$[-]19.6$ $[-]19.55$ $+6.8$ 739.5 max $= -16.9$ W Neb $= 10$ snowing				9 pm	-18.5 -18.5 $+14.0$ 736.6 87.2% calm Neb $= 10$

It snowed quite a lot until 8 am. From yesterday midnight till 7:30 am this morning, 3.25 of snow has fallen. It continued snowing the rest of the day, but in a very small amount. At 2 pm, 0.65 mm had fallen. The ground is completely white. Only the places with 90° slopes have no snow. It is the first time we have seen so much snow on the ground. Today, a little exercise is done on the "ski." Sliding on the snow from the top of the hills. Today, the Swedish people celebrate *"Midsommar dag"* [Midsummer's Day] with lunch and extraordinary food (table-cloth on the table). Also, the gramophone plays its repertoire. The courses for the lunch are: first, cold cuts (tongue, fried onion, and raw fish, with bread and butter); second, canned soup; third, penguin cooked with dried, stewed vegetables. And for dessert: a chocolate pudding.

5.8.19   *June 25, [1903]*

7 am	$[-]14.35$ $[-]14.4$ max $= -6.9$ $+5.4$ 734.5 91.5% SSW 2 m Neb $= 10$				2 pm	$[-]21.75$ $[-]21.8$ $+18.2$ 737.35 min $= -23.6$ SSW 13 m Neb $= 10$				9 pm	$[-]18.5$ $[-]18.6$ $+14.8$ 738.8 SSW 10 m Neb $= 9$

The SSW began to blow at 7 am. At 9 am, it had a speed of 13 m, and a lot of snow was flying around, blown by the wind. The SW continued until 11 pm, and the sky cleared up completely at midnight. The wind has blown away most of the snow that there was on the ground. The maximum speed of the wind occurred at 9 am and at 2 pm.

5.8.20   *June 26, [1903]*

<table>
<tr><td rowspan="8">7 am</td><td>−23.9</td><td rowspan="8">2 pm</td><td>[−]26.45</td><td rowspan="8">9 pm</td><td>[−]24.95</td></tr>
<tr><td>−23.9</td><td>[−]26.45</td><td>[−]24.95</td></tr>
<tr><td>+4.8</td><td>+17.0</td><td>+12.0</td></tr>
<tr><td>735.2</td><td>736.45</td><td>739.95</td></tr>
<tr><td>Max = −13.0</td><td>min = −28.0</td><td>82%</td></tr>
<tr><td>83.7%</td><td>81.5%</td><td>SSW 15.0 m</td></tr>
<tr><td>Calm</td><td>SSW 14.0 m</td><td>neb = 10</td></tr>
<tr><td>Neb = 0.5</td><td>neb = 10</td><td>snöfyk</td></tr>
</table>

Until 12 am, nebulosity was between 0.5 and 1. Fog from the NE started to advance at 11 am, at the same time when the wind was blowing weakly, successively from the NW, N, NNE, NE, and ENE. The thermometer started to rise, and the barometer dropped slowly. At 11:25, after some moments of calm, the SSW started to blow. The thermometer was still rising, and a few minutes later, the fog closed, and a huge amount of snöfyk came. An hour later, it was blowing with more than 11 m of strength, and the thermometer started to drop as quickly as it had risen. At 4 pm, 6, and 9 pm, the wind had a speed of 15 m, which was its maximum, but between 1 pm and 10 pm, the speed wasn't below 14 m. I don't know when the barometer began to rise.

5.8.21   *June 27, [1903]*

<table>
<tr><td rowspan="8">7 am</td><td>[−]24.65</td><td rowspan="8">2 pm</td><td>[−]25.55</td><td rowspan="8">9 pm</td><td>[−]27.6</td></tr>
<tr><td>[−]24.65</td><td>[−]25.6</td><td>[−]27.6</td></tr>
<tr><td>736.9</td><td>+15.2</td><td>+12.7</td></tr>
<tr><td>max = [−]19.6</td><td>737.8</td><td>737.5</td></tr>
<tr><td>83%</td><td>min = −26.6</td><td>calm</td></tr>
<tr><td>SW 7 m</td><td>SSW 5.5 m</td><td>Neb = 10</td></tr>
<tr><td>Neb = 10</td><td>Neb = 8</td><td></td></tr>
</table>

During the day, the wind remained between 5 and 10 m. The maximum was 10 m at 1 am. It calmed at 7:35. Nebulosity began to diminish at 1 pm; at 5 pm, it was only 2; and at 6, the sky was completely clear. At night, after 9 pm, the thermometer started to drop very slightly; at 11 pm, it was at −30.5°.

5.8.22 *June 28, [1903]*

$$
7\,\text{am}
\begin{bmatrix}
[-]25.6 \\
[-]25.6 \\
+3.4 \\
733.75 \\
\text{Max} = -23.9 \\
\text{Calm} \\
\text{neb} = 1
\end{bmatrix}
\quad
2\,\text{pm}
\begin{bmatrix}
[-]23.0 \\
[-]23.0 \\
+12.7 \\
731.75 \\
\min = -30.6 \\
84.4\% \\
\text{calm} \\
\text{neb} = 10
\end{bmatrix}
\quad
9\,\text{pm}
\begin{bmatrix}
-8.0 \\
-8.5 \\
+12.0 \\
725.6 \\
85\% \\
\text{ENE}\,10\,\text{m} \\
\text{neb} = 9
\end{bmatrix}
$$

ENE blew from 1 to 5. From this last hour until 11 am, it was calm, and from 11 to 5, variable winds from the NE, ENE, and SSW, alternating with calm, their intensity very little. At 6 pm, it started to blow from the NE, sometimes moving to the ENE. One of the threads of the anemometer broke, so wind speed was only observed at a certain hour. The maximum observed was at 9 pm. At 12 at night, there was complete calm.

5.8.23 *June 29, [1903]*

$$
7\,\text{am}
\begin{bmatrix}
-9.7 \\
-10.0 \\
+5.5 \\
733.0 \\
72.8\% \\
\max = +0.4 \\
\text{calm} \\
\text{neb} = 3
\end{bmatrix}
\quad
2\,\text{pm}
\begin{bmatrix}
-3.6 \\
-5.9 \\
+10.2 \\
735.1 \\
54.5\% \\
\min = -24.0 \\
\text{calm} \\
\text{neb} = 9
\end{bmatrix}
\quad
9\,\text{pm}
\begin{bmatrix}
-3.0 \\
-3.3 \\
+15.6 \\
733.6 \\
\text{NE}\,9\,\text{m} \\
\text{neb} = 9
\end{bmatrix}
$$

At 1 am, it was calm. When I got up to make the observations, the SW was blowing at a speed of 12 m per second. At 5 am, it was calm again, and it remained that way, sometimes interrupted by weak variable winds until 2:10 pm, when the NE started to blow, quickly increasing in strength: At 6 pm, the speed was 12.5 m, which was the maximum. The average wind speed from 2 pm to 9 pm was 9 m. The temperature, which yesterday at 3:30 pm was −24°, started to rise swiftly: At 8 pm, it was −7.2°, and it remained between −7° and −9° until 1:30 am (today), when it rose to −5°. At 2 am, it was −3°, and it dropped again to −7°, but at 3 am, just when the SSW was blowing, it rose again, it reached up to +0.3° at 5 am, and then, it dropped again.

At 7 am, it was −10°. During the day, at ____ [blank], it rose again to + ____ [blank], and it hasn't dropped below −5. The barometer got to its minimum at 10 pm (yesterday). At 11, it started to rise, and the NE calmed at that hour. For a moment today, humidity was around 45%.

5.8.24   *June 30, [1903]*

$$
7\,\text{am}
\begin{bmatrix}
-1.0 \\
-2.45 \\
+9.0 \\
730.9 \\
\text{Max} = +0.5 \\
73.8\% \\
\text{Var}\,4\,\text{m} \\
\text{Neb} = 6
\end{bmatrix}
\qquad
2\,\text{pm}
\begin{bmatrix}
-18.25 \\
-18.4 \\
+15.0 \\
734.0 \\
\text{min} = -18.9 \\
80.1\% \\
\text{SSW}\,15.5\,\text{m} \\
\text{Neb} = 3
\end{bmatrix}
\qquad
9\,\text{pm}
\begin{bmatrix}
-19.5 \\
-19.45 \\
+12.0 \\
736.45 \\
92\% \\
\text{ENE}\,0.1\,\text{m} \\
\text{Neb} = 10
\end{bmatrix}
$$

During the first hours of the morning until 6 am, the wind blew from the NE and ENE. Then, the wind blew in several gusts which came from different directions. It would calm from the SSW, and after some seconds, the NE would blow with the same intensity (around 4 m). Then, it would move to the SE and S. At 7:20, variable gusts calmed, and at 7:35, after 15 min of calm, it started to blow from the SW. Some seconds later, it moved to the SSW. At 9, the SSW had calmed, and it was blowing from the SE, but at 9:05, it started blowing from the SW again. The maximum strength of the wind was at 2 pm; at that time, it started to diminish while the sky was clearing up. At 5 pm, the sky was completely clear, and at 7 pm, the calm was complete. At 9 pm, the sky became overcast with clouds, and the NE came at the same time. (When the SSW blew, the temperature dropped swiftly to $-18°$.)

5.8.25   *July 1, [1903]*

$$
7\,\text{am}
\begin{bmatrix}
-16.5 \\
- \\
+5.75 \\
733.6 \\
\text{Max} = -10 \\
56.5\% \\
\text{Var}\,2\,\text{m} \\
\text{Neb} = 1
\end{bmatrix}
\qquad
2\,\text{pm}
\begin{bmatrix}
[-]20.5 \\
[-]20.6 \\
+12.4 \\
736.25 \\
\text{min} = -25.8 \\
\text{SSE}\,1\,\text{m} \\
\text{Neb} = 1
\end{bmatrix}
\qquad
9\,\text{pm}
\begin{bmatrix}
-22.5 \\
-22.5 \\
+9.5 \\
737.2 \\
\text{calm} \\
\text{Neb} = 1
\end{bmatrix}
$$

At 1 and 2 am, variable winds. At 3, 4, and 5, wind from the SW and at 6 SE. The rest of the day, calm and very weak variable winds. Nebulosity was mostly Ci and Ci-S that came from the W. The afternoon was quite cold. Today, we're making magnetic observations.

5.8.26   *July 2, [1903]*

$$
7\,\text{am}
\begin{bmatrix}
-10.05 \\
-11.0 \\
+7.4 \\
737.85 \\
\text{SSE 4 m} \\
\text{Neb} = 9
\end{bmatrix}
\quad
2\,\text{pm}
\begin{bmatrix}
-12.55 \\
-12.6 \\
+18.2 \\
723 \\
91.3\% \\
\text{min} = -29.0 \\
\text{ESE 7 m} \\
\text{Neb} = 9 \\
\text{snowing}
\end{bmatrix}
\quad
9\,\text{pm}
\begin{bmatrix}
-20.05 \\
-20.1 \\
+15.9 \\
724.15 \\
85.8\% \\
\text{SSW 15 m} \\
\text{Neb} = 9 \\
\text{snowing}
\end{bmatrix}
$$

At 1 am and 2 am, completely calm. At 3 and 4, weak NE; at 5, 6, 7, 8, and 9, SSE. Since 9 am, the wind has been varying between ESE and ENE, and a lot of snow has been falling since 11 pm. And at 5:15, the SSW; SW at 11 pm. The speed of the wind has been increasing at a rate of 2 m per hour; the maximum of the day was 21 m at 12 at night. I don't know how to classify the wind changes that take place after a storm from the NE and before starting one from the SW. Today, for example, the wind blew from ENE, ESE, and SE and sometimes from the SSW in isolated gusts. In between these gusts, there were some seconds of complete calm. I don't think that the reason for this is the precipitations caused by the high land that we have so close by, because we have had wind blowing from that direction continuously and regularly.

5.8.27   *July 3, [1903]*

$$
7\,\text{am}
\begin{bmatrix}
[-]25.4 \\
[-]25.35 \\
+3.8 \\
727.8 \\
\text{max} = -8.0 \\
\text{SW 23 m} \\
\text{Neb} = 10 \\
\text{snöfyk}
\end{bmatrix}
\quad
2\,\text{pm}
\begin{bmatrix}
[-]24.8 \\
[-]25.0 \\
+7.6 \\
732.0 \\
\text{min} = [-]24.8 \\
\text{SW 24.3 m} \\
\text{Neb} = 10 \\
\text{snöfyk}
\end{bmatrix}
\quad
9\,\text{pm}
\begin{bmatrix}
[-]24.3 \\
[-]24.5 \\
+8.5 \\
736.8 \\
\text{SW 22 m} \\
\text{Neb} = 8 \\
\text{snöfyk}
\end{bmatrix}
$$

SW blowing all day. It would sometimes move to the WSW. The maximum speed was 25 m at 11 am, and the minimum was 20 m at 7 pm. The barometer rises as usual.

5.8.28   *July 4, [1903]*

$$
7\,\text{am}\begin{bmatrix}[-]19.8\\ [-]20.0\\ +2.5\\ 738.55\\ 75.5\%\\ \text{Max}=-19.0\\ \text{WSW}\,10\,\text{m}\\ \text{Neb}=3\end{bmatrix}\qquad
2\,\text{pm}\begin{bmatrix}[-]19.55\\ [-]19.75\\ +14.0\\ 739.4\\ 78.9\%\\ \text{SSW}\,2\,\text{m}\\ \text{neb}=2\end{bmatrix}\qquad
9\,\text{pm}\begin{bmatrix}[-]21.95\\ [-]22.2\\ +13.2\\ 739.1\\ 62.2\%\\ \text{ENE}\,2\,\text{m}\end{bmatrix}
$$

Wind from the WSW for the first 5 h of the morning. The maximum strength was 26 m at 2 am, which wasn't only the maximum of the day, but also the maximum of this storm. At 2 am, the intensity of the wind started to diminish around 2 m per hour; at 8 am, it moved to the SSW, and the strength was only 6 m per second. Between 1 pm and 3 pm, the strength was 2 m on average, and it was completely calm at times. After 3, it blew with 5 m until 7:30, when it calmed. From 8:10 to 12 at night, the wind would vary between SSE and ENE, the latter being predominant. In the morning, nebulosity decreased together with the wind: At 10 am, it was 2, not being completely clear until 7 pm.

5.8.29   *July 5, [1903]*

$$
7\,\text{am}\begin{bmatrix}[-]28.3\\ [-]28.5\\ +4.8\\ 740.3\\ \text{SSW}\\ \text{Neb}=0\end{bmatrix}\qquad
2\,\text{pm}\begin{bmatrix}[-]28.8\\ -\\ +11.7\\ 743.0\\ \text{min}=-32.8\\ \text{SSW}\,11.5\,\text{m}\\ \text{Neb}=0\end{bmatrix}\qquad
9\,\text{pm}\begin{bmatrix}[-]30.3\\ [-]30.6\\ +7.8\\ 742.2\\ \text{calm}\\ \text{Neb}=0\end{bmatrix}
$$

The sky has been almost clear during the day. Only some clouds would appear now and then, but not for long. The only thing that hindered the sun from shining clearly was the snow that the SSW lifted from the ice. The SSW started to blow at 3 am and calmed at 5:30 pm. The maximum speed was 13 m at 10 am. The temperature has been quite low today: According to the data recorder, it was around $-33.5°$ at 8 am.

5.8.30 *July 6, [1903]*

$$
7\,\text{am} \begin{bmatrix} [-]24.2 \\ [-]24.0 \\ 735.25 \\ \text{Max} = -20.1 \\ \text{Neb} = 7 \\ \text{Calm} \end{bmatrix}
\quad
2\,\text{pm} \begin{bmatrix} [-]19.5 \\ [-]19.5 \\ +12.6 \\ 733.4 \\ \text{min} = -30.5 \\ -19.0 \\ 71.4\% \\ \text{calm} \\ \text{neb} = 10 \end{bmatrix}
\quad
9\,\text{pm} \begin{bmatrix} +14.2 \\ 733.9 \\ -16.8 \\ 76.2\% \\ \text{calm} \\ \text{neb} = 10 \end{bmatrix}
$$

It can be said that it has been calm all day, except between 5 pm and 8 pm, and between 10 and 12 pm, when some breezes were felt, but the strongest one barely reached 0.5 m. The sky has been overcast almost all day.

5.8.31 *July 7, [1903]*

$$
7\,\text{am} \begin{bmatrix} -10.8 \\ -11.6 \\ +4.7 \\ 730.0 \\ 70\% \\ \text{Var}\,2\,\text{m} \\ \text{Neb} = 2 \end{bmatrix}
\quad
2\,\text{pm} \begin{bmatrix} -5.0 \\ -6.5 \\ +14.3 \\ 730.5 \\ \text{min} = -20.5 \\ 5\,2 \\ \text{S}\,10\,\text{m} \\ \text{Neb} = 2 \end{bmatrix}
\quad
9\,\text{pm} \begin{bmatrix} -18.1 \\ -18.2 \\ +15.0 \\ 736.55 \\ \text{S}\,9\,\text{m} \\ \text{Neb} = 2 \end{bmatrix}
$$

Until midday, variable wind and overcast sky. The temperature [is] high, and it has risen above 0°. At 12:30, the SSW and the S started to blow alternately until midnight. When the S started to blow, the sky began to clear up on that side, the barometer began to rise, and the thermometer dropped (although slowly). At 3 pm, nebulosity was 3, completely clearing up at midnight.

The temperature rose to −0.6° according to the thermometer for maximum and minimum and to 0° according to the data recorder (8 am).

My diary, it may be said, is nothing more than meteorological, because there truly isn't anything else to observe. Nevertheless, even though I have already done it, I am going to describe our life; that is, I'm going to describe what we do during a day, and it applies to all the rest [of the days]. In general, we wake up at 9:30, sometimes close to 10. Half an hour before, Åkerlundh goes around the rooms yelling "*Fruckosten är färdig*" which means "Lunch is ready!" [Sic—Normally, "frukost" is breakfast.] At the same time, he fills the washbowls with water for personal washing.

At 10, we have "*frukost.*" After the frukost, one usually goes to bed again to read or to do any other thing, if that is possible. Otherwise, we work in the dining room. In the morning, the temperature drops to +1°, +2°, or +3°. Åkerlundh is woken up at 7 am, and he gets up at 7:30 when the fire is lit in the kitchen, and they start to cook. Temperature with the heat of the kitchen [fire] starts to rise in the dining room, but very gradually. But by frukost time, it is never below 10°, and sometimes, it reaches 18°. When the temperature is between 10° and 13°, the heater is lit. From 2:30 to 3 pm, we have *middag* [dinner], and then, B, E, and N [Bodman, Ekelöf, and Nordenskjöld] go to sleep, and I begin studying Swedish while walking to do exercise, and I do gymnastics. After 3:30, usually, the fires are put out in the kitchen; at 5 pm, the coffee that has been left over from the morning is reheated, and Åkerlundh walks around the rooms yelling "*Kaffet är färdig*" [coffee is ready]. After having coffee (which is anything but coffee), we work on whatever needs to be done, or we read. At 8, water is heated and some food, and we have *Kväll* [evening meal] at 9. After that, we get together and play cards until 11 pm. At this hour, everyone is there except for whoever is on guard duty. Guard duty shifts at night are three: 10 pm to 1 am, 3 am, and from 5 am to 9 am. Whoever has to get up at 3 is woken up by an alarm clock, and after making the 3 am observations, goes to bed, setting the alarm clock for whoever has to get up at 5. Night guard shifts are performed by Nordenskjöld, Bodman, Ekelöf, and me. In this way, every three nights one of us is fresh, that is, he has "*sofnau*" [sic—sleep], as we call it. Daytime guard shifts are from 9 am to 9 pm, and Bodman and I take these, except for the days when there are magnetic observations during the daytime, in which case Nordenskjöld takes part. When there are magnetic observations at night, the same procedure is followed. The only difference is that the one who worked the first guard duty shift is awake until 2:30, and then, the one who works the second shift gets up at 3 am and goes to bed at 4:30 am. For the third guard shift, there's no difference. As seen, in general, there are only 2 hours when the weather is not observed—at 2 am and at 4 am. On the days of magnetic observations, these are done at every hour. Another task that we have is astronomical observations. These consist of observations of latitude, longitude, and time. I make the first one, and Bodman and I make the last two. The tools used are transit theodolite and a _______ [unreadable—possibly formula] circle. Meals are composed of: First, *frukost* is coffee, bread with butter, and penguin or seal steaks; second, *middag* is soup and penguin with dried vegetables, and bread with no butter. On Sundays, instead of penguin, we have canned food, and dessert is added. Third, the *Kväll* is tea, bread with butter, and penguin or seal steaks (half the portion of the morning); sometimes, instead of penguin or seal steak, [we have] a little bit of stew made of the same material [penguin or seal], or some canned food like sardine, or some cooked penguin.

Sobral is very self-conscious about his diary. He realizes that the past several entries have all been meteorological data, and he makes the effort to illustrate life at Snow Hill, telling the story of the men's daily routine. The last page of this entry

*is only half filled—a rarity for Sobral, to leave half a page empty in his diary.
Perhaps, he had intended to fill the remaining space with more details about daily
life in Antarctica. Much of this information he would later use in his speech given
for the Naval Center (Centro Naval) in Buenos Aires (Sobral 1903a).*

5.8.32 July 8, [1903]

$$
7\,\text{am}\begin{bmatrix} -24.55 \\ - \\ +4.3 \\ 740.2 \\ \text{Max} = -1.2 \\ -24.5 \\ 82\% \\ \text{ENE 1 m} \\ \text{Neb} = 1 \end{bmatrix}
\quad
2\,\text{pm}\begin{bmatrix} -22.5 \\ -22.55 \\ +14.0 \\ 741.85 \\ \text{min} = -29.0 \\ -22.5 \\ 82\% \\ \text{ENE 3.5 m} \\ \text{Neb} = 10 \end{bmatrix}
\quad
9\,\text{pm}\begin{bmatrix} -21.5 \\ - \\ +13.0 \\ 741.4 \\ 87.2\% \\ \text{calm} \\ \text{Neb} = 10 \end{bmatrix}
$$

Variable winds and calm during the day, but between 10 am and 6 pm, it blew
from the ENE, E, and ESW at a speed of 2, 3, and 4 m. Between 6 pm and 12 pm,
the night was completely calm. The barometer has started to drop.

5.8.33 July 9, [1903]

$$
7\,\text{am}\begin{bmatrix} [-]19.25 \\ [-]19.15? \\ +5.0 \\ 738.25 \\ 87\% \\ \text{Calm} \\ \text{Neb} = 10 \end{bmatrix}
\quad
2\,\text{pm}\begin{bmatrix} -12.1 \\ -11.8 \\ +15.6 \\ 734.6 \\ \text{min} = -22.5 \\ [-]11.8 \\ \text{calm} \\ \text{Neb} = 10 \end{bmatrix}
\quad
9\,\text{pm}\begin{bmatrix} -4.8 \\ -5.3 \\ +12.9 \\ 727.75 \\ \text{NE 11 m} \\ \text{neb} = 8 \end{bmatrix}
$$

During the entire morning until 2 pm, calm and overcast sky. At 3, it started to
blow from the NE at a speed of 5 m. At 7 pm, the speed was 13 m and the wind
was from the NNE (it was from the NNE between 5 and 8 pm); that was the
maximum. Nebulosity was 9, and at 11 pm, the wind moved to the ENE, and
nebulosity was only 2, wind speed 7 m. At 11:45, the wind calmed completely, and
at 11:57, the SSW started at a speed of 6 m per second. The barometer, which had

been dropping all day, started to rise with the SSW; the thermometer, which had been high all day and indicated $-2.5°$ at 11:57, started to drop quickly at that hour, and it indicated -14.5 at 12.30.

5.8.34   *July 10, [1903]*

7 am
$$\begin{bmatrix} [-]24.95 \\ [-]24.95 \\ +7.0 \\ 721.6 \\ \text{Max} = -2.0 \\ \text{SSW } 14\,\text{m} \\ \text{Neb} = 1 \end{bmatrix}$$
2 pm
$$\begin{bmatrix} [-]22.3 \\ [\text{crossed out}] \\ +15.3 \\ 738.5 \\ \text{min} = -25.8 \\ 66.1\% \\ \text{var } 3\,\text{m} \\ \text{Neb} = 0.5 \end{bmatrix}$$
9 pm
$$\begin{bmatrix} [-]23.0 \\ [-]23.3 \\ +12.8 \\ 739.45 \\ 76\% \\ \text{NE } 0.5\,\text{m} \\ \text{Neb} = 10 \end{bmatrix}$$

The barometer is rising, and the thermometer is dropping. From 6 am to 6 pm, nebulosity was almost nonexistent, but at 7, the sky was completely overcast, although only with cirrus and Ci-S. Today, I observed the most splendid lunar corona, with three perfectly straight rings. The maximum strength of the wind was 16 m at 5 am. Since that time, it started to diminish quickly, and at 1:30 pm, the SSW ceased, varying to the E. At 4, the NE began to blow.

5.8.35   *July 11, [1903]*

7 am
$$\begin{bmatrix} [-]20.95 \\ [-]21.0 \\ +3.9 \\ 735.7 \\ 85\% \\ \text{NE } 1\,\text{m} \\ \text{Neb} = 10 \\ \text{snöfyk} \end{bmatrix}$$
2 pm
$$\begin{bmatrix} [-]26.3 \\ [-]26.5 \\ +12.8 \\ 733.9 \\ \text{min} = -27.6 \\ [-]26.2 \\ \text{SSW } 12\,\text{m} \\ \text{Neb} = 10 \\ \text{snöfyk} \end{bmatrix}$$
9 pm
$$\begin{bmatrix} [-]28.8 \\ [-]28.9 \\ +12.2 \\ 732.8 \\ \text{SSW } 11\,\text{m} \\ \text{Neb} = 10 \\ \text{snöfyk} \end{bmatrix}$$

Wind from the ENE and NE all morning, except for 5 am, when it blew from the SSW. At 11:45, it started to blow from the SSW, and it kept blowing for the rest of the day. Nebulosity has been 10 all day. The maximum wind speed was 14 m at 10 pm.

5.8.36   *July 12, [1903]*

$$
7\,\text{am}
\begin{bmatrix}
[-]28.6 \\
[-]28.7 \\
+3.0 \\
735.45 \\
79\% \\
\text{Max} = -19.9 \\
\text{SSW 5 m} \\
\text{Neb} = 10 \text{ Stratus}
\end{bmatrix}
\quad
2\,\text{pm}
\begin{bmatrix}
[-]28.7 \\
[-]28.8 \\
+14.2 \\
737.4 \\
\min = -29.5 \\
81.8\% \\
\text{SW 6 m} \\
\text{Neb} = 10 \\
\text{snöfyk}
\end{bmatrix}
\quad
9\,\text{pm}
\begin{bmatrix}
[-]29.5 \\
[-]29.5 \\
+7.9 \\
738.15 \\
81.2\% \\
\text{SSW 9 m} \\
\text{Neb} = 10 \\
\text{snöfyk}
\end{bmatrix}
$$

Overcast sky and SSW blowing all day. In the morning, the wind diminished until midday, when it started to increase in intensity. The maximum was 15.5 m at 12 at night.

5.9   I Keep Losing Weight

5.9.1   *July 13, [1903]*

$$
7\,\text{am}
\begin{bmatrix}
[-]31.0 \\
[-]31.0 \\
+1.5 \\
735.05 \\
\text{Max} = -27.3 \,(?) \\
\text{SW 20 m} \\
\text{Neb} = 10
\end{bmatrix}
\quad
2\,\text{pm}
\begin{bmatrix}
[-]29.2 \\
[-]29.45 \\
+9.6 \\
732.25 \\
\min = -32.55 \\
\text{SW 16 m} \\
\text{Neb} = 1
\end{bmatrix}
\quad
9\,\text{pm}
\begin{bmatrix}
[-]28.8 \\
[-]28.9 \\
+9.6 \\
725.3 \\
\text{NE 2 m} \\
\text{Neb} = 0
\end{bmatrix}
$$

The wind from the SW increased in intensity until 6 am, at which time it hit a maximum of 22 m; then, it diminished. At 3 pm, it moved to the SSW, and at 6:30 pm, it was completely calm, but the barometer dropped swiftly, and the NE started at 7 pm. At 10 pm, the wind kept varying between the ENE and the ESE. Nebulosity has been nonexistent since midday; at 10, it was 2, and great Cu-Ni and high cumulus would come from the SW. At the same time, moon halos could be observed, and upper tangent arches. The barometer began to rise at 11:15. At 11 pm, the wind was from the SE, and suddenly, without calming, it moved to the SSW. At the same time, the sky became overcast. A quarter of an hour later, nebulosity was 10. From 10:55 to 11 pm, it was calm. Today, we have weighed ourselves. I keep losing weight and have gotten down to the minimum of 142 lb. In April of 1902, I weighed 158 lb.

In the extreme cold temperatures of the Polar regions, it is important to increase caloric intake so as to be able to function in the extreme cold. The appetite increases exponentially with the enormous amount of energy expended in the

Antarctic. Sobral had reason to worry about not eating enough and about losing weight.

5.9.2   *July 14, [1903]*

$$
7\,\text{am}
\begin{bmatrix}
[-]25.8 \\
[-]26.35 \\
+\,2.0 \\
731.4 \\
\text{Max} = -21.5 \\
\text{SSW}\,3.5\,\text{m} \\
\text{Neb} = 0
\end{bmatrix}
\quad
2\,\text{pm}
\begin{bmatrix}
[-]28.0 \\
[-]28.2 \\
+\,11.0 \\
736.8 \\
\text{min} = -30.0 \\
73.8\% \\
\text{var}\,1\,\text{m} \\
\text{Neb} = 2
\end{bmatrix}
\quad
9\,\text{pm}
\begin{bmatrix}
[-]22.0 \\
[-]22.0 \\
+\,14.0 \\
740.35 \\
75\% \\
\text{ENE}\,1.5\,\text{m} \\
\text{Neb} = 2
\end{bmatrix}
$$

The SSW blew until 7 am. It calmed at 8. Between 9 am and 9 pm, variable winds. At 9 ENE, and then ESE and E until 12. At 12, it had a speed of 9 m. From 3 am to 10 am, nebulosity was 0; from that time until 9, it was between 1 and 3. At 11, overcast sky.

5.9.3   *July 15, [1903]*

$$
7\,\text{am}
\begin{bmatrix}
[-]20.6 \\
[-]20.65 \\
+\,1.8 \\
733.1 \\
81\% \\
\text{Max} = -20.2 \\
\text{E}\,4\,\text{m} \\
\text{Neb} = 4 \\
\text{snöfyk}
\end{bmatrix}
\quad
2\,\text{pm}
\begin{bmatrix}
[-]28.0 \\
[-]28.2 \\
+\,9.7 \\
733.5 \\
\text{min} = -29.1 \\
\text{SSW}\,18.5\,\text{m} \\
\text{Neb} = 8 \\
\text{snöfyk}
\end{bmatrix}
\quad
9\,\text{pm}
\begin{bmatrix}
[-]29.5 \\
[-]29.6 \\
+\,10.3 \\
734.75 \\
\text{SW}\,18\,\text{m} \\
\text{Neb} = 10
\end{bmatrix}
$$

From 1 am to 7:55, extraordinarily strong wind from the E (between 1 am and 8 am, [it was] 8 m). The SSW started at 8. The sky was cloudy almost all day. At 5 in the afternoon, the wind moved to the SSW. The maximum wind speed was 21 m at 11 pm.

5.9.4   July 16, [1903]

$$
7\,\text{am}
\begin{bmatrix}
[-]29.75 \\
- \\
+2.6 \\
737.0 \\
76.8\% \\
\text{Max} = -19.9 \\
\text{SW } 19\,\text{m} \\
\text{Neb} = 10 \\
\text{snöfyk}
\end{bmatrix}
\qquad
2\,\text{pm}
\begin{bmatrix}
[-]28.3 \\
[-]28.5 \\
+15.6 \\
742.8 \\
\text{min} = -30.0 \\
75\% \\
\text{SSW } 10\,\text{m} \\
\text{Neb} = 3
\end{bmatrix}
\qquad
9\,\text{pm}
\begin{bmatrix}
[-]27.6 \\
[-]28.05 \\
+11.6 \\
746.25 \\
64\% \\
\text{calm} \\
\text{Neb} = 0
\end{bmatrix}
$$

The SW continued until 10 am, with overcast sky, reaching a maximum of 20 m at 1 am and a minimum of 18.5 m at 10 am. At that hour, the wind began to ease, and it moved to the SSW, clearing the sky. The sky was completely clear at 5 pm, and the wind calmed at 9 pm.

5.9.5   July 17, [1903]

$$
7\,\text{am}
\begin{bmatrix}
[-]24.15 \\
[-]24.45 \\
+3.2 \\
741.2 \\
81\% \\
\text{NE} \\
\text{Neb} = 10
\end{bmatrix}
\qquad
2\,\text{pm}
\begin{bmatrix}
-2.55 \\
-4.0 \\
+15.4 \\
741.2 \\
68.5\% \\
\text{min} = -31.8 \\
\text{ENE} \\
\text{Neb} = 9
\end{bmatrix}
\qquad
9\,\text{pm}
\begin{bmatrix}
-0.9 \\
- \\
+14.2 \\
739.75 \\
82\% \\
\text{NE} \\
\text{Neb} = 9
\end{bmatrix}
$$

Overcast sky all day starting at 3 am. From 1 am to 1 pm, weak variable winds alternated with calm. At 1:45, the NE began at a speed of 5 m. After 7 pm, it would blow sometimes from the ENE and sometimes from the N. The temperature started to rise swiftly beginning at 9 am. At 11, it dropped again, but it steadily rose at midday, and at 5, it was above 0°. At 10 pm, it was +4.3°, which is the maximum of the day, and at 12 at night, it was still above 0°. At this hour, the NNE keeps blowing, and it is clearing up between the SE and the WSW, a sign that the SW will soon come.

5.9.6   *July 18, [1903]*

$$
7\,\text{am}
\begin{bmatrix}
+0.3 \\
-0.6 \\
+8.2 \\
739.6 \\
82\% \\
\text{Max} = +3.8 \\
\text{NNE}\,9\,\text{m} \\
\text{Neb} = 9
\end{bmatrix}
\quad
2\,\text{pm}
\begin{bmatrix}
+1.5 \\
-0.05 \\
+15.2 \\
740.1 \\
73.6\% \\
-4.2 = \min \\
\text{NNE}\,9\,\text{m} \\
\text{Neb} = 10
\end{bmatrix}
\quad
9\,\text{pm}
\begin{bmatrix}
+0.4 \\
-0.3 \\
+14.0 \\
738.9 \\
86\% \\
\text{NNE}\,10\,\text{m} \\
\text{Neb} = 10
\end{bmatrix}
$$

NNE blowing all day, with temperatures above 0° most of the time. The barometer is dropping, but very slowly. Today, some water clearings have been seen.

Inside the house, it rains, one could say, especially on the beds. Thanks to pieces of rugs and blankets that have been nailed to the ceiling, I keep my bed a little bit dry. The ice, which in great amounts had formed on the glass, walls, floors, etc., melts, and so the house becomes a puddle. The doors are open all day, because, if you close them when the kitchen [fire] is lit, the temperature rises above 20° quite rapidly.

5.9.7   *July 19, [1903]*

$$
7\,\text{am}
\begin{bmatrix}
+2.1 \\
+0.6 \\
+9.1 \\
739.05 \\
77\% \\
\text{Max} = +2.6 \\
\text{N}\,12\,\text{m} \\
\text{Neb} = 8
\end{bmatrix}
\quad
2\,\text{pm}
\begin{bmatrix}
-0.6 \\
-1.25 \\
+15.7 \\
739.7 \\
91\% \\
\min = -1.55 \\
\text{NE}\,13\,\text{m} \\
\text{Neb} = 10
\end{bmatrix}
\quad
9\,\text{pm}
\begin{bmatrix}
-1.7 \\
-2.5 \\
+13.2 \\
736.0 \\
\text{NNE}\,8\,\text{m} \\
\text{Neb} = 6
\end{bmatrix}
$$

The NNE continued all day, with almost the same temperatures. The barometer started to drop swiftly in the afternoon. The minimum wind speed was 6 m at 8 pm, and the maximum was 16 m at 10 am. But between 1 am and 6 pm, the wind has not blown below 10 m.

It doesn't rain in the house anymore, but the floor is still a puddle. We all climbed up to the top of the basalt at midday, and we could see that the ice-free water channel that extends up to the latitude of Snow Hill, around two miles from the coast to the ENE, has 4 or 5 miles of width; at Cape Seymour, it takes a northerly direction, and it passes some miles through Cockburn. In that part (the one in the direction of Cockburn), we couldn't see the limit of the ice-free channel, but we were unable to determine whether or not it continued to extend over a distance from there, because there was a bit of fog that would not let us see too far. Since

12 pm, the thermometer has remained below zero. It looks like the warmer weather we had in September has anticipated [come early], because, if we use last year as a guide, this is completely abnormal. Despite the fact that during this recent time (June, July) we have been constantly looking for seals, we can't find them. We have seen some [seal] holes and their [seal] tracks on the snow, but nothing more. Probably, these animals come out only on calm and warm days, but in these conditions, none have been seen by us. The poor dogs eat whatever they can find, and it's not much. The skeletons of the seals no longer have any meat, and the only thing they [the dogs] have to eat is the hide with the fat.

5.9.8   *July 20, [1903]*

$$
7\,\text{am}
\begin{bmatrix}
-0.5 \\
-0.75 \\
+9.6 \\
729.7 \\
\text{Max} = +2.5? \\
89.2\% \\
12\,\text{m} \\
\text{Neb} = 10
\end{bmatrix}
\qquad
2\,\text{pm}
\begin{bmatrix}
+17.4 \\
732.0 \\
-1.9 \\
-2.45 \\
73.2\% \\
\text{min} = -3.6 \\
\text{NE}\,5\,\text{m} \\
\text{Neb} = 10 \\
\text{snowing}
\end{bmatrix}
\qquad
9\,\text{pm}
\begin{bmatrix}
-19.3 \\
-19.45 \\
+13.8 \\
733.85 \\
73.6\% \\
\text{SW}\,19\,\text{m} \\
\text{neb}\,10 \\
\text{snowing}
\end{bmatrix}
$$

The NNE blew all day until 7 pm. At 7:15, the SSW started, with a lot of snow. At that same hour, the barometer began to rise quickly. The maximum wind speed was at 9 pm, and it started to ease right away. The maximum speed from the NNE was 10 m at 1 am, but during the whole day, it didn't drop below 5 m. The temperature dropped swiftly when the SW came. At 11 pm, the sky started to clear up, and at midnight, nebulosity was just 3.

5.9.9   *July 21, [1903]*

$$
7\,\text{am}
\begin{bmatrix}
-19.95 \\
-20.1 \\
+5.4 \\
738.35 \\
\text{Max} = -0.25 \\
70\% \\
\text{Calm} \\
\text{Neb} = 3
\end{bmatrix}
\qquad
2\,\text{pm}
\begin{bmatrix}
-7.7 \\
-9.3 \\
+17.0 \\
736.5 \\
\text{min} = -20.6 \\
51.5\% \\
\text{NE} \\
\text{Neb} = 9
\end{bmatrix}
\qquad
9\,\text{pm}
\begin{bmatrix}
-4.8 \\
-5.2 \\
+13.5 \\
732.35 \\
\text{SW} \\
\text{Neb} = 10
\end{bmatrix}
$$

The SSW calmed at 4:30 am, and at 5 am, the NE was already blowing; the sky was clear. From 5 am to 11 am, complete calm and few clouds. Thermometer rising and barometer dropping. At midday, the ENE started, which lasted until 8:30 pm. At 1 pm, the sky became overcast with Ni, and snow started to fall at 6. At 9 pm, wind started to blow, varying between the SSW and WSW. At 10, it was from the SW, but although the barometer started to rise at 9, the thermometer didn't drop, and at 12 at night, it is still around $-5°$. Today, we have seen much less free water than the day before yesterday. The day before yesterday, a giant petrel and a snow petrel were seen; today, a snow petrel.

5.9.10 July 22, [1903]

$$
7\,\text{am}
\begin{bmatrix}
-10.3 \\
-11.3 \\
+8.0 \\
742.3 \\
\text{Max} = -3.0 \\
\text{Calm} \\
\text{Neb} = 1
\end{bmatrix}
\quad
2\,\text{pm}
\begin{bmatrix}
-2.95 \\
-5.3 \\
+16.0 \\
742.35 \\
\text{min} = -11 \\
56.7\% \\
\text{NNE}\,5.5\,\text{m} \\
\text{Neb} = 10
\end{bmatrix}
\quad
9\,\text{pm}
\begin{bmatrix}
-3.0 \\
-3.1 \\
+15.3 \\
733.5 \\
89\% \\
\text{ENE}\,10.5\,\text{m} \\
\text{neb} = 10 \\
\text{snowing}
\end{bmatrix}
$$

Clear sky until 7 am, when the wind from the S calmed (at 4 am it moved from the SSW to the S). The NE started at 9, with almost overcast sky. At 2 pm, it moved to the NNE, starting to snow at 5 pm. Between 7 and 9 pm, wind from the NNE, with a lot of fog. At 10, it was from the NE, with little snow, and at 10:50, at the same time that the barometer started to rise, and with only some seconds of calm, the SSW began. Magnetic observations are being made today.

5.9.11 July 23, [1903]

$$
7\,\text{am}
\begin{bmatrix}
[-]20.0 \\
[-]20.05 \\
+9.9 \\
724.0 \\
86\% \\
\text{Max} = -0.9 \\
\text{SSW}\,12\,\text{m} \\
\text{Neb} = 10
\end{bmatrix}
\quad
2\,\text{pm}
\begin{bmatrix}
[-]18.5 \\
[-]18.7 \\
+13.2 \\
727.95 \\
\text{min} = -20.6 \\
\text{SW}\,10\,\text{m} \\
\text{Neb} = 3
\end{bmatrix}
\quad
9\,\text{pm}
\begin{bmatrix}
[-]14.8 \\
[-]15.25 \\
+14.0 \\
737.8 \\
\text{WSW}\,3\,\text{m} \\
\text{Neb} = 2
\end{bmatrix}
$$

Overcast sky until 11 am, with wind from the SSW. At that hour, it started to clear, and the wind moved to the SW. Until 9 pm, nebulosity was between 2 and 4, with alternating winds from the SSW and SW. There was a period of calm at 9, but a while later, the SSW started again. The temperature, which was around $-20°$ in the morning, kept rising all day, reaching $-6.5°$. The barometer is rising.

5.9.12   *July 24, [1903]*

$$
7\,\text{am}\,\begin{bmatrix} -10.9 \\ -12.25 \\ +6.0 \\ 739.55 \\ 52\% \\ \text{Max} = -6.3 \\ \text{SSW}\,8\,\text{m} \\ \text{Neb} = 0 \end{bmatrix}
\quad
2\,\text{pm}\,\begin{bmatrix} -10.5 \\ -12.0 \\ +16.1 \\ 742.85 \\ 44.1\% \\ \text{min} = -18.0 \\ \text{var} = 0.2 \\ \text{Neb} = 9 \end{bmatrix}
\quad
9\,\text{pm}\,\begin{bmatrix} -4.3 \\ -5.75 \\ +14.3 \\ 741.05 \\ 67.8\% \\ \text{ENE}\,4\,\text{m} \\ \text{Neb} = 9 \end{bmatrix}
$$

Very nice day, mild and extraordinarily dry, especially in the morning. Clear sky before midday, and cloudy since 2 pm. In the morning, the temperature remained around $-10°$, sometimes dropping to $-14°$, but it started to rise after midday, and it reached $-3°$ at night.

5.9.13   *July 25, [1903]*

$$
7\,\text{am}\,\begin{bmatrix} -4.35 \\ -5.45 \\ +9.0 \\ 737.6 \\ 74\% \\ \text{Max} = -1.7 \\ \text{ENE}\,2.5\,\text{m} \\ \text{Neb} = 10 \end{bmatrix}
\quad
2\,\text{pm}\,\begin{bmatrix} -5.25 \\ -6.4 \\ +17.6 \\ 739.0 \\ 69.7\% \\ \text{min} = -12.65 \\ \text{calm} \\ \text{Neb} = 8 \end{bmatrix}
\quad
9\,\text{pm}\,\begin{bmatrix} -12.1 \\ -12.8 \\ +14.8 \\ 738.3 \\ 88\% \\ \text{calm} \\ \text{Neb} = 0 \end{bmatrix}
$$

In the morning, variable winds and overcast sky, temperature remains high (around $-4°$, $-5°$). After 1, it started to clear up, and there was complete calm. At night, the temperature started to drop, but not by much. The barometer doesn't drop, and it's around the average. The weather this month is really curious. Today, for example, the day couldn't have been more beautiful, especially after midday: clear sky, temperature $-5°$, the barometer indicates the average, without dropping. Complete calm. Needless to say, with these temperatures, we don't have to light the heater, the house being so warm that, every so often, we have to open the doors in order to let the too-hot air escape. I've worked all morning pulling off the ice that had formed on the walls of my room. In many parts, it was over 10 cm thick. I had to use an axe, a hammer, and a lot of patience to remove it. When it's cold outside, it [the ice] does no harm, but during these days of relatively high temperatures, it melts, and the room turns into a lake during the entire day. According to Nordenskjöld, today there is as much ice-free water as there was in September of last year.

Yesterday, I took four photographs: one of our W.C.; the second of the house, looking at it from the N, including the anemometer, W.C., and meteorological casings; "Peridota" and her son "Kalle" eating; and one of the glens.

Yesterday, they brought the skin of a seal that I had slaughtered on April 4. Inside the skin I put these pieces of meat that are of great use to us today. That meat and the skin weren't picked up before because Jonassen didn't feel like it, and for that same reason, we have lost a lot of seals.

Ever mindful of waste, and seeing his waistline receding, Sobral criticizes—for the first time—Jonassen, who "did not feel like" fetching the seal that Sobral had painstakingly killed over three months ago. The fact was that the men by this time were very short on fresh protein, and even as their diet suffered, the physical reality surrounding them was deteriorating as well: Their house was becoming an ice-encrusted, sticky soot-filled, moldy abode; their clothes were disintegrating on their backs (and feet); and the dogs among them were becoming weak with hunger. And yet, as can be seen from his diary, Sobral continued his methodical meteorological work, observing the atmospheric elements as well as the state of the water, ice, and terrain, documenting the wintering station and its surroundings photographically, and making psychological observations about the men as well.

5.9.14   *July 26, [1903]*

$$
7\,\text{am}
\begin{bmatrix}
-3.9 \\
-6.4 \\
+7.7 \\
731.3 \\
\text{Max} = -1.5 \\
47\% \\
\text{NE } 3.5\,\text{m} \\
\text{Neb} = 10
\end{bmatrix}
\quad
2\,\text{pm}
\begin{bmatrix}
-3.5 \\
-4.3 \\
+15.0 \\
726.2 \\
85\% \\
\text{min} = -16.0 \\
\text{NNE } 9\,\text{m} \\
\text{Neb} = 10
\end{bmatrix}
\quad
9\,\text{pm}
\begin{bmatrix}
-3.5 \\
-3.8 \\
+14.4 \\
716.8 \\
88.0\% \\
\text{NNE } 7.5\,\text{m} \\
\text{Neb} = 10
\end{bmatrix}
$$

All day, wind from the NNE. Sometimes, it would move to the NE, ENE, NW, or NNE, but that was the general direction from 1 pm to 9 pm. Until 5 pm, sky overcast with Ci and Ni, Ni-Cu. At 8:30, it cleared up almost completely, but the nebulosity started to increase, and the sky was overcast at midnight. The maximum speed from the NNE was 10 m at 7 pm. The barometer dropped swiftly (around 2 mm per hour), and it reached its minimum at 10 (0.5 mm lower than at 9). Since 6 am, the thermometer has been varying between 0° and −5°, and relative humidity between 40 and 100%. The SSW started to blow at 9:50, and the barometer remained without any variations until 10:30, when it began to rise slowly. When the SSW started, the thermometer was indicating −3°, and it dropped to −13° at midnight.

5.9.15   *July 27, [1903]*

	7 am		2 pm		9 pm
	−21.75		−23.8		−24.8
	−21.90		−24.35		−25.0
	+5.4		+14.6		+11.8
	720.5		723.55		722.9
	Max = 0		50%		calm
	68%		min = −24.0		Neb = 0
	SW 17.5 m		SWS 4 m		
	Neb = 4		Neb = 0		

SW from 1 am to 7 am; SSW until 1 pm; and until 5 pm, SSW, SW, and WSW. Clear sky since 2 pm. The SSW calmed completely at 5 pm, remaining calm for the rest of the day, with completely clear sky. The temperature keeps dropping at night.

Yesterday, one of Amager's puppies appeared dead, killed by some of the other dogs. I don't know who had the idea of cooking a little dog meat in the form of steaks, which was brought to the table. Only Bodman and I ate it. It is a tough meat, and it has the consistency of rubber, but it's not disagreeable.

No one else would partake of the perished puppy, whom Sobral and Bodman eat for lack of other fresh meat. This page of the diary has bright orange stains that look to be of oil.

5.10   To Experience the Feeling of Such Beauty, You Have to See It

5.10.1   *July 28, [1903]*

	7 am		2 pm		9 pm
	−31.0		−14.0		−16.05
	−31.0		−14.0		−16.2
	74.0		+14.2		+12.8
	726.55		732.55		741.35
	69%		81%		79%
	Max = −22.0		SSW 17 m		WSW 24.5 m
	Neb = 2		Neb = 10		Neb = 10
			Snöfyk = 3		Snöfyk = 2

Between 1 and 6 am, clear sky and complete calm. Outside, everything is silence, except for the noise that the ice makes in its cracking. The thermometer is around −30°. So the master of these regions is the only one that raises his voice;

even for someone who's used to it, it's a surprise to suddenly hear that loud noise.[*]
But despite the cold, how beautiful is nature in these moments! The sky completely
full of stars, and the white of the frozen plain forms a contrast with the black of the
basalt cliffs. At 7 pm, you can see the pinkish tinge of the aurora to the east, and
toward Mt. Haddington, on the enormous snowdrift, like a reflection more beautiful
than the image itself, you can see the transitions of the delicate tints between red
and yellow. Futile will be the attempt to paint or describe these colors and this
harmony; to experience the feeling of such beauty, you have to see it.

But to this very beautiful picture, and so as to form a contrast, this happens: The
barometer is very low, at 8:30 the wind from SSW starts to blow, and the sky
rapidly becomes overcast, with enormous snow whirlwinds flying in the air. In the
same instant, the thermometer drops to $-32.8°$ and it rises to $-19°$. It is as though
there is a fight between the temperature, the pressure, and the wind, and once the
latter has made the others vary degree by degree, and millimeter by millimeter, with
the expense of titanic efforts, it wins.

[*]During the summer, or when the temperature is high, sea ice is very flexible; therefore, the
movements that take place in its mass do not make any noise, whereas, with great cold, its
flexibility disappears, and, due to the acquired rigidity, whenever it moves because of wind
or tides, it cracks with a noise probably proportional to the cold, caused by the rigidity. The
ice of glaciers and snowdrifts is completely different; being the product of the compression
of snow, it has the nature of ice resulting from the freezing of freshwater, always having, no
matter what the temperature is, the property of being fragile and glassy, so to speak.
Obviously, the state of the glassiness is relative and is subject to being augmented or
diminished by changes in temperature.

*Sobral's poetic description of the sky and atmosphere is accompanied by a
scientific explanation of the different types of ice and the various sounds they make.
It seems he is indeed trying to paint a picture and relay an orchestration of what he
is seeing and hearing. As he gives voice to his poetic tendencies, the scientist in him
is emerging.*

5.10.2   *July 29, [1903]*

7 am	2 pm	9 pm
-17.2	-18.25	-20.2
-17.4	-18.35	-20.4
$+14.7$	$+18.2$	$+12.5$
744.65	min $= -18.1$	753.25
Max $= -12.5$	SSW 14 m	WSW 20 m
SW 24 m	Neb $= 10$	Neb $= 10$
Neb $= 10$	Snöfyk $= 3$	Snöfyk $= 2$
Snökyk $= 2$		

The wind blowing from the SSW, SW, and WSW, the SW predominating, and a
lot of snow. The maximum wind speed has been 28.1 m at 3 am, and the minimum
was 13.5 m at 5 pm, but it started to rise again at that time. The 24-hour average

(from 1 am to 12 pm) is 23.0. At 12 at night, it has a speed of 22.8 m. The barometer is high and it keeps rising, while the thermometer is dropping. Inside the house, the temperature is most pleasant. At night, it starts to drop; at midnight, it is $-10°$, but it will probably be around +3 or +4° tomorrow at 6 or 7 am.

5.10.3 July 30, [1903]

$$
7\,\text{am}
\begin{bmatrix}
-21.5 \\
-21.6 \\
+6.0 \\
753.3 \\
\text{Max} = -16.6 \\
\text{SW } 21.9\,\text{m} \\
\text{Neb} = 10 \\
\text{snöfyk}
\end{bmatrix}
\quad
2\,\text{pm}
\begin{bmatrix}
-21.9 \\
-22.1 \\
+17.5 \\
755.6 \\
70.7\% \\
-21.7 = \text{min} \\
(-21.7) \\
\text{SW } 22\,\text{m} \\
\text{snöfyk} \\
\text{neb} < 2
\end{bmatrix}
\quad
9\,\text{pm}
\begin{bmatrix}
-21.6 \\
-22.05 \\
+11.8 \\
755.95 \\
64.5\% \\
\text{SSW } 18.4\,\text{m} \\
\text{neb} = 0 \\
\text{snöfyk}
\end{bmatrix}
$$

SW and SSW blowing, with a lot of snow blown by the wind. The sky has been very clear in the afternoon. The maximum wind speed was 24.7 m at 5 am, and the minimum was 14 m at 12 at night. The 24-hour average wind speed was 20 m.

5.11 I Find Myself Isolated, With No Friends

5.11.1 July 31, [1903]

$$
7\,\text{am}
\begin{bmatrix}
-19.0 \\
+3.1 \\
754.1 \\
\text{Max} = -16.2 \\
44\% \\
\text{SW } 20\,\text{m} \\
\text{Neb} = 0 \\
\text{snöfyk}
\end{bmatrix}
\quad
2\,\text{pm}
\begin{bmatrix}
-10.15 \\
-12.5 \\
+14.7 \\
754.25 \\
\text{min} = -21.9 \\
29.5\% \\
\text{SW } 2\,\text{m} \\
\text{Neb} = 0
\end{bmatrix}
\quad
9\,\text{pm}
\begin{bmatrix}
-18.3 \\
-18.8 \\
+15.7 \\
753.15 \\
62.5\% \\
\text{calm} \\
\text{neb} = 0
\end{bmatrix}
$$

SW and SSW blowing until 5 pm, then complete calm. Clear sky all day, and extremely dry air. At 1 pm, the relative humidity was 26%. The wind diminished its intensity from 1 am to 5 am, but at that hour it suddenly started to increase until it reached its maximum, and then, it decreased again; 24-hour average wind speed was 7 m.

Another month has concluded. I have never wanted time to pass as quickly as I do now. This month of July seems to have been exceedingly long. Six long months are yet to pass before we have any chances of getting out of here. I feel and suffer so much, not physically, but morally. I find myself isolated, with no friends, without having someone to talk to, I would say, because I do exchange some words with these people, but they are only the absolutely necessary ones. This month has been extraordinary, it seems, from a climatological point of view: It's around two degrees warmer than the month of May and close to four degrees warmer than the month of June of 1903. The first half of the month of July had an average temperature of around $-22°$, while in the second half it was $-13.1°$. Except for this last storm (from the 27th through the 31st of July), all the other storms have been of short duration and relatively weak. This last storm is the only one that makes us remember the ones of last winter, even though it has not been that cold. In July of last year, the maximum the thermometer reached was $-1.4°$, that month being the only month when the temperature did not rise above $0°$, while in this month of July, the temperature has been above $0°$ on the 18th and the 19th.

5.11.2   August 1, [1903]

Very beautiful day, high temperature, very dry, and calm air and clear sky. Magnetic observations are made. Today, they have set off to look for seals, but in vain. I think that they don't find seals because they don't look for them well. Rather than harnessing the dogs to the sledge, and having four people go in a different direction each for around 4 or 6 h, I'm positive that they would find seals if, instead, one man went out on foot, searched for one or 2 hours, and then came back with the news that seals were not seen [in that direction]. Furthermore, when they're not found on the ice, they have to be hunted in their element with a harpoon, like the Eskimos do.

7 am	2 pm	9 pm
-15.05	-12.55	-14.3
-15.00	-11.3	-14.55
$+4.8$	$+12.8$	$+13.0$
748.3	747.6	748.2
89.2%	min $= -22.5$	Hy $= 65\%$
Max $= -10.0$	(-11.4)	Calm
(-15.0)	49.5%	Neb $= 4$
Calm	calm	
Neb $= 0$	Neb $= 1$	
	Ci and S	

5.11.3   August 2, [1903]

$$
7\,\text{am}\begin{bmatrix}
-20.65 \\
-21.30 \\
+9.1 \\
751.1 \\
64\% \\
\text{Max} = -10.3 \\
\text{SSW}\,0.3 \\
\text{Neb} = 0
\end{bmatrix}
\quad
2\,\text{pm}\begin{bmatrix}
-9.05 \\
-9.9 \\
+19.1 \\
750.85 \\
60\% \\
\text{min} = -22.45 \\
(-9.9) \\
\text{ENE}\,4\,\text{m} \\
\text{Neb} = 0
\end{bmatrix}
\quad
9\,\text{pm}\begin{bmatrix}
-14.3 \\
- \\
+14.4 \\
747.3 \\
\text{S} \\
\text{Calm} \\
\text{Neb} = 0
\end{bmatrix}
$$

Weak and variable winds, except between 2 and 6, when NE and ENE blew with a strength of 3 or 4 m. The maximum of the day was 5 m at 4 pm. The temperature, which had been quite low during the first hours of the morning, rose and remained between 9 and 10 degrees from midday until 3 pm, at which time it started to drop. At 12 at night, it was −25°.

After lunch, feeling even more hungry than usual, and motivated by that feeling, I decided to go out and hunt seals. At 10:30, I went in the direction of the little strait that separates Snow Hill from Seymour Island, followed by "Kalle" and "Kvik." The ice presented a very resistant snow surface because it had been hardened by the wind, so one could walk without sinking. If, to this good condition of the route, you add a complete calm of the air and a temperature of −10°, one, who knows what walking on ice is, understands that this exercise in such circumstances was really nice. Not one cloud broke the endless continuity of the blue sky. The sun was shining in all its glory. This, which would have been very unpleasant in spring due to the risk of ophthalmia, now, since the sun is so low, doesn't cause any discomfort. Half an hour after leaving the station, I turned around and saw five of the other dogs coming toward me. When "Kvik" saw them, he became scared and ran away toward the coast, going up the mountain. When that pack of starving dogs got closer to me, I continued walking, and by the time I reached the strait, the number had increased to seven, and they went away, heading toward Seymour Island. When I got out of a labyrinth of ice and snow that I had entered looking for some seals, I couldn't see them anymore. Congratulating myself to a certain extent for that, since _______ [unreadable] the animals _______ [unreadable] the happiness of slaughtering some amphibious was enough for me, they wouldn't leave me alone. And then, after touring every corner and looking around the crevasses and snow mounds, I headed toward the strait to explore the ice on the E side of Seymour Island. Here, the ice is completely flat, and the only signs of its movements are rows of crevasses more or less parallel to the coast. In some parts, some of them are so even that they resemble the cut of a knife. At this moment, I can't think of anything more similar. But I'll say that the walls were perfectly vertical and flat. The east coast of Seymour Island in this part is steep with cliffs of about 20–30 m high that fall directly to the sea. But in between these cliff areas, which are the majority in this portion of the coast, you can see some long and narrow canyons that come from inland to the sea—some of them completely full of snow and ice, they are true glaciers. In some parts of the ravines and also inland, you can see interesting little mountains of sand that probably have some geological meaning. After two-and-a-half hours of hiking and not seeing any seals around, I decided to cross Seymour Island from E to W and then take the path back. With this goal, I climbed up one of the small snowdrifts with some difficulty, but suddenly I regretted taking that land trail. Deep glens cut the island from approximately S to W and N to E; in general, all of them with a lot of snow, but very difficult to cross because of the steepness of the flanks. Despite the abundant snow that covers the land, I could see some small snail fossils, like the ones in Snow Hill. I picked up some kind of shell, but it broke in my pocket (I keep the fragments). These fossils are at a height of 30 or 40 m, but in Snow Hill you can find them at the same or even higher altitudes. Land toward the W decreases continuously, forming ________ [unreadable] steps, kind of a terrace until it finally ends in a completely flat plain but with a slight slope toward the sea. Of about three miles long, and a quarter mile of width, this plain goes down to the sea with a very sweet slope. Here, the ice is completely irregular; large and numerous mounds, resulting from pressure, rise across it, making it very difficult to walk. The reason for this irregular ice is probably the ebb tide that comes from approximately WSW and hits against this coast. From here, I took the direction toward the station, always looking anywhere I could to find prey, and with no seals, and hungrier than when I had left, and tired after 6 h of continuous walking, I sat at the table, where I found something to slightly ease the desires of the stomach.

The dogs have not come back, so, very likely, they, happier than me, have found prey.

5.11.4   August 3, [1903]

$$
7\,\text{am}
\begin{bmatrix}
-5.6 \\
-8.1 \\
+6.0 \\
742.2 \\
\text{Max} = -5.9 \\
47\% \\
\text{S}\,2\,\text{m} \\
\text{Neb} = 1
\end{bmatrix}
\quad
2\,\text{pm}
\begin{bmatrix}
-15.55 \\
-16.0 \\
+18.1 \\
740.25 \\
81.9\% \\
\text{min} = -22.9 \\
-15.55 \\
\text{NE}\,1\,\text{m} \\
\text{Neb} = 9
\end{bmatrix}
\quad
9\,\text{pm}
\begin{bmatrix}
-8.1 \\
-9.5 \\
+14.2 \\
741.65 \\
59\% \\
\text{SSW}\,3\,\text{m} \\
\text{Neb} = 0
\end{bmatrix}
$$

Between 1 am and 9 am, warm winds from the S and SSW. At 12:30 at night, the thermometer indicated −22.9°, and it started to rise swiftly with the south wind. The increase began at 1 am, and it was from −22.9 to −9.2 (all at once). Between 1 pm and 3 pm, NNE, NE, and ENE; from 4 pm to 12 midnight, winds from the S and SSW. Nebulosity was nonexistent before midday and very variable in the afternoon (between 3 and 10). Today, I went to look for the dogs, without finding them. We suppose that they have gone to the rocky part of Seymour Island, where there is a great quantity of penguin remains. At nightfall, the dogs that are still left at the station killed one of the puppies belonging to Amaguer [Amager].

5.11.5   *August 4, [1903]*

$$
7\,\text{am} \begin{bmatrix} -8.95 \\ -10.25 \\ +7.3 \\ 743.1 \\ 56\% \\ \text{Max} = -2.5 \\ \text{Calm} \\ \text{Neb} = 9 \end{bmatrix} \quad 2\,\text{pm} \begin{bmatrix} -0.3 \\ -3.0 \\ +17.2 \\ 736.7 \\ 55.7\% \\ \text{WSW}\,0.5\,\text{m} \\ \text{neb} = 9 \end{bmatrix} \quad 9\,\text{pm} \begin{bmatrix} -0.25 \\ -2.3 \\ +14.4 \\ 730.2 \\ \text{ENE}\,3\,\text{m} \\ \text{neb} = 9 \end{bmatrix}
$$

During the first hours of the morning, it blew from the south at a speed between 5 and 8 m. Variable winds for the rest of the day. Overcast sky almost all day. The dogs that are supposed to be the authors of Qvik's [Kvik's] death have been given a huge beating this morning. How barbaric! As if the animals will understand that it was because of what they did yesterday. It was Jonassen's idea, and, naturally, Nordenskjöld approves. In addition, the dogs have been tied to a pole, and every time someone walks by them, they receive terrible kicks. According to what they say, the dogs will be tied and go without eating, for a week. At mealtime, Ekelöf, who is to my left, serves himself first and then passes to me what's left, so that I do the same, but he always accompanies the action of passing things with deep sighs, and this is every day. And many times, after I have served myself, he points out that there is little food. He always argues! He talks about these things, and it's very natural that I think that his sighs are because I always eat too much. So after lunch, I interrupted him and told him that if I really ate too much, he should tell me what my portion is. He got upset, telling me that I was insulting him, and that instead of me being thankful for how well he had treated me, I was making him uncomfortable. The thing is that now, at the table, he doesn't sigh any more. I don't know how he has the courage to lie to me in my face and tell me that he has treated me well.

Sobral's desperation to find food, as seen by his extensive efforts two days prior in trekking all over the Island to hunt seal, is exacerbated by Ekelöf's behavior at the table. Once again, Sobral finds he must confront the temperamental doctor regarding his treatment of the earnest under-lieutenant. To make matters worse, the dogs that had not followed him on his trek two days ago had, on the previous day, killed Amager's puppy, whom Sobral names "Qvik" in this entry. "Qvik" must be the same as "Kvik" who had followed Sobral two days prior and had run home after seeing the other five, then seven, hungry dogs join the hunt. Unfortunately, Kvik was then killed at home by some of the dogs remaining at the station.

Jonassen's brutal corporal punishment of those dogs, to which Nordenskjöld acquiesces, disgusts Sobral, who understands that, in the minds of the dogs, there will be no cause-and-effect connection between the murder of yesterday and the punishment of today. Meanwhile, the dogs who had gone hunting with Sobral still have not returned.

In his handwritten notes for his speeches given at the end of 1903, Sobral would later discuss some of the illnesses that the men suffered during their winter habitation, including the ophthalmia which he mentions in the August 2nd entry—the condition of inflamed eyes that results from exposure to the brightness of the light and snow, also as in snow blindness. He also mentions dysentery affecting the men during the last months at Snow Hill, the cruel irony of this being that, while they did not have enough to eat to fill their stomachs, their intestines also became inflamed and infected. He attributed the dysentery to "the leavening that was used to make bread" (Sobral 1903b). Nordenskjöld corroborated this statement, saying in his book that it was none other than Dr. Ekelöf who had discovered a fungus that he cultivated as a yeast, which the men then used in the dough to leaven the bread (Nordenskjöld et al. 1905). It was this fungus-related leavening that caused the stomach problems. When mentioning the argumentative doctor later in his notes for his speeches, Sobral would only say that Ekelöf kept "interesting data about the changes that our blood experienced during the second year, as well as significant observations on the effects of snow and light on the visual organs" (Sobral 1903b).

5.11.6   August 5, [1903]

$$7\,\text{am} \begin{bmatrix} +3.8 \\ +1.4 \\ +10.3 \\ 721.9 \\ 64.7\% \\ \text{Max} = +9.3 \\ \text{SSW } 9.5\,\text{m} \\ \text{Neb} = 10 \end{bmatrix} \qquad 2\,\text{pm} \begin{bmatrix} +3.75 \\ +0.95 \\ +14.5 \\ 718.5 \\ 57\% \\ \text{min} = -4.0 \\ \text{ENE } 30\,\text{m} \\ \text{neb} = 9 \end{bmatrix} \qquad 9\,\text{pm} \begin{bmatrix} -1.0 \\ -2.5 \\ +15.0 \\ 716.5 \\ 74.5\% \\ \text{var } 2.5\,\text{m} \\ \text{neb} = 9 \end{bmatrix}$$

A very cloudy day. Between 3 am and 6, it blew from the NNE, N, and NNW, with a strength of 5 or 6 m. After 10 min of calm, the SSW started to blow at 6:10, but it calmed at 8. In general, these winds start from the WSW, and the temperature drops, but if it then moves to the SSW and S, the thermometer rises again. The rest of the day, variable winds and calm, but with predominance of winds from the ENE, E, and ESE. At 5 pm, the E reached a speed of 9 m.

Today, they went to look for seals, with a negative result. At 3 pm, the dogs that had left the other day came back. It looks as if they have not been to Seymour Island, because they were hungry. Poor "Kalle," the smallest of all of them, was so tired that he could barely stand. They were probably around one of those seal holes

that they saw, waiting for it [a seal] to come out of the water. "Baskin" came around
2 h later. Who knows what he was doing.

Today is Bodman's birthday. So lunch is a true banquet. Canned vegetable soup,
bites of meat, skua, and giant petrel with rice. Some kind of cake of something
indefinable. Asparagus and candy and port wine. At night, Nordenskjöld presents
the smokers with two pipes of tobacco for each one of them. A very valuable gift,
because they are reduced to smoking very little tobacco. A lot of tea and coffee.
Despite what they had promised, today they have freed the alleged assassins of
"Qvik." This morning, the temperature inside the house is the same as outside.

*As Bodman's birthday feast engenders good will among the men, the prisoner
dogs are released from their beatings; more good news arrives with the return of
the dogs who had gone seal hunting, including the small Kalle who, together with
the now deceased Kvik, had initially followed Sobral on the hunt. It must be that
both of these puppies are part of Amager's litter. It is Sobral's tendency toward
thorough documentation, and empathy for the dogs, that compel him to note these
events; it is also probable that his mentor Nordenskjöld took pity on the dogs. The
men, however, are still operating on very little food, as no seals have been found to
sustain them.*

5.11.7   *August 6, [1903]*

7 am	$\begin{bmatrix} -19.6 \\ -19.75 \\ +9.9 \\ 720.75 \\ \text{Max} = +5.5 \\ \text{SW } 18 \text{ m} \\ \text{Neb} = 10 \end{bmatrix}$	2 pm	$\begin{bmatrix} -21.65 \\ -21.95 \\ +10.4 \\ +10.4 \\ 742.2 \\ 79\% \\ \text{min} = -21.65 \\ \text{SW } 21.8 \text{ m} \\ \text{Neb} = 8 \end{bmatrix}$	9 pm $\begin{bmatrix} -19.05 \\ -19.7 \\ +10.6 \\ 731.0 \\ 64\% \\ \text{SSW } 5 \text{ m} \\ \text{Neb} = 1 \end{bmatrix}$

At 12:30 at night, the SSW started to blow with a sudden temperature drop, and
the barometer rose. At 4, it moved to the SW and increased its intensity. Overcast
sky until 7 am. At 11 am, nebulosity was 1, but it again became overcast at midday.
At 6 pm, it diminished again, being between 3 and 1 until 11, when it got overcast
again. During the whole morning, the wind blew at a speed of 18 m, but at midday
it diminished to 12.5 m, only to increase again, the maximum of the day occurring
at 2 pm and the minimum of 5 m being at 12 at night. The thermometer has been
rising since 2 pm, and the wind has diminished its intensity. At midnight, the
temperature is $-17°$. This storm had the peculiarity of being without snow. At
night, beautiful coronas are seen. What will this weather lead to? Everything makes
me believe that we won't have any more cold. Toward Cockburn, but much farther,

the sky is of a very dark color, which makes one believe in the presence of a respectable quantity of [ice-] free water.

5.11.8   August 7, [1903]

$$7\,\text{am} \begin{bmatrix} -18.0 \\ -18.5 \\ +5.0 \\ 740.0 \\ \text{Max} = +15.8 \\ \text{SW}\,13\,\text{m} \\ \text{Neb} = 2 \end{bmatrix} \quad 2\,\text{pm} \begin{bmatrix} -14.45 \\ -15.55 \\ +15.9 \\ 744.5 \\ 53.8\% \\ \text{min} = -21.5 \\ \text{SSW}\,4.5 \\ \text{Neb} = 0 \end{bmatrix} \quad 9\,\text{pm} \begin{bmatrix} -17.4 \\ -17.9 \\ +14.7 \\ 745.4 \\ 78.8\% \\ \text{E}\,4\,\text{m} \\ \text{Neb} = 10 \end{bmatrix}$$

The SSW blew until 2 pm, with clear sky since 7 am. At 5 pm, it began to blow between the ESE and the ENE, starting firmly from the E at 9 pm. Sky overcast with Cirro-S since 8, and magnificent halos and coronas can be seen. In the afternoon, the temperature dropped to −22°, but it rose again with the E. The 24-hour average wind speed was 7 m. The maximum was 13 m at 9 pm and the minimum 1 m at 4 pm.

5.11.9   August 8, [1903]

$$7\,\text{am} \begin{bmatrix} +10.7 \\ +10.9 \\ +4.3 \\ 739.9 \\ 94\% \\ \text{Max} = -10.2 \\ \text{ENE}\,10\,\text{m} \\ \text{Neb} = 10 \end{bmatrix} \quad 2\,\text{pm} \begin{bmatrix} -13.65 \\ -13.55 \\ +15.7 \\ 739.4 \\ 90.5\% \\ \text{min} = -22.0 \\ \text{NNE}\,0.5\,\text{m} \\ \text{neb} = 10 \end{bmatrix} \quad 9\,\text{pm} \begin{bmatrix} -22.05 \\ -22.15 \\ +12.6 \\ 738.45 \\ \text{SW}\,3.5\,\text{m} \\ \text{neb} = 10 \end{bmatrix}$$

Wind from the ENE until 9 pm; NNE until 5 pm, with some intervals of calm and variable; 6 ENE; at 6:45, the SSW started. Overcast sky all day. Snow falling since 1:30. The 24-hour average wind speed was 5 m. The maximum from the ENE was 11 m at 8 am, and the maximum from the SSW was 12 m at 12 at night. At 4 pm, it was completely calm. The day before yesterday, I saw that they had thrown away an enormous quantity of beans and oatmeal. I can't explain why they would do this, in the circumstances that we are in. The other day (on the 5th, in the morning), at the moment that I was going to observe the land thermometers, I saw Åkerlundh throwing some pieces of cookies to the dogs. I went and took them from them [the dogs] and I ate them. I'm very hungry; when I finish breakfast, I'm already thinking about lunch. After that ends, during the afternoon coffee, the food is very little. If, instead of taking it to the table for each one to serve themselves, they would ration it out, I wouldn't be so hungry, because I would at least get my portion. But not being that way, when I see these tremendous eyes observing what I am serving myself, and hear heartbreaking sighs, I can't take what should be designated for me, and so I eat the least amount possible.

Sobral's hunger overtakes him, as he takes food away from the dogs—food that was thrown to them. He is still highly sensitive to the "looks" from Nordenskjöld and the "sighs" from Ekelöf, both of which inhibit his food intake. He is hungry, but afraid to eat for fear of social retribution.

5.11.10   August 9, [1903]

7 am		2 pm		9 pm	
	−25.2		−24.5		−22.6
	−25.4		24.55		−22.9
	+3.7		+11.5		+9.8
	725.75		737.05		738.65
	Max = −7.1		min = −24.5		71.5%
	82.8%		79.8%		SSW 9.5 m
	SSW 15 m		SSW 14.5 m		Neb = 10
	Neb = 10		Neb = 10		
	Snöfyk = 2		Snöfyk = 2		

SSW blowing all day, with a 24-hour average strength of 12.5 m. The maximum speed was at 7 am and the minimum 8 m at 12 pm. Overcast sky all day, except for 6 and 7 pm, when it cleared up a little on the zenith. Snow fell in the morning.

5.12   I am Getting Thinner, in an Alarming Way

5.12.1   August 10, [1903]

7 am		2 pm		9 pm	
	−20.95		−20.5		−19.6
	−21.7		−20.8		−19.85
	+4.8		+13.1		+12.8
	738.3		738.65		738.6
	64%		min = −24.25		SSW 6.5 m
	SSW 5 m		SSW 8 m		Neb = 0
	Neb = 7		Neb = 3		

SSW and S blowing all day, with temperatures rising slowly, and the barometer at around the same height. From midday until 4 pm, nebulosity was 3, then it was 1 until 6 pm, and clear sky until 9 pm. At 11 pm, the SSW calmed, and it is blowing a little from the NE, the W, and the WSW, and the sky is starting to get overcast to the N. The 24-hour average wind speed is 6 m.

I am getting thinner, in an alarming way. Naturally, the cause is the poor diet, and also because I spend all day in bed. But this I cannot remedy. I would like to be a million miles away from this station. Here, one suffers not only in morale, but also physically, the latter actually being avoidable to a great extent. Here, everything is dirty: muddy water on the floor, grease, and everything possible on it, when, with a little bit of care, we could have this place like a jewel. One can't get water to wash oneself, and when it is possible, the water is with pieces of food or garbage in it. Now, every day, I melt ice in a can (which is my washbasin). My weight has diminished to 138 lb.

At 11, a wind from the W and SW came, lowering the thermometer to $-25°$. What happens with the seals is truly curious: They don't come out to lie down on the ice in this season, even though the temperature is many times above $0°$.

Fastidious by nature and as a result of his naval training, Sobral does not abide by dirt well, nor does he appreciate sloppy surroundings. His dismay at the state of the wintering house is visceral, as is his anxiety about his severe shedding of weight. He fears that he is turning into nothing more than skin and bones. Photographs of Sobral before, during, and after the expedition reveal a composed young man in fine form, carefully dressed, and stoically posed, with a trim yet muscled physique. Both his posture and his expression convey a sense of his confidence in his own being, a comfort in his own skin. For a young man who prides himself on his physical strength and prowess, as well as his mental acumen, the loss of weight—and thus of physical stature—is a demoralizing situation in the midst of his isolation.

5.12.2   *August 11, [1903]*

$$
7\,\text{am}
\begin{bmatrix}
-22.5 \\
-22.6 \\
+2.9 \\
735.25 \\
\text{Max} = -18.8 \\
\text{Calm} \\
\text{Neb} = 9
\end{bmatrix}
\quad
2\,\text{pm}
\begin{bmatrix}
-18.95 \\
-19.1 \\
+14.0 \\
734.5 \\
\text{min} = -24.2 \\
77.5\% \\
\text{SSW}\,6\,\text{m} \\
\text{Neb} = 10
\end{bmatrix}
\quad
9\,\text{pm}
\begin{bmatrix}
-20.1 \\
-20.25 \\
+14.0 \\
733.95 \\
80\% \\
\text{SW}\,3.5\,\text{m} \\
\text{neb} = 4
\end{bmatrix}
$$

Calm until 11 am, at which time the WSW began to blow, moving to the SW at 12 am, and to the SSW at 1. Overcast sky until 5 pm, when it cleared up. Snow fell from 9 am to 3 pm. The SSW calmed at 10 pm. The barometer has dropped all day, although slowly. The temperature remains around $-20°$. The 24-hour average wind speed was 3.5 m; the maximum was 7.5 m at 4 pm.

5.12.3 August 12, [1903]

$$
7\,\text{am}
\begin{bmatrix}
-19.9 \\
-19.8 \\
+3.3 \\
731.65 \\
\text{Max} = -17.2 \\
\text{E } 1.5\,\text{m} \\
\text{Neb} = 7
\end{bmatrix}
\quad
2\,\text{pm}
\begin{bmatrix}
-11.6 \\
-12.2 \\
+10.8 \\
730.75 \\
84.5\% \\
\text{min} = -27.0 \\
\text{E } 1\,\text{m} \\
\text{Neb} = 4
\end{bmatrix}
\quad
9\,\text{pm}
\begin{bmatrix}
-2.6 \\
-3.4 \\
+16.2 \\
728.4 \\
87\% \\
\text{N } 8\,\text{m} \\
\text{Neb} = 6
\end{bmatrix}
$$

The average wind speed during the day was 2 m; the maximum was at 9 pm. It was calm until midday; then, the NE started, which—moving sometimes to the N, NNW—kept blowing until 12 at night. The temperature rising all day, and the thermometer dropping.

5.12.4 August 13, [1903]

$$
7\,\text{am}
\begin{bmatrix}
-8.3 \\
-9.4 \\
+6.3 \\
733.8 \\
\text{Max} = -0.2 \\
\text{Hy } 68\% \\
\text{NE } 7\,\text{m} \\
\text{Neb} = 10
\end{bmatrix}
\quad
2\,\text{pm}
\begin{bmatrix}
+1.1 \\
-1.7 \\
+12.0 \\
733.75 \\
\text{min} = -18.0 \\
54\% \\
\text{NE } 2\,\text{m} \\
\text{Neb} = 10
\end{bmatrix}
\quad
9\,\text{pm}
\begin{bmatrix}
+5.25 \\
+1.45 \\
+14.3 \\
735.4 \\
51.8\% \\
\text{var } 7\,\text{m} \\
\text{Neb} = 10
\end{bmatrix}
$$

Very warm day, and NE blowing almost all day. Sky overcast with Ci-Ci-S and Cu-Ni.

At midday, J [Jonassen] finally found a seal, and it was a Weddell. With fresh meat, not only we, but also the dogs, are celebrating. It's been a year minus 2 days that the first seal was killed. It was like today, a very hot day.

All day an ossifraga [giant petrel] has been flying around here. Yesterday, "Peridota" began. A lot of snow petrels are seen.

During the whole afternoon and until 11 pm, very violent gusts have blown from all directions, with periods of calm in between them. These gusts have been mainly from the SSW, S, SW, and NNE. The temperature is very high at night, having reached around +7°. The average wind speed during the day was 6 m.

5.12.5   *August 14, [1903]*

$$
7\,\text{am}\begin{bmatrix}
-0.5 \\
-1.6 \\
+9.4 \\
734.0 \\
\text{Max} = +6.8 \\
88\% \\
\text{NE}\,13\,\text{m} \\
\text{Neb} = 10
\end{bmatrix}
\quad
2\,\text{pm}\begin{bmatrix}
+0.3 \\
-0.15 \\
+18.0 \\
728.0 \\
\text{min} = -1.4 \\
91\% \\
\text{NNE}\,10\,\text{m} \\
\text{Neb} = 10
\end{bmatrix}
\quad
9\,\text{pm}\begin{bmatrix}
-18.05 \\
-18.02 \\
+15.7 \\
726.1 \\
\text{SSW}\,13\,\text{m} \\
\text{Neb} = 5
\end{bmatrix}
$$

From 1 am to 7 am, variable winds. From 7 to 10, NNE and N. The rest of the day, NNE and N, but sometimes moving to the ENE and NE. The maximum speed of the wind was 18 m at 8 am—the maximum speed at which the NNE wind has blown. Until 7 pm, sky overcast with Ni and Cu-Ni. At 7:30, the SW started to blow, and the temperature dropped swiftly to $-18°$. The wind calmed completely at 11:30, but with a clear sky. In the morning, the ENE blew close to the ground, while the N and NNW blew at the height of the Ni. The amount of water free of ice is enormous. From 8 pm to 12 pm, half the sky was overcast.

5.12.6   *August 15, [1903]*

$$
7\,\text{am}\begin{bmatrix}
+1.7 \\
+0.5 \\
+10.8 \\
728.65 \\
\text{Max} = +1.9 \\
\text{N} \\
\text{Neb} = 9
\end{bmatrix}
\quad
2\,\text{pm}\begin{bmatrix}
-9.5 \\
-10.2 \\
+18.1 \\
726.35 \\
81.7\% \\
\text{min} = -18.7 \\
\text{WSW} \\
\text{Neb} = 10 \\
\text{snowing}
\end{bmatrix}
\quad
9\,\text{pm}\begin{bmatrix}
-22.6 \\
-23.0 \\
-13.9 \\
733.45 \\
64.8\% \\
\text{SSW} \\
\text{Neb} = 1
\end{bmatrix}
$$

Last night's cold, as can be seen from the observations, was of short duration. NNE and N blew all morning, with a maximum strength of 13 m at 9 pm. Some snow and raindrops falling. At 1 pm, the wind moved to the ENE, and at 1:40, after some moments of calm, it began to blow from the WSW with snow, and the temperature dropped swiftly. As can be seen from the observations, the barometer rose around 8 mm between 2 pm and 9 pm, but this storm didn't last long either. At 10:35, the SSW calmed completely, and at 11:20, the NE started, and the temperature began to rise. The maximum strength of the SSW was 21 m at 7 pm. The 24-hour average wind speed was 9.4 m.

5.13   I Don't Understand How This Man Can Be So Shameless

5.13.1   August 16, [1903]

$$
7\text{am}
\begin{bmatrix}
-4.0 \\
-5.05 \\
+9.8 \\
735.0 \\
\text{Max} = +1.4 \\
\text{NE } 7.5\,\text{m} \\
\text{Neb} = 4
\end{bmatrix}
\quad
2\text{pm}
\begin{bmatrix}
-1.1 \\
-2.6 \\
+15.4 \\
736.1 \\
\text{min} = -23.2 \\
70.5\% \\
\text{E } 5.5\,\text{m} \\
\text{Neb} = 8
\end{bmatrix}
\quad
9\text{pm}
\begin{bmatrix}
-3.3 \\
-3.9 \\
+14.6 \\
734.8 \\
86.5\% \\
\text{NE } 9\,\text{m} \\
\text{Neb} = 6
\end{bmatrix}
$$

Winds from the NE and NNE blowing all day; the latter sometimes moving to the N, and the former to the E. The maximum strength was 9.5 m at 7 pm, and the minimum was 3 m at midday.

In general, the thermometer has been around −3°, and the barometer began to drop in the afternoon (but it rose until around 2 pm). The 24-hour average wind speed was 6.5 m. Nebulosity has been very variable, but, in general, half the sky has been overcast [he includes a drawing].

At 12 at midday, I went out for a walk on the ice to see if I could find any seals. After I had walked 2 or 3 km, Nordenskjöld caught up with me, telling me that he had to talk to me. He asked me if I was thinking about going on the sledge trips that were to be made. I answered yes, but with the condition that Jonassen would not be on the expedition that I would take part in. Right away he told me that, in 2 or 3 days, he's planning to go to Cape "Depósito" [the provisions depot], to get the food that is there, and that in mid-September, he is thinking of going to "Paulet" Island, following the coast of Louis Filippe Land with the objective of seeing if any ship has left news or anything for us (!!)—that he's planning to spend a great part of the summer on Seymour Island, making observations, and that to all these places he will go with Jonassen alone. But that in mid-October, he plans on making an expedition to go explore the west of Cape Foster, and that he will go there with me. I don't understand how this man can be so shameless, after promising the government that I would take part in every expedition that left the station, whatever its objective, and other things; he is not honoring everything that he promised.

This is the severest criticism of Nordenskjöld by Sobral, and the only time the young under-lieutenant accuses the scientist of being without shame. At all other times, Sobral had defended the expedition leader, even when some of the other Swedish scientists had criticized him. Sobral is sensitive to the fact that Nordenskjöld made a pact with Perito Moreno and the Argentine government, and that he must live up to his promise to include Sobral in all the Antarctic scientific expeditions and overland exploratory journeys. Sobral's stated terms that Jonassen not accompany them on the sledging expeditions may be due to his observations of Jonassen's brutal treatment of the dogs, as well as Jonassen's failure to retrieve the seals that Sobral had killed for food—two complaints he had previously made in his diary. For his part, Nordenskjöld would later write a justification of his decision to take only Jonassen on these sledge journeys. In his book, he cites the dogs as the reason for his decision. Not wanting to split up his six dogs into two teams of three in order to pull two sledges, which he said would be necessary for taking three people, and not wanting to burden the dogs with carrying the additional clothes, food, equipment, and sleeping gear for a third person, whom he saw as having minimal value, he chose to take only one sledge and one person and that person was Jonassen. Without specifically mentioning Sobral by name, Nordenskjöld states that he felt that the "inconvenience" and "delay" caused by taking a third person would outweigh the good achieved (Nordenskjöld et al. 1905: 296). With Jonassen, he would pursue geological findings as well as geographical discoveries and make his way to Paulet Island as well as to Seymour Island. Of course, Nordenskjöld has no idea whatsoever that Captain Larsen, the crew, and the scientists who were aboard the Antarctic *were at this very moment shipwrecked on Paulet Island.*

5.13.2   August 17, [1903]

$$
7\,\text{am}\begin{bmatrix} -2.7 \\ -3.9 \\ +8.5 \\ 727.5 \\ 73\% \\ \text{Max} = 0^{\circ} \\ \text{NE}\,9\,\text{m} \\ \text{Neb} = 9 \end{bmatrix}
\quad
2\,\text{pm}\begin{bmatrix} -22.0 \\ -22.1 \\ +17.0 \\ 730.05 \\ 76\% \\ -22.7 \\ \text{SSW}\,11.5\,\text{m} \\ \text{Neb} = 10 \\ \text{Fog} \end{bmatrix}
\quad
9\,\text{pm}\begin{bmatrix} -23.45 \\ -23.5 \\ +16.1 \\ 729.5 \\ 76.5\% \\ \text{SW}\,16.3\,\text{m} \\ \text{Neb} = 10 \end{bmatrix}
$$

Wind from the NNE and NE blew until 8:30, with sky overcast with Cu-Ni, Ci-S, and S-Cu. At 8:45, after 7 min of calm, the SSW started at a speed of around 10 m. The thermometer, which was around $-3°$, suddenly dropped to $-18°$, and the barometer started to rise. The rest of the day there was fog, and it blew from the SW. At 9 pm, it moved to the SW, increasing its intensity, and snow began to fall. The maximum was 18 m at 10 and 11 pm. The average of the day was 12 m, and the minimum was 7.5 m at 8 am.

5.13.3 August 18, [1903]

$$
7\text{ am}
\begin{bmatrix}
-19.35 \\
-19.5 \\
+6.0 \\
732.2 \\
78.5\% \\
\text{Max} = -2.3 \\
\text{SSW } 17\,\text{m} \\
\text{Neb} = 10 \\
\text{snöfyk}
\end{bmatrix}
\quad
2\text{ pm}
\begin{bmatrix}
-13.95 \\
-14.5 \\
+11.4 \\
737.35 \\
\text{min} = -23.5 \\
\text{SSW } 10\,\text{m} \\
\text{Neb} = 9
\end{bmatrix}
\quad
9\text{ pm}
\begin{bmatrix}
-11.8 \\
-12.8 \\
+14.0 \\
741.9 \\
\text{SSW } 11\,\text{m} \\
\text{neb} = 3
\end{bmatrix}
$$

The SW kept blowing until 6 am, and from that time until 10 pm, SSW with rising temperature. Overcast sky until 1 pm; at that time, it started to clear up. At 11, the wind was from the SW, and at 12 at night, it was WSW, with the temperature dropping swiftly. It rose until 9:30 pm, up to $-11°$, $-11.5°$. At 12, it is around $-17°$. The barometer rising all day. The 24-hour average wind speed was 12.9; the maximum was 19 m at 1 am, and the minimum was 10 m at 11 am.

5.13.4 August 19, [1903]

$$
7\text{am}
\begin{bmatrix}
-15.8 \\
-16.5 \\
+6.4 \\
749.7 \\
\text{Max} = -11.4 \\
60\% \\
\text{SSW } 10.5\,\text{m} \\
\text{Neb} = 1
\end{bmatrix}
\quad
2\text{ pm}
\begin{bmatrix}
-12.5 \\
-13.45 \\
+17.0 \\
754.35 \\
53.7\% \\
\text{min} = -20.4 \\
\text{SSW } 10\,\text{m} \\
\text{Neb} = 0
\end{bmatrix}
\quad
9\text{ pm}
\begin{bmatrix}
-16.1 \\
-17.05 \\
+12.8 \\
756.9 \\
50\% \\
\text{SSW } 11\,\text{m} \\
\text{Neb} = 0
\end{bmatrix}
$$

All day, SSW blowing with a completely clear sky. In the morning, the wind wasn't too strong, but after midday, it started to increase in intensity. The temperature relatively very high, around $-10°$ and $-16°$. From the station, you can't see signs of water. I had the intention of climbing up to the top of the basalt, but because of the observations, I couldn't do it. N [Nordenskjöld] is making preparations for a sledge excursion to Cape Depósito with the objective of getting the provisions that are there; he's going with J [Jonassen], and I think they're leaving tomorrow. The barometer is high, and it keeps rising. At 10, the wind moved to the SW, and at 10:30, it was from the WSW. If it keeps blowing from that direction all night, I wouldn't be surprised if the temperature dropped a lot. The 24-hour average wind speed is 10 m. The maximum was 13 m at 1 am and 10 pm.

5.13.5   *August 20, [1903]*

$$
7\,\text{am}
\begin{bmatrix}
-15.95 \\
-16.7 \\
+3.8 \\
758.95 \\
53\% \\
\text{Max} = -11.9 \\
\text{SSW}\,14\,\text{m} \\
\text{Neb} = 0
\end{bmatrix}
\quad
2\,\text{pm}
\begin{bmatrix}
-10.7 \\
-12.3 \\
+11.5 \\
762.2 \\
72.3\% \\
\text{min} = -18.7 \\
\text{S}\,4\,\text{m} \\
\text{Neb} = 3
\end{bmatrix}
\quad
9\,\text{pm}
\begin{bmatrix}
-25.0 \\
-25.15 \\
+12.0 \\
7862.75 \\
\text{calm} \\
\text{Neb} = 0
\end{bmatrix}
$$

SSW blowing until midday, with a clear sky and dry air, which caused us to feel a lot of cold with a temperature of around $-15°$. Between midday and 2 pm, it blew from the S. At 2:10, winds blew variably from the E, ENE, and NW. Variable and very weak breezes blew for the rest of the day, alternating with calm. The 24-hour average wind speed was 6 m; the maximum was 16 m at 6 am. At 11:30, Nordenskjöld sets off with the laborer on the sledge excursion; he is taking six dogs, and he plans to be out there for 2 or 3 days.

Labeling him a "laborer," Sobral is indirectly insulting Jonassen, not even mentioning him by name, and perhaps even being a bit classist in his displeasure. He is still upset with Nordenskjöld's decision not to take Sobral on the sledge journey.

5.13.6   *August 21, [1903]*

$$
7\,\text{am}
\begin{bmatrix}
-18.5 \\
-18.55 \\
+3.3 \\
757.9 \\
83\% \\
\text{max} = -10.3 \\
\text{calm} \\
\text{neb} = 0
\end{bmatrix}
\quad
2\,\text{pm}
\begin{bmatrix}
-9.45 \\
-9.95 \\
+15.7 \\
754.25 \\
79.3\% \\
\text{min} = -25.6 \\
\text{NE} \\
\text{neb} = 10
\end{bmatrix}
\quad
9\,\text{pm}
\begin{bmatrix}
-5.6 \\
-6.0 \\
+16.2 \\
749.75 \\
89.5\% \\
\text{NE} \\
\text{neb} = 0
\end{bmatrix}
$$

At 9 am, it began to blow from the NE, and it kept blowing all day, except for 5 and 6 pm, when it blew from the NNE. The temperature rose quickly in the morning, but then it dropped. Barometer decreasing. Snowing since 6 pm. The 24-hour average wind speed was 4 m; the maximum was 7.5 m at 1 pm.

5.13.7　August 22, [1903]

$$7\text{ am}\begin{bmatrix} -3.75 \\ -4.0 \\ +8.7 \\ 743.35 \\ \text{NNE 8 m} \\ \text{Neb} = 10 \end{bmatrix} \quad 2\text{ pm}\begin{bmatrix} -8.0 \\ -8.35 \\ +15.16 \\ 741.6 \\ \text{min} = -10.5 \\ \text{calm} \\ \text{Neb} = 10 \end{bmatrix} \quad 9\text{ pm}\begin{bmatrix} -17.8 \\ -18.05 \\ +14.4 \\ 739.35 \\ \text{S 17 m} \\ \text{Neb} = 8 \end{bmatrix}$$

Overcast sky all day. NNE blowing all morning, and in the afternoon, until 6, NNE and calm. The SSW started at 6:30. The maximum wind speed (S) was at 9 pm, and from the NNE, it was 13 m at 2 am. The 24-hour average wind speed was 7.5 m.

5.13.8　August 23, [1903]

$$7\text{ am}\begin{bmatrix} -20.3 \\ -20.5 \\ +6.0 \\ 738.6 \\ 49.1\% \\ \text{Max} = -3.2 \\ \text{SSW 19 m} \\ \text{Neb} = 10 \\ \text{snöfyk} \end{bmatrix} \quad 2\text{ pm}\begin{bmatrix} -23.6 \\ -23.8 \\ +13.0 \\ 740.7 \\ 73.5\% \\ -24.0 \\ \text{SSW 20 m} \\ \text{Neb} = 10 \\ \text{snöfyk} \end{bmatrix} \quad 9\text{ pm}\begin{bmatrix} -21.25 \\ -21.50 \\ +11.9 \\ 741.2 \\ 74\% \\ \text{SW 25 m} \\ \text{Neb} = 10 \\ \text{snöfyk} \end{bmatrix}$$

The SSW blew until 2 pm. The rest of the day, SW. Overcast sky and snow.
Wind speed increased since midday, reaching its maximum (?) of 29.5 m at 11 pm. The anemometer broke at that hour, but the wind kept increasing. We haven't had such strong winds since the first days of June. The minimum wind speed was 15 m at 1 am, and the 24-hour average was 20.5 m.

5.13.9   August 24, [1903]

$$
\text{7 am} \begin{bmatrix} -23.45 \\ -23.55 \\ +1.8 \\ 741.95 \\ 75\% \\ \text{Max} = -18.5 \\ \text{SW 23 m} \\ \text{Neb} = 10 \\ \text{snöfyk} \end{bmatrix}
\qquad
\text{2 pm} \begin{bmatrix} -22.75 \\ -23.2 \\ +13.2 \\ 743.6 \\ \text{WSW 23 m} \\ \text{Neb} = 10 \\ \text{snöfyk} \end{bmatrix}
\qquad
\text{9 pm} \begin{bmatrix} -21.9 \\ -22.5 \\ +9.9 \\ 743.4 \\ \text{SW 25 m} \\ \text{Neb} = 8 \\ \text{snöfyk} \end{bmatrix}
$$

The average wind speed between midday and midnight (13 h.) was 24 m. The recording anemometer remained broken until midday, so it is not possible to know what the maximum wind speed was, not observing this every 3 hours. With the pocket an [anemometer], I observed 32.4 m at 3 am. The minimum was 21.5 m at 3 pm.

5.13.10   August 25, [1903]

$$
\text{7 am} \begin{bmatrix} -22.25 \\ -22.7 \\ +4.2 \\ 742.7 \\ 71.5\% \\ \text{SW 29 m} \\ \text{Neb} = 9 \\ \text{Snöfyk} \end{bmatrix}
\qquad
\text{2 pm} \begin{bmatrix} -19.1 \\ -21.5\,(\text{obs}) \\ +10.2 \\ 744.75 \\ 73\% \\ \text{SW 24 m} \\ \text{Neb} = 10 \\ \text{Snöfyk} = 2 \end{bmatrix}
\qquad
\text{9 pm} \begin{bmatrix} -18.1 \\ 20.4\,(\text{obs}) \\ +9.7 \\ 744.15 \\ 71.5\% \\ \text{SW 23 m} \\ \text{Neb} = 10 \end{bmatrix}
$$

The sky started to clear up at 5 pm, and wind from the WSW and SW was blowing all day. The average wind speed was 25 m. The maximum was at 7 am, and the minimum was 22.2 m at 5 pm. The axis of the anemograph broke at that time. If I don't recall incorrectly, today is the day that the wind has blown the strongest this year. In the afternoon, the temperature rose to $-17°$ or $-16°$, dropping again at night. The barometer isn't rising, but it is continuously showing small undulations.

Fía started.

In addition to keeping track of the wind, weather, temperatures, and celestial activities, it seems that Sobral was also keeping track of the female dogs' reproductive schedules and recording who was in heat at which time.

5.13.11   August 26, [1903]

7 am $\begin{bmatrix} -19.8 \\ -22.25\,(\text{obs.}) \\ +2.3 \\ 744.3 \\ 64\% \\ \text{Max} = -16.0 \\ \text{WSW } 20\,\text{m} \\ \text{Neb} = 1 \end{bmatrix}$ 2 pm $\begin{bmatrix} -21.4 \\ -23.8\,(\text{obs.}) \\ +12.8 \\ 745.5 \\ \text{WSW } 17\,\text{m} \\ \text{neb} = 1 \end{bmatrix}$ 9 pm $\begin{bmatrix} -22.1 \\ -24.5 \\ +13.6 \\ 745.6 \\ \text{SW } 15.3\,\text{m} \\ \text{neb} = 1 \end{bmatrix}$

In the morning, I saw a flock of eight snow petrels flying over the gully. Quite a clear day. The maximum nebulosity was 4.

The anemograph is still broken. Wind speed is observed every 4 h with a pocket anemometer. The maximum wind speed was 22.5 at 1 am, the average was 18.9 m, and the minimum was at 9 pm. This morning, while I was looking for the spheres of the anemometer that had been blown away by the wind yesterday afternoon when the axis had broken, the tip of my nose froze. It is remarkable how, with relatively high temperatures, one freezes so easily. And it's true, too, that the wind was blowing at more than 15 m. I found the spheres buried in the snow around 50 m away from where the anemometer was located.

5.13.12   August 27, [1903]

7 am $\begin{bmatrix} -22.25 \\ -24.65 \\ +4.0 \\ 743.9 \\ \text{Max} = -19.6 \\ \text{SW } 13\,\text{m} \\ \text{Neb} = 8 \\ \text{snöfyk} \end{bmatrix}$ 2 pm $\begin{bmatrix} -23.1 \\ -25.5 \\ +11.9 \\ 744.7 \\ 68\% \\ \text{min} = -23.4 \\ \text{SSW } 14\,\text{m} \\ \text{Neb} = 2 \end{bmatrix}$ 9 pm $\begin{bmatrix} -20.9 \\ -23.45 \\ +9.8 \\ 746.3 \\ 62\% \\ \text{SW } 3.4\,\text{m} \\ \text{Neb} = 3 \end{bmatrix}$

During the first hours of the morning, it blew from the SW, but at 9 am, it moved to the SSW, and it remained like that until 8 pm, when it moved to the SW. But at midnight, that wind came back again. Between 9:30 and 10 pm, it was calm, and the wind blew variably from the NW, N, etc., for some moments. The same happened between 10:30 and 11, when the thermometer dropped to −26°. But at 11, the SSW started again, and the temperature rose. The maximum wind speed observed was 14 m at 1 pm, and the minimum was 2 m at 10 pm. The average wind speed for the day was 9.5 m. Between 5 am and 9 am, pretty cloudy sky, but it was very clear the rest of the day, the maximum nebulosity being 3 at those same hours.

5.13.13   *August 28, [1903]*

$$
7\,\text{am}
\begin{bmatrix}
-27.5 \\
-29.65 \\
+2.9 \\
747.5 \\
\text{Max} = [-]19.4 \\
71\% \\
\text{Calm} \\
\text{Neb} = 1
\end{bmatrix}
\quad
2\,\text{pm}
\begin{bmatrix}
-23.5 \\
-26.0 \\
+15.7 \\
750.0 \\
\text{min} = -28.0 \\
68.5\% \\
\text{calm} \\
\text{Neb} = 6
\end{bmatrix}
\quad
9\,\text{pm}
\begin{bmatrix}
-25.25 \\
- \\
+11.2 \\
752.2 \\
\text{SW}\,0.5\,\text{m} \\
\text{neb} = 3
\end{bmatrix}
$$

Quite a clear sky in the morning, but in the afternoon, nebulosity varied between 5 and 9. The first three hours of the morning, wind from the SSW; the rest of the day was completely calm. At 9 pm, the SW started to blow. The maximum wind speed was 8 m at 1 am. The 24-hour average wind speed was 1 m.

5.13.14   *August 29, [1903]*

$$
7\,\text{am}
\begin{bmatrix}
-28.1 \\
-27.65(?) \\
+4.6 \\
753.8 \\
69.5\% \\
\text{Max} = -21.3 \\
\text{SSW}\,11\,\text{m} \\
\text{Neb} = 2
\end{bmatrix}
\quad
2\,\text{pm}
\begin{bmatrix}
-27.05 \\
-26.8 \\
+11.9 \\
754.95 \\
-29.25 = \text{min} \\
\text{SSW}\,8.5\,\text{m} \\
\text{Neb} = 10
\end{bmatrix}
\quad
9\,\text{pm}
\begin{bmatrix}
-28.9 \\
-28.5(?) \\
+10.0 \\
755.05 \\
68.7\% \\
\text{SSW}\,8\,\text{m} \\
\text{Neb} = 3
\end{bmatrix}
$$

SSW blowing all day. Between 8 am and 10 pm, nebulosity was 10, but before 10 am and after 9 pm, it was lower than 5. The average wind speed was 9.4 m. The maximum was 13 m at 5 am and the minimum at 9 and 10 pm.

Today, for the first time this month, the temperature has dropped to −30 degrees.

5.13.15   *August 30, [1903]*

$$
7\,\text{am}
\begin{bmatrix}
-30.5 \\
-30.0(?) \\
+1.7 \\
754.55 \\
\text{Max} = -26.3 \\
67.8\% \\
\text{SSW}\,2\,\text{m} \\
\text{Neb} = 1
\end{bmatrix}
\quad
2\,\text{pm}
\begin{bmatrix}
-28.5 \\
-28.45 \\
+13.2 \\
755.0 \\
\text{SW}\,2\,\text{m} \\
\text{Neb} = 1
\end{bmatrix}
\quad
9\,\text{pm}
\begin{bmatrix}
-30.0 \\
-29.5 \\
+10.4 \\
751.4 \\
\text{calm} \\
\text{Neb} = 1
\end{bmatrix}
$$

Wind from the SSW, SW, and WSW until 4 pm. From 8 pm to 11 pm, weak breezes from the ENE, and the rest of the day, calm. Very clear sky all day. In general, nebulosity was 1 (a bank of clouds to the NE). This cold coincides with last year's cold during the same season. Min. wind speed = 3.5 m. Max. = 10 m at 3 am.

5.13.16   August 31, [1903]

$$
7\,\text{am}\begin{bmatrix} -26.6 \\ -25.9(?) \\ +2.0 \\ 745.7 \\ \text{Max} = -25.8 \\ \text{SE} \\ \text{Neb} = 8 \end{bmatrix}
\quad
2\,\text{pm}\begin{bmatrix} -33.05 \\ -32.55(?) \\ +15.2 \\ 750.8 \\ \text{SSW} \\ \text{Neb} = 0 \end{bmatrix}
\quad
9\,\text{pm}\begin{bmatrix} -30.6 \\ -30.4(?) \\ +9.6 \\ 752.6 \\ \text{SW} \\ \text{Neb} = 10 \end{bmatrix}
$$

Since 10 am, wind from the SSW and SW with fog. Between 1 am and 10 am, wind from the ENE, E, and ESE. Overcast sky most of the day. The 24-hour average wind speed was 10 m; the maximum was 20 m at 8 pm, and the minimum was 0 at 9 am.

Another month has concluded! What joy! At night, we drink Tody [hot toddy] to bid farewell to it. And since I don't use spirituous beverages, I content myself with lemonade.

Spring begins, it brings us light and heat, and also the spirit becomes fortified with new hopes, because the season when the ship will come is drawing closer. These last 15 days of August have been very cold. And since the 15th, more or less, we haven't seen any [ice-] free water.

The end of August and the coming of Spring mark a great change in Sobral's psyche. His spirits are up, despite his refusal to drink spirits. His handwriting becomes much neater and smaller in this entry. He is looking forward to finding warmth in the coming sun and transportation away from the snow-and-ice-clad hill that is his temporary home.

5.13.17   September 1, [1903]

$$
7\,\text{am}\begin{bmatrix} -29.65 \\ -29.45 \\ +1.5 \\ 751.7 \\ 61.7\% \\ \text{Max} = -24.0 \\ \text{SSW}\,9\,\text{m} \\ \text{Neb} = 8 \end{bmatrix}
\quad
2\,\text{pm}\begin{bmatrix} -26.5 \\ - \\ +12.9 \\ 750.05 \\ 66.5\% \\ \text{min} = -32.0 \\ \text{NE}\,3\,\text{m} \\ \text{Neb} = 8 \end{bmatrix}
\quad
9\,\text{pm}\begin{bmatrix} -27.4 \\ -27.0 \\ +12.0 \\ 746.4 \\ \text{calm} \\ \text{Neb} = 0 \end{bmatrix}
$$

The SSW blew until 9 am; overcast sky from 1 am to 7:30 am. The rest of the day, nebulosity was lower than 1, and winds from the ENE, NE, and ESE blew. The average wind speed was 4.5 m. The maximum wind speed was 15.5 m at 2 am, and at 8, 9, and 12 pm, it was completely calm. Today was a very nice day. Magnetic observations were made.

5.13.18 September 2, [1903]

$$
7\,\text{am}
\begin{bmatrix}
-30.1 \\
-31.6 \\
+6.0 \\
745.0 \\
\text{Max} = -24.9 \\
71.5\% \\
\text{Calm} \\
\text{Neb} = 0
\end{bmatrix}
\quad
2\,\text{pm}
\begin{bmatrix}
-17.5 \\
-17.9 \\
+14.6 \\
744.6 \\
\min = -32.4 \\
61.2\% \\
\text{calm} \\
\text{neb} = 6 \\
\text{Ci-S}
\end{bmatrix}
\quad
9\,\text{pm}
\begin{bmatrix}
-21.5 \\
-21.0 \\
+16.5 \\
743.55 \\
74\% \\
\text{calm} \\
\text{neb} = 0 \\
\text{Ci-S, halo, corona}
\end{bmatrix}
$$

Between 1 am and 10 am, nebulosity was lower than 3. The rest of the day, the sky was almost completely overcast with Ci-S. At night, very nice halos and rings. The average wind speed during the day was 0.7 m. The barometer has dropped all day, but slowly. At 10:30 pm, the SSW started to blow, and at 11, it had a speed of 10 m, but the temperature was rising.

Today has been a beautiful day. Despite the bank of Ci-S that covered all the sky, the sun was heating significantly. The ice on the glass windows disappears even with temperatures of $-30°$. The snow that's in contact with dark bodies—for example, the house, which is covered with black paper on the outside—melts. But at 4 in the afternoon, when the sun hides behind the gully, the window glass gets covered again with beautiful crystallization, and the house crunches, for the wood contracts when it gets colder. With the winds and the cold of these last days, the small or big snowdrifts found here and there have acquired great consistency, so that one walks without sinking. And also on the snow, you see deep furrows, whose immaculate whiteness resembles marble. Now, we have 12 h of light from 6 am to 6 pm; that is, at 6 it is still difficult to read, but it's very clear.

In a major lecture that Sobral would later give in Buenos Aires in December 1903 for the Naval Center, the young under-lieutenant would call upon future generations to "tear down the huge snowdrifts" in Antarctica and "raise in their place the marble blocks that will serve as a pedestal to the glory of Argentina" (Sobral 1903a: 487). This futuristic vision and seemingly patriotic statement, it seems, had its origin in these observations written in this diary entry, in which he likens the snowdrift and snow furrows to immaculate white marble.

5.13.19   September 3, [1903]

$$
7\,\text{am}\begin{bmatrix} -19.0 \\ -19.4 \\ +4.3 \\ 748.55 \\ \text{Max} = -10.8 \\ 52.1\%\,\text{SSW}\,8\,\text{m} \\ \text{Neb} = 0 \end{bmatrix}
\quad
2\,\text{pm}\begin{bmatrix} -13.9 \\ -14.85 \\ +14.8 \\ 751.3 \\ \text{min} = -24.6 \\ 44\% \\ \text{S}\,4\,\text{m} \\ \text{Neb} = 0 \end{bmatrix}
\quad
9\,\text{pm}\begin{bmatrix} -21.8 \\ -21.6 \\ +13.8 \\ 750.9 \\ \text{calm} \\ \text{Neb} = 0 \end{bmatrix}
$$

Until 9 am, the SSW blew, with a clear sky. Between 10 am and 4 pm, variable winds from the S, WSW, and NNE. Clear sky almost all day. The highest nebulosity was 2 Ci-S. The 24-hour average wind speed was 4.5 m; the maximum was 13 m at 2 am, and from 4:30 pm to 12 at night, it was complete calm. The day couldn't have been more beautiful, and the night more splendid. Longitude observations were made.

5.13.20   September 4, [1903]

$$
7\,\text{am}\begin{bmatrix} -23.25 \\ 23.4 \\ +5.0 \\ 747.55 \\ 74.5\% \\ 14.5\% \\ \text{Calm} \\ \text{Neb} = 1 \end{bmatrix}
\quad
2\,\text{pm}\begin{bmatrix} -11.5 \\ -12.5 \\ +15.0 \\ 747.0 \\ \text{min} = -25.0 \\ 57\% \\ \text{calm} \\ \text{neb} = 9 \\ (\text{Ci-S}) \end{bmatrix}
\quad
9\,\text{pm}\begin{bmatrix} -9.3 \\ -11.2 \\ +15.2 \\ 746.45 \\ 46.8\% \\ \text{calm} \\ \text{neb} = 10 \\ \text{Ci-S} \\ \text{A-Cu corona} \end{bmatrix}
$$

Almost all day, sky overcast with Ci and Ci-S, and at night, A-Cu. Almost complete calm. Some very weak breezes from several directions. At 2 pm, I could observe the direction of some ________ [unreadable] S 40 W, with a lot of speed. At night, coronas are seen continuously.

5.13.21   *September 5, [1903]*

$$
7\,\text{am}
\begin{bmatrix}
-4.35 \\
-5.55 \\
+7.3 \\
745.6 \\
\text{Max} = +3.3 \\
55\% \\
\text{Calm} \\
\text{neb} = 6
\end{bmatrix}
\qquad
2\,\text{pm}
\begin{bmatrix}
+3.05 \\
-4.4 \\
+11.3 \\
743.85 \\
60\% \\
\text{NNE 1 m} \\
\text{neb} = 5
\end{bmatrix}
\qquad
9\,\text{pm}
\begin{bmatrix}
-8.7 \\
-9.75 \\
+16.1 \\
741.75 \\
69.5\% \\
\text{calm} \\
\text{neb} = 3
\end{bmatrix}
$$

24-hour average wind speed = 0.4; max. = 2 m at 4 pm.

Weak and variable breezes from the SW and NW (mainly), all day. Sky quite overcast with Ci, Ci-S, and A-Cu.

After the *frukost*, I took my camera and I went to the *"iceberg"* with the objective of taking some photographs and searching for seals. The ice is hard as you walk. One doesn't sink, but it tires me a lot; it's hard to walk because of my heavy boots. I walked across all the floes close to the icebergs, but without finding any signs of seals; I took a picture of the *"iceberg"* looking at it from the north. When I went back to the station, it was 3 pm; the sun is shining in all its splendor, and temperature is above 0°; you can hear the happy sound of the water running down the plains of the hills to the sea.

5.14   I Listen to the Sound of the Water, which is the Only Thing that Gives Life to this Nature of Ours, and that Makes Us Happy and Cheers Us Up, Because It's the Sign of the Beginning of a New Season, of Happier Days, Warmer Days that Melt the Ice, Giving Us Freedom, and Perhaps Saving Our Lives

5.14.1   *September 6, [1903]*

$$
7\,\text{am}
\begin{bmatrix}
-0.25 \\
-2.0 \\
+9.1 \\
740.5 \\
65\% \\
\text{Max} = +5.8 \\
\text{SSW 11 m} \\
\text{Neb} = 6
\end{bmatrix}
\qquad
2\,\text{pm}
\begin{bmatrix}
+4.1 \\
+2.7 \\
44.6\% \\
+16.0 \\
745.75 \\
\text{min} = -16.5 \\
\text{SSW 3 m} \\
\text{Neb} = 2
\end{bmatrix}
\qquad
9\,\text{pm}
\begin{bmatrix}
-5.95 \\
-7.0 \\
+16.0 \\
750.85 \\
72\% \\
\text{NE 1 m} \\
\text{Neb} = 2
\end{bmatrix}
$$

At 8 am, overcast sky. It started to clear up at that hour, and nebulosity was 1 at 10 am. Between 10 am and 9 pm, the maximum nebulosity was 2, but a bank of

A-Cu came from the NE, and at 10 pm, the sky was overcast. Wind from the S and SSW almost all day. The average wind speed was 4 m. The maximum was at 7 am, and the minimum was 0.5 m at 10 pm. Today, the temperature has also been some degrees above 0, and water ran along in many areas. After six months, yesterday was the first day that we heard that merry noise. Yesterday, when I came back from my walk, I stopped for some moments to listen to the sound of the water, which is the only thing that gives life to this nature of ours, and that makes us happy and cheers us up, because it's the sign of the beginning of a new season, of happier days, warmer days that melt the ice, giving us freedom, and perhaps saving our lives.

Fortunately, today J [Jonassen] found a seal, and even though it is small, we'll have meat for around 2 weeks. And the poor dogs, who haven't had anything to eat, swallowed until they were completely stuffed. It was cheerful to see that bunch of beasts latched onto the seal, eating until they could eat no more, to finally lie down completely stretched out in the sun, to digest their splendid and favorite feast.

Almost all day, we've had wind from the S and SW, with very high temperatures. At 8:30, the NE started to blow weakly, and the temperature started to drop. At 10:30 pm, the SSW continued again, with decreasing temperature.

The sounds of running water revive the morale and spirits of Sobral, who equates these sounds of melting snow and ice with the possibility of a rescue. Once again, he also shows his empathy for the sled dogs, commiserating with them in their hunger, and exalting in their finally being able to feast on a seal that Jonassen has killed.

5.14.2   *September 7, [1903]*

$$
7\,\text{am} \begin{bmatrix} -15.8 \\ -15.8 \\ +8.2 \\ 757.85 \\ 81\% \\ \text{Max} = +4.0 \\ \text{SSW } 8.5\,\text{m} \\ \text{Neb} = 10 \end{bmatrix} \quad 2\,\text{pm} \begin{bmatrix} -14.95 \\ -15.2 \\ 17.3 \\ 762.3 \\ 72.5\% \\ \text{min} = -17.4 \\ \text{var} = 1 \\ \text{Neb} = 7 \end{bmatrix} \quad 9\,\text{pm} \begin{bmatrix} -15.5 \\ -15.1 \\ +13.2 \\ 762.3 \\ \text{E } 3\,\text{m} \\ \text{Neb} = 10 \end{bmatrix}
$$

From 1 am to 1 pm, wind from the SSW, with overcast sky. Some snow fell between 3 and 6 am. Between 2 and 5, partially overcast sky, but after 6 pm, nebulosity was 10. Between 2 pm and 12 at night, winds from the NE and SE. A Larus [gull] was seen. All day, the temperature was around - ________ [unreadable]. The average wind speed was 5 m; the maximum was 10 m at 8 am and the minimum 0.2 m at 4 pm.

5.14.3   *September 8, [1903]*

$$
7\text{ am}
\begin{bmatrix}
-14.9 \\
-14.4 \\
+5.7 \\
757.8 \\
93.5\% \\
\text{Max} = -13.7 \\
\text{Calm} \\
\text{Neb} = 10
\end{bmatrix}
\quad
2\text{ pm}
\begin{bmatrix}
-8.8 \\
-9.05 \\
+14.6 \\
756.0 \\
\min = -16.4 \\
\text{calm} \\
\text{Neb} = 5 \\
\text{Fog}
\end{bmatrix}
\quad
9\text{ pm}
\begin{bmatrix}
-8.4 \\
-8.4 \\
+14.5 \\
750.85 \\
88\% \\
\text{E 1 m} \\
\text{Neb} = 4 \\
\text{Ci-S}
\end{bmatrix}
$$

Average wind speed = 0.5 m. Maximum wind speed = 2.5 m. In the morning, fog and very weak winds from the NE and ENE. From 3 to 9 am, sky partially overcast with Ci-S. Neb. < 4 and after 10 pm neb. = 10, with halos and moon coronas. Today, we've had the fortune of finding another seal. A Larus [gull] was flying around near the station, and at 12 at night, I heard a Sterna [tern] screech. Almost all day, temperature above −10°; barometer dropping.

5.14.4   *September 9, [1903]*

$$
7\text{am}
\begin{bmatrix}
-1.9 \\
-3.15 \\
+7.8 \\
742.5 \\
\text{Max} = -1.7 \\
73.6\% \\
\text{NE 8.5 m} \\
\text{Neb} = 8
\end{bmatrix}
\quad
2\text{pm}
\begin{bmatrix}
-2.5 \\
-3.0 \\
+17.4 \\
737.8 \\
\min = -10.4 \\
\text{NE 13.5 m} \\
\text{Neb} = 6
\end{bmatrix}
\quad
9\text{pm}
\begin{bmatrix}
-1.5 \\
-2.05 \\
+15.7 \\
732.0 \\
\text{ENE 6 m} \\
\text{Neb} = 10
\end{bmatrix}
$$

Average wind speed during the day = 8 m. The maximum = 13.5 m at 2 pm. There have been 4 h of calm. The minimum speed was at 2 pm = 6. Almost all day, the temperature remained below 0° but very high; at 10 pm, it rose to +1.2° and it dropped again at 11:30. Snöfyk [snow] and wind from the NE almost all day.

5.14.5 *September 10, [1903]*

$$
7\,\mathrm{am}
\begin{bmatrix}
-4.65 \\
-5.4 \\
+9.75 \\
730.65 \\
\mathrm{Max} = +0.6 \\
79\% \\
\mathrm{Calm} \\
\mathrm{Neb} = 2
\end{bmatrix}
\quad
2\,\mathrm{pm}
\begin{bmatrix}
+2.5 \\
-0.25 \\
+15.2 \\
732.75 \\
\min = -8.5 \\
55\% \\
\mathrm{SSW}\,8.5\,\mathrm{m} \\
\mathrm{neb} = 6
\end{bmatrix}
\quad
9\,\mathrm{pm}
\begin{bmatrix}
+2.2 \\
-0.5 \\
+17.0 \\
736.45 \\
45.6\% \\
\mathrm{SSW}\,3\,\mathrm{m} \\
\mathrm{neb} = 2
\end{bmatrix}
$$

The average wind speed was 2 m; the maximum = 10 m at 3 pm, and there was 13 h of calm. Fully overcast sky between 9 am and 7 pm. At 1:35 pm, the SW started to blow until 9 pm, with temperatures above 0° during the entire time. The maximum temperature was around +3.2°. The thermometer has been above 0 between midday and 9 pm. The barometer rose with the SSW, but it started to drop again before midnight. The direction of the cirrus is W, as usual.

Today, a seal was seen on the east side of "*Snow Hill.*" The large amount of snow that is observed everywhere is remarkable. The boat is underneath 3 and a half meters of snow. This is probably due to the relatively slow speed of the wind, because I don't think there has been more precipitation this year than last year. Conversely, I think more snow fell last year than this year. A few birds are seen.

5.14.6 *September 11, [1903]*

$$
7\,\mathrm{am}
\begin{bmatrix}
-4.25 \\
-4.4 \\
+10.1 \\
733.45 \\
\mathrm{Max} = +3.2 \\
92.2\% \\
\text{Very weak and variable} \\
\text{wind, and almost calm} \\
\mathrm{Neb} = 10
\end{bmatrix}
\quad
2\,\mathrm{pm}
\begin{bmatrix}
-9.3 \\
-9.3 \\
-20.2 \\
732.8 \\
\min = -9.7 \\
\mathrm{SW}\,12\,\mathrm{m} \\
\mathrm{neb} = 10
\end{bmatrix}
\quad
9\,\mathrm{pm}
\begin{bmatrix}
-14.4 \\
-14.25 \\
+18.1 \\
734.25 \\
\mathrm{SSW}\,14.5\,\mathrm{m} \\
\mathrm{neb} = 10
\end{bmatrix}
$$

Since the morning until 1 pm, it blew from the NE with a lot of strength and snöfyk. At 1:15 (I think the NE had been blowing for just a second), it started to blow from the WSW and then from the SW and SSW, with a lot of strength and snow.

Jonassen, who had left on a sledge to look for a seal that was near the coast on the east side of Snow Hill, was caught by the storm. At 4, he came back, bringing a seal, a big female seal, which was pregnant. The tapichi [fetus] was very big (from what I heard), and from what Jonassen said, the little animal came shooting out (!!!) from her mother's womb as soon as he opened it.

At night, we ate a little seal stew. It was delicious; it can be compared to the best of meats. Before the SW began (i.e., when the NE was blowing), a lot of snow fell.

According to what they say, great areas of ice-free water could be seen to the E of Snow Hill and to the NE of Cape Seymour this morning. The average wind speed = 7.5 m; the maximum = 16 m at 7 pm. The first seven hours of the day were calm. From 7:30 am to 1:15 pm, wind from the E blew, with snow since midday. The rest of the day, SW and SSW.

A tapichi *is an unborn baby animal or calf taken from the womb of the slaughtered mother. Other expedition diaries—for example, those of the Norwegian Antarctic Expedition who reached the South Pole in 1911—speak of live seal fetuses that shot out from their slaughtered mothers and startled the men, who would then proceed to kill the unhappily born baby. At Snow Hill, during the spring of 1903, a seal was a welcome sight to the starving men and dogs of the Swedish Expedition. Presumably, the seal stew Sobral and his companions ate on this night included parts of the fetus, as Sobral later mentions in his speech notes that the smaller baby seals offered the more tender meat. Unfortunately, it was only through the demise of mother and baby seals, and male seals and both sexes of penguins, that the lives of the men—and the dogs—at Snow Hill were saved.*

5.14.7   *September 12, [1903]*

$$7\,\text{am}\begin{bmatrix} -15.5 \\ -15.5 \\ +7.5 \\ 734.05 \\ 81\% \\ \text{Max} = +2.4 \\ \text{SW } 13\,\text{m} \\ \text{Neb} = 8 \\ \text{snöfyk} \end{bmatrix} \quad 2\,\text{pm}\begin{bmatrix} -15.25 \\ -15.2 \\ +18.0 \\ 736.3 \\ 81.3\% \\ \text{min} = -15.75 \\ \text{SW } 17\,\text{m} \\ \text{neb} = 10 \\ \text{snöfyk} \end{bmatrix} \quad 9\,\text{pm}\begin{bmatrix} -20.2 \\ -20.0 \\ +11.9 \\ 737.9 \\ 80.5\% \\ \text{SW } 20\,\text{m} \\ \text{neb} = 8 \\ \text{snöfyk} \end{bmatrix}$$

All day, SW blowing, with snow and overcast sky. Sometimes, A-Cu could be seen at the zenith. Average wind speed during the day = 17 m. The maximum = 21.3 m at 7 pm and the minimum 12 m at 3 am.

Today is Ekelöf's birthday. To celebrate it, we are having a splendid meal, almost a banquet. Cold cut meat, dried penguin, bread with butter, natural sheep tongue, liver, noodle soup, canned meat, seal steak with beans, and finally, a splendid cake with sweet strawberry jam. We all were satisfied with the meal and got up from the table saying that every day there should be a birthday. I forgot to say that port wine was served, and they served candies and corncob with butter.

The effect that a good meal makes on the human spirit is remarkable. One feels inclined toward goodness and tolerance; sees black things with vivid and brilliant colors; and allows everything. That old saying, "full belly, happy heart."

Until 2 pm, the temperature remained around −15 or −16°, but at that time, it started to drop. At 12 at night, it was −22°.

5.14.8   *September 13, [1903]*

$$
7\,\text{am}
\begin{bmatrix}
-23.0 \\
-22.85 \\
+3.2 \\
739.5 \\
\text{Max} = [-]14.2 \\
\text{SW } 19\,\text{m} \\
\text{Neb} = 10 \\
\text{snöfyk}
\end{bmatrix}
\qquad
2\,\text{pm}
\begin{bmatrix}
-23.0 \\
-22.9 \\
+15.6 \\
742.5 \\
\text{min} = [-]23.2 \\
\text{SW } 17\,\text{m} \\
\text{Neb} = 10 \\
\text{snöfyk}
\end{bmatrix}
\qquad
9\,\text{pm}
\begin{bmatrix}
-23.0 \\
-23.2 \\
+14.2 \\
744.35 \\
\text{SSW } 10\,\text{m} \\
\text{Neb} = 3
\end{bmatrix}
$$

24-hour average wind speed = 15.6 m.
24-hour max. wind speed = 23.4 m.
24-hour min. wind speed = 6.
Temperature around −23° all day.

5.14.9   *September 14, [1903]*

$$
7\,\text{am}
\begin{bmatrix}
-22.5 \\
-22.4 \\
+5.5 \\
743.45 \\
71\% \\
\text{Max} = [-]21.4 \\
\text{Calm} \\
\text{Neb} = 8
\end{bmatrix}
\qquad
2\,\text{pm}
\begin{bmatrix}
-18.8 \\
-18.8 \\
+15.0 \\
743.7 \\
74.2\% \\
\text{min} = -23.6 \\
\text{calm} \\
\text{neb} = 10
\end{bmatrix}
\qquad
9\,\text{pm}
\begin{bmatrix}
-20.0 \\
-19.9 \\
+17.2 \\
742.95 \\
78.5\% \\
\text{calm} \\
\text{neb} = 10 \\
\text{snowing(very little)}
\end{bmatrix}
$$

Average wind speed = 2 m.
Maximum wind speed = 8 m at 1 am.
There was 8 h of calm. All day, overcast sky with variable winds, predominating the SSW and the SW.

5.14.10   *September 15, [1903]*

$$
7\,\text{am}
\begin{bmatrix}
-18.15 \\
-18.0 \\
+6.5 \\
739.25 \\
80.5\% \\
\text{E}\,4\,\text{m} \\
\text{Neb} = 10
\end{bmatrix}
\quad
2\,\text{pm}
\begin{bmatrix}
-15.0 \\
-14.9 \\
+16.7 \\
740.3 \\
89.5\% \\
\text{min} = -20.7 \\
\text{ENE}\,2\ \text{m} \\
\text{Neb} = 10
\end{bmatrix}
\quad
9\,\text{pm}
\begin{bmatrix}
-13.85 \\
-13.65 \\
+13.3 \\
740.4 \\
\text{E}\,1.5\,\text{m} \\
\text{Neb} = 10
\end{bmatrix}
$$

Average wind speed during the day = 3 m.
Maximum wind speed during the day = 4.5 m.
There was an hour of calm. All day, sky overcast with S. E and ENE blowing.
Temperature rising very slowly. Magnetic observations are made.
An ossifraga [giant petrel] was seen yesterday.

5.14.11   *September 16, [1903]*

$$
7\,\text{am}
\begin{bmatrix}
-14.95 \\
-14.7 \\
+7.0 \\
741.3 \\
\text{Max} = \quad 13.5 \\
89.2\% \\
\text{E} \\
\text{Neb} = 10\ \text{Fog} \\
\text{Frost}
\end{bmatrix}
\quad
2\,\text{pm}
\begin{bmatrix}
-17.8 \\
-17.55 \\
+19.2 \\
745.4 \\
\text{min} = -18.0 \\
87.2 \\
\text{SW} \\
\text{Neb} = 10 \\
\text{Stratus}
\end{bmatrix}
\quad
9\,\text{pm}
\begin{bmatrix}
-16.5 \\
-16.5 \\
+15.9 \\
747.3 \\
87.3 \\
\text{calm} \\
\text{Neb} = 10
\end{bmatrix}
$$

Average wind speed = 2.5 m.
Maximum wind speed = 5.5 m.
There was 5 h of calm.
Wind from the E and ENE until 7 am. From 9 am to 8 pm, WSW, SW, and SSW,
and NE the rest of the day. Fog with frost all day.

5.14.12 September 17, [1903]

$$
7\,\text{am}\begin{bmatrix} -17.3 \\ -17.1 \\ +5.0 \\ 743.4 \\ \text{ENE}\,5\,\text{m} \\ \text{Neb} = 6 \end{bmatrix}
\qquad
2\,\text{pm}\begin{bmatrix} -9.0 \\ -8.85 \\ +17.2 \\ 737.65 \\ 95\% \\ \text{min} = -18.5 \\ \text{NNE}\,13\,\text{m} \end{bmatrix}
\qquad
9\,\text{pm}\begin{bmatrix} -22.1 \\ -22.0 \\ +12.3 \\ 736.1 \\ \text{SSW}\,18.5\,\text{m} \\ \text{Neb} = 10 \end{bmatrix}
$$

Average wind speed = 14.5 m.
Maximum wind speed = 18 m at 8 pm (SW).
Minimum wind speed = 1 m at 1 am (ENE).
The maximum speed from the NNE was 13 m at 2 pm. ENE blowing until 8 am.
From 8 am until midday, NE, and from that last hour until 4 pm, it blew from the
NNE. At 4:45, after some seconds of calm, the SW started.

5.14.13 September 18, [1903]

$$
7\,\text{am}\begin{bmatrix} -24.4 \\ -24.15 \\ +2.8 \\ 734.4 \\ \text{Max} = -6.0 \\ \text{SW}\,8\,\text{m} \\ \text{Neb} = 10 \end{bmatrix}
\qquad
2\,\text{pm}\begin{bmatrix} -23.45 \\ -23.2 \\ +15.0 \\ 737.25 \\ \text{min} = -25.1 \\ 80\% \\ \text{SW}\,15\,\text{m} \\ \text{Neb} = 10 \\ \text{snöfyk} \end{bmatrix}
\qquad
9\,\text{pm}\begin{bmatrix} -22.55 \\ -22.55 \\ +10.9 \\ 739.4 \\ 70.8\% \\ \text{NE almost calm} \\ \text{Neb} = 2 \end{bmatrix}
$$

Until 3 pm, completely overcast sky. From that time until 3 pm, nebulosity was
1, 2, 3, but at midnight it was 10 again.

Average wind speed = 8 m.
Max. wind speed = 15 m at 2 pm.
There was 3 h of calm.

5.14.14   September 19, [1903]

$$
7\,\text{am}\begin{bmatrix} -14.2 \\ -14.2 \\ +4.9 \\ 730.65 \\ 86.5\% \\ \text{ENE 6 m} \\ \text{Neb} = 10 \end{bmatrix}
\quad
2\,\text{pm}\begin{bmatrix} -21.25 \\ -21.05 \\ +15.5 \\ 730.35 \\ \text{min} = -23.6 \\ \text{SW 21 m} \\ \text{Neb} = 10 \end{bmatrix}
\quad
9\,\text{pm}\begin{bmatrix} -22.1 \\ -22.05 \\ +11.2 \\ 734.4 \\ \text{SW 16 m} \\ \text{Neb} = 10 \end{bmatrix}
$$

Overcast sky until 3 pm. At that time, it started to clear up, and at 9 pm, nebulosity was = 0. From 1 am to 8 am, it blew from the ENE. The SW started at that same hour, and the temperature rose to $-12.5°$, but it swiftly dropped again with the SW.

Average wind speed = 15 m; the maximum = 22 m at 4 pm. The minimum = 3.5 m at 1 am, the wind being from the ENE.

Nordenskjöld is waiting for the weather to clear up so he can go out on a sledge excursion, I believe to Paulet Island. The barometer has been rising since the morning, but very irregularly.

5.14.15   September 20, [1903]

$$
7\,\text{am}\begin{bmatrix} -23.5 \\ -23.3 \\ +2.8 \\ \text{bar} = 740.1 \\ 72\% \\ \text{Max} = -12.5 \\ \text{Neb} = 1 \\ \text{SSW 8 m} \end{bmatrix}
\quad
2\,\text{pm}\begin{bmatrix} -20.55 \\ -20.55 \\ +13.6 \\ 745.5 \\ 70\% \\ \text{min} = -27.0 \\ \text{WSW} \\ \text{Neb} = 1 \end{bmatrix}
\quad
9\,\text{pm}\begin{bmatrix} -24.4 \\ -23.8 \\ +15.9 \\ 747.35 \\ 76\% \\ \text{calm} \\ \text{Neb} = 1 \end{bmatrix}
$$

Until 6 am, completely clear sky. At 7, nebulosity = 1; there was a bank of clouds to the N. At 2 am, the wind moved to the SSW, but at 5 would blow from the WSW, and temperature would drop quickly. It continued like that until 6:50, the thermometer reaching around $-27°$. At 7, the wind moved to the SSW, and the thermometer rose again. While I was going to read the thermometers, I saw four pagodromas [snow petrels] flying to the S. Average wind speed was 6 m; the maximum was 18 m at 1 am. There was 6 h of calm (from 7 pm to 12).

5.14.16 September 21, [1903]

$$
7\,\text{am}
\begin{bmatrix}
-20.5 \\
-20.3 \\
+4.6 \\
745.9 \\
75\% \\
\text{Max} = -18.9 \\
\text{NE 1 m} \\
\text{Neb} = 3
\end{bmatrix}
\qquad
2\,\text{pm}
\begin{bmatrix}
-17.8 \\
-18.0 \\
+14,2 \\
746.55 \\
66\% \\
\text{NE 4 m} \\
\text{Neb} = 10
\end{bmatrix}
\qquad
9\,\text{pm}
\begin{bmatrix}
-18.05 \\
-18.1 \\
+14.4 \\
747.2 \\
\text{ENE 3 m} \\
\text{Neb} = 10
\end{bmatrix}
$$

During the day, the temperature has remained between −17° and −18°, generally closer to −18°.

Barometer rising, overcast sky, wind from the NE until 3 pm, and the rest of the day from the ENE, E, and ESE. Several cormorants were seen in the morning, I think four, a Larus [gull], and an ossifraga [giant petrel], and some pagodromas [snow petrels], too.

Average wind speed 3 m. The maximum was 5 m at 5 pm, and there were some moments of calm. Nordenskjöld told me that as soon as he thinks that the good weather has " _______ " [unreadable], he'll set off on a sledge trip with Jonassen. He's planning to go through the west of Cape Foster to come out at Sydney Herbert Bay, that is, around Mount Haddington, because he thinks that Mount Haddington is an island (I think so too, at least last year in November, when we were coming back from King Oscar Land, the slope of Mount Haddington to the west was very quick, and everything made us believe that there was a strait; probably, due to its smallness, it's covered with ice from the glacier, but, nevertheless, I don't think it will represent a big inconvenience for a sledge trip). If he manages to pass through Sydney Herbert Bay, from there he will continue toward Paulet [Island], where he plans on posting a sign and leaving a letter.

5.14.17 September 22, [1903]

$$
7\,\text{am}
\begin{bmatrix}
-17.35 \\
-17.45 \\
+3.5 \\
746.35 \\
\text{Max} = -16.9 \\
71.8\% \\
\text{ESE 3 m} \\
\text{Neb} = 10
\end{bmatrix}
\qquad
2\,\text{pm}
\begin{bmatrix}
-16.0 \\
-16.3 \\
+15.8 \\
747.3 \\
68\% \\
\text{min} = -18.7 \\
\text{ENE 3 m} \\
\text{Neb} = 10
\end{bmatrix}
\qquad
9\,\text{pm}
\begin{bmatrix}
-20.0 \\
-19.8 \\
+17.4 \\
747.75\,(747.15) \\
\text{SSW 7.5 m} \\
76.7\% \\
\text{Neb} = 10
\end{bmatrix}
$$

Overcast sky all day; some snow fell between 3 pm and 9 pm. Average wind speed during the day was = 3.5 m. The maximum was 8 m at 12 at night, and the minimum was 1 m. From 1 am to 6 pm, winds from the E, ESE, and ENE; from

4 pm to 12 pm, SSW. Today, a Larus dominicanus [kelp gull] was seen flying around the skeleton of a seal. I grabbed a gun, but, when I came back, it had disappeared.

5.14.18   September 23, [1903]

$$
7\,\text{am}
\begin{bmatrix}
-20.25 \\
-20.1 \\
+5.0 \\
745.1 \\
77\% \\
\text{Max} = -15.4 \\
\text{SSW} \\
\text{Neb} = 10 \\
\text{snöfyk}
\end{bmatrix}
\quad
2\,\text{pm}
\begin{bmatrix}
-20.0 \\
-19.8 \\
+16.2 \\
748.5 \\
\text{min} = -20.45 \\
\text{SSW} \\
\text{Neb} = 10 \\
\text{snöfyk}
\end{bmatrix}
\quad
9\,\text{pm}
\begin{bmatrix}
-18.5 \\
-18.3 \\
+14.2 \\
751.7 \\
\text{SSW} \\
\text{Neb} = 10
\end{bmatrix}
$$

Overcast sky all day, and SSW blowing. Average wind speed was 10 m. The maximum was 14 m at 1 pm, and the minimum was 6 m at 12 at night.

People are ready for the sledge trip, but Nordenskjöld says he'll wait for better weather. If Mount Haddington is not an island, and if they can't get to Sydney Herbert Bay by passing through over the glacier, he plans on coming back to the station in order to start the trip directly from here to Paulet [Island]. In the letter that he'll leave on Paulet, which is written in English, it says that we are all well, but that provisions are very scarce, and that if the ship that finds the letter can't go to the station (it provides its geographical location) to pick us up, they should leave a store of provisions for six people to last over one year, and they should set [indicate] a place where these six people would go to look for the ship next summer. I think he also plans on asking them [those who find the letter] to leave a boat, because ours (the big one) was wrecked by a storm.

5.14.19   September 24, [1903]

$$
7\,\text{am}
\begin{bmatrix}
-17.6 \\
-17.5 \\
+4.1 \\
751.45 \\
80.5\% \\
\text{Max} = -16.5 \\
\text{ENE}\ 1.5\ \text{m} \\
\text{Neb} = 10 \\
\text{Snowing}
\end{bmatrix}
\quad
2\,\text{pm}
\begin{bmatrix}
-18.0 \\
-18.0 \\
+16.0 \\
752.15 \\
74.1\% \\
\text{min} = -20.5 \\
\text{SSW}\ 12.5\ \text{m} \\
\text{Neb} = 10 \\
\text{Snowing}
\end{bmatrix}
\quad
9\,\text{pm}
\begin{bmatrix}
-20.5 \\
-20.4 \\
+16.9 \\
750.05 \\
80\% \\
\text{SSW}\ 13\ \text{m} \\
\text{Neb} = 10 \\
\text{Snowing}
\end{bmatrix}
$$

All day overcast sky, and snowing almost all day, although little. At 1:15 am, the SSW calmed, and the SE started to blow at 3; then, E and ENE. At 9 am, the SSW started again, and it continued for the rest of the day. Average wind speed was 6.5 m, the maximum was 18 m at 12 at night, and there was an hour of calm (2 am). The barometer dropped and so did the temperature, and the wind has been increasing since 5 pm. A year ago today, a storm started from the SW and lasted until the 29th. Will it be the same for the one that is starting today?

5.14.20   September 25, [1903]

$$
7\,\text{am}\begin{bmatrix} -24.45 \\ -24.0 \\ +5.2 \\ 737.1 \\ 80.5\% \\ \text{SW } 25.8\,\text{m} \\ \text{Neb} = 10 \end{bmatrix}
\quad
2\,\text{pm}\begin{bmatrix} -24.7 \\ -24.5 \\ +16.3 \\ 737.7 \\ \min = -25.2 \\ \text{WSW } 27.7 \\ \text{Neb} = 10 \end{bmatrix}
\quad
9\,\text{pm}\begin{bmatrix} -24.3 \\ - \\ +13.0 \\ 737.95 \\ \text{SW } 24.7\,\text{m} \\ \text{Neb} = 10 \end{bmatrix}
$$

All day, blowing between the SW and the WSW, with a lot of snow and overcast sky. Average wind speed was 25.6 m; the maximum (hourly average) was 28.7 m at midday, and the minimum was 18.5 m at 1 am. There have been moments between 10 am and 1 pm when the speed of the wind has definitely exceeded 30 m. Rocks were flying during these moments and broke one of the glass windows of the kitchen. On September 28 of last year, the same thing happened and almost at the same time. It must be truly enormous, the effect that these strong winds have on these lands, even on the hardest rocks. I think that the door of the shelter for the data recorder broke, because I saw that they had tied it with a rope, and the screws were missing.

5.14.21   September 26, [1903]

$$
7\,\text{am}\begin{bmatrix} -23.8 \\ -23.5 \\ +2.3 \\ 739.3 \\ \text{Max} = -17.2 \\ \text{SW } 17\,\text{m} \\ \text{Neb} = 10 \\ \text{snöfyk} \end{bmatrix}
\quad
2\,\text{pm}\begin{bmatrix} -19.0 \\ -19.1 \\ +16.8 \\ 742.6 \\ \min = -23.8 \\ \text{SSW } 6.5\,\text{m} \\ \text{Neb} = 3 \end{bmatrix}
\quad
9\,\text{pm}\begin{bmatrix} -18.8 \\ -18.6 \\ +16.8 \\ 741.7 \\ 81.2\% \\ \text{E } 46\,\text{m[sic]} \\ \text{Neb} = 2 \end{bmatrix}
$$

Average wind speed during the day was 11 m. The maximum was 24 m at 1 am, and there were a couple of hours of calm (4 pm). The SSW blew until 3:20 pm, and the ENE started at 5:30 and continued for the rest of the day. The weather seemed to have improved, but the barometer started to drop at 4 pm, and the temperature started to rise, too. After 9, the sky started to become overcast little by little. A stone blown by the wind broke the glass of the window of the little magnetic house, even though there was wire mesh on the outside to prevent this from happening.

5.14.22   September 27, [1903]

$$
7\,\text{am}
\begin{bmatrix}
-14.75 \\
-14.05 \\
+5.4 \\
733.9 \\
\text{Max} = -13.1 \\
83.5\% \\
\text{ENE}\,9\,\text{m} \\
\text{Neb} = 10
\end{bmatrix}
\quad
2\,\text{pm}
\begin{bmatrix}
-20.7 \\
-20.55 \\
+17.4 \\
735.2 \\
\min = -21.3 \\
79;5 \\
\text{S}\,11\,\text{m} \\
\text{Neb} = 10
\end{bmatrix}
\quad
9\,\text{pm}
\begin{bmatrix}
[-]22.0 \\
[-]21.8 \\
+13.7 \\
734.95 \\
\text{S}\,11.5\,\text{m} \\
\text{Neb} = 9
\end{bmatrix}
$$

Average wind speed = 9.5 m.
Maximum wind speed = 11.5 m.
Minimum wind speed = 4 m.

The ENE and the NE blew until 11:30, with snow. Barometer dropping and temperature rising, and after having variable winds for an hour, the SSW started at 11:30 and continued for the rest of the day, with the barometer dropping and the thermometer below −20°. At 10:30, the temperature rose to −10.5°, but then it dropped swiftly.

The last of the puppies of "Amager" was found dead this morning. And since the alleged criminal is "Nemö," he was cruelly disfigured by being hit with a stick over his head, eyes, and snout.

5.14.23   September 28, [1903]

$$
7\,\text{am}
\begin{bmatrix}
-20.8 \\
-20.5 \\
+3.4 \\
732.35 \\
80.1\% \\
\text{Max} = -10.5 \\
\text{SSW}\,14\,\text{m} \\
\text{Neb} = 10 \\
\text{Snowing}
\end{bmatrix}
\quad
2\,\text{pm}
\begin{bmatrix}
-20.2 \\
-20.1 \\
+14.5 \\
+14.5 \\
735.1 \\
79.3\% \\
\min = -21.7 \\
\text{WSW}\,18.5 \\
\text{Neb} = 10
\end{bmatrix}
\quad
9\,\text{pm}
\begin{bmatrix}
-23.55 \\
-23.45 \\
+14.6 \\
739.5 \\
76.5\% \\
\text{SSW}\,12\,\text{m} \\
\text{Neb} = 3
\end{bmatrix}
$$

Average wind speed = 14.5 [m].
Maximum wind speed = 19.5.
Minimum wind speed = 9.5.

Between 11 am and 6 pm, SW and WSW blowing. The rest of the day, SSW. Until 6 pm, completely overcast sky. At 7, it started to clear up, Ci being the only clouds that could be seen. At night, the sky became overcast with Ci, and there was a moon halo. The barometer has risen swiftly all day.

5.14.24   September 29, [1903]

$$
7\,\text{am}
\begin{bmatrix}
-24.45 \\
-24.55 \\
+4.7 \\
741.05 \\
70\% \\
\text{Max} = -19.5 \\
\text{ENE}\,1\,\text{m} \\
\text{Neb} = 1
\end{bmatrix}
\qquad
2\,\text{pm}
\begin{bmatrix}
-20.15 \\
-20.05 \\
+19.4 \\
742.0 \\
\text{min} = -25.8 \\
79.5\% \\
\text{NE}\,2\,\text{m} \\
\text{Neb} = 10
\end{bmatrix}
\qquad
9\,\text{pm}
\begin{bmatrix}
-24.6 \\
-24.3 \\
+11.2 \\
739.85 \\
\text{SSW}\,8\,\text{m} \\
\text{Neb} = 10
\end{bmatrix}
$$

Average wind speed = 5.0.
Maximum wind speed = 12.2.
There was 3 h of calm.

Until 12 midday, sky with few clouds; to the NE, a bank of Ni and A-S could be seen coming, but since 2 in the afternoon, nebulosity was 10. The SSW calmed at 6 am. Between 7 am and 6 pm, the wind was from the NE, NNE, and NNW, the barometer dropping. At 6 h. 5 min. pm, it started to blow from the SW, and then, it moved to the SSW, and a lot of snow fell.

Nordenskjöld decided to set off today, despite the fact that the weather does not promise anything good. Everything was prepared, and at 12:30, he took off into motion, leaving instructions with Bodman. The sledge is pulled by six good dogs.

Sobral's observations, instincts, and words that "the weather does not promise anything good" are prophetic, as Nordenskjöld would indeed meet with challenges on this sledge journey. Aside from the fact that the expedition leader had decided not to take Sobral with him on this scientific excursion, despite his solemn promise to the Argentine government, Nordenskjöld also selected a time to leave when the weather conditions would work against him. He and Jonassen also forgot to pack some much-needed provisions in their 30-day supply, and one of their trusty dogs, Suggen, sported a foot injury that would reopen and leave tracks of blood along the snow during their journey (Nordenskjöld et al. 1905.)

Ironically, this day that Nordenskjöld selects to depart from Snow Hill in search of Paulet Island, where, unbeknownst to him, Captain Larsen and the crew and passengers of the Antarctic *are at that very moment taking refuge after the sinking of their ship, is also the exact same day that Johan Gunnar Andersson and his two companions set off on foot from Hope Bay, leaving their makeshift stone hut in the hopes of crossing the ice and reaching Snow Hill. It seems that both Nordenskjöld and Andersson had a sense that this was the day to travel, although with different respective results.*

5.14.25   *September 30, [1903]*

$$
7\,\text{am}
\begin{bmatrix}
-25.95 \\
- \\
+3.2 \\
735.85 \\
81\% \\
\text{SW} \\
\text{Neb} = 10 \\
\text{Snöfyk} = 2
\end{bmatrix}
\quad
2\,\text{pm}
\begin{bmatrix}
-25.25 \\
-25.0 \\
+14.4 \\
735.8 \\
81.5\% \\
\text{WSW} \\
\text{neb} = 10 \\
\text{Snöfyk} = 2
\end{bmatrix}
\quad
9\,\text{am}
\begin{bmatrix}
-26.4 \\
-26.0 \\
+12.8 \\
737.5 \\
\text{SW} \\
\text{neb} = 10 \\
\text{Snöfyk} = 2
\end{bmatrix}
$$

SW and WSW blowing all day with a lot of snow.
Average wind speed = 21.5 m.
Max. wind speed = 25.3 m at 2 pm.
Min. wind speed = 14.1 m at 1 am.

The wind has been increasing its intensity since yesterday, and it moved to the SW at 1 am. In general, it seems that the wind will increase until 2 and 3 pm, to diminish at midnight; maybe its speed increases together with the temperature (?).

5.14.26   *October 1, [1903]*

$$
7\,\text{am}
\begin{bmatrix}
-25.7 \\
- \\
+1.8 \\
737.0 \\
\text{Max} = -24.5 \\
\text{WSW} \\
\text{Neb} = 10 \\
\text{Snöfyk} = 2
\end{bmatrix}
\quad
2\,\text{pm}
\begin{bmatrix}
-24.8 \\
-24.6 \\
+13.8 \\
742.7 \\
\text{WSW} \\
\text{Neb} = 9 \\
\text{Snöfyk} = 2
\end{bmatrix}
\quad
9\,\text{pm}
\begin{bmatrix}
-22.8 \\
-22.6 \\
+15.1 \\
746.25 \\
\text{WSW} \\
\text{Neb} = 10 \\
\text{Snöfyk} = 2
\end{bmatrix}
$$

SW and WSW blowing all day, with a lot of snow.

Average wind speed = 23.3.
Max. wind speed = 26.1 at midday.
Min. wind speed = 17 m at 3 pm.
Magnetic observations were made today.

5.14.27 October 2, [1903]

$$
7\,\text{am}
\begin{bmatrix}
-21.95 \\
-21.85 \\
+5.6 \\
751.5 \\
80\% \\
\text{Max} = -21.2 \\
\text{SW} \\
\text{Neb} = 10 \\
\text{Snöfyk} = 1
\end{bmatrix}
\qquad
2\,\text{pm}
\begin{bmatrix}
-14.6 \\
-15.25 \\
+17.0 \\
756.05 \\
63.1\% \\
\text{SSW} \\
\text{neb} = 0.5
\end{bmatrix}
\qquad
9\,\text{pm}
\begin{bmatrix}
-18.2 \\
-18.0 \\
+11.4 \\
756.45 \\
80\% \\
\text{NE} \\
\text{neb} = 0.5
\end{bmatrix}
$$

The barometer rose until 7 pm; at 8 pm, it began to drop. Wind from the SSW, SW, or S blew until 4:30 pm. At 5, it blew from the N and at 7 from the ENE, and it remained this way (sometimes moving to the NE) until the end of the day.

Average wind speed = 12.8 m; maximum = 20 m at 1 am, and the minimum = 1 meter at 5 pm. The day was clear and the sun shone, but the wind blew quite strongly in the morning.

At 6 pm, those who had set off on the sledge trip appeared again. The day they had left, they had reached Lockyer Island, camping leeward of it, but because of the strong wind and whirlwinds, their tent broke, and that is why they have come back. The day before yesterday, speaking with Bodman, I predicted this. I think it's an error to camp leeward of small islands or mountains where the wind can swirl and come from the sides. I believe that camping amidst the ice is much better, in order to receive the constant power of the wind and not to have [intermittent] moments of calm and then furious gusts, which is what happens when one is close to glens or valleys or peaks where the wind blows to the sides with enormous strength. Before the sledge arrived, I climbed up the mountain to see if I could spot any seals, but with no result. I could see the Land of Louis Filippe very clearly, as well as the one that is to the south of Mount Haddington. I think that if there is a strait on the other side of Cape Foster, it does not lead to Sydney Herbert Bay but to Cape Corry.

5.14.28   October 3, [1903]

$$
7\,\text{am}
\begin{bmatrix}
-13.0 \\
-13.05 \\
+4.2 \\
752.2 \\
79.9\% \\
\text{Max} = -11.9 \\
\text{NE}\,7.5\,\text{m} \\
\text{Neb} = 10
\end{bmatrix}
\quad
2\,\text{pm}
\begin{bmatrix}
-5.7 \\
-6.1 \\
+17.2 \\
751.8 \\
\min = -19.4 \\
\text{NE}\,9\,\text{m} \\
\text{Neb} = 10
\end{bmatrix}
\quad
9\,\text{pm}
\begin{bmatrix}
-3.5 \\
-4.0 \\
+14.8 \\
749.0 \\
\text{NE}\,10\,\text{m} \\
\text{Neb} = 9
\end{bmatrix}
$$

NE blowing all day, sometimes with a lot of *snöfyk* [snow cover]. Completely overcast sky. The temperature has been rising since last night, when it was around −18°. After midday, it remained between −4 and −5°. Today, several large water clearings have been sighted (some 2 or 3 miles long) to the east side of Snow Hill. Today, I saw the tent. It seems that what broke isn't such a big deal; 2 or three scratches of little importance, at least from what I saw. It looks like we're condemned to not have a nice day. Ever since the first days of September, we haven't had a single day of relative calm and with sun. Yesterday, Nordenskjöld said that if the tent was fixed today, he would plan on leaving tomorrow, depending on the weather being clear, of course, but I think that, immediately after this Northeaster, a Southeaster is coming. The poor dogs are very bad in terms of food, since there are only two or three bare skeletons of seals left. The average speed of the wind = 9 m, the max. = 13.5 m, and the min. = 6 m.

5.14.29   October 4, [1903]

$$
7\,\text{am}
\begin{bmatrix}
-3.35 \\
-3.6 \\
+7.5 \\
746.0 \\
\text{Max} = -2.9 \\
\text{NE}\,11.0\,\text{m} \\
\text{Neb} = 7 \\
\text{snöfyk}
\end{bmatrix}
\quad
2\,\text{pm}
\begin{bmatrix}
-0.45 \\
-1.3 \\
+16.5 \\
745.5 \\
81\% \\
\min = -6.25 \\
\text{NE}\,7\,\text{m} \\
\text{Neb} = 10
\end{bmatrix}
\quad
9\,\text{pm}
\begin{bmatrix}
-1.95 \\
-2.4 \\
+15.6 \\
742.8 \\
88\% \\
\text{NNE}\,10.5\,\text{m} \\
\text{Neb} = 10
\end{bmatrix}
$$

Average wind speed = 10.5 m.
Max. wind speed = 13.5 m.
Min. wind speed = 5.6 at 1 pm.

From 1 am to 3 pm, it blew from the NE. Then, from 3 pm through the rest of the day, the NNE blew. Nebulosity began to diminish at 5 am, and at 10 am, only some Cu could be seen to the NE, but a bank of Ci was coming from the NW. At midday, the sky was overcast with Ci with SW-NE radiation. But then the Ci turned into Ci-S, and Cu-Ni started to appear.

Nordenskjöld set off again at midday; from what Bodman said, the main objective of this excursion is the supposed strait and then, as a secondary spot, Paulet [Island].

Bodman killed a seal. We're lucky, not only for us, but for the poor dogs that haven't had anything to eat. The spaces of free water that there are to the E of Snow Hill are considerable, according to Bodman. One extends from Snow Hill to as far as the eye can see in the direction of Seymour Island. Today, I took a photograph of the grooves that the wind leaves in the snow. Since the temperature has been so high, there has been a lot of water running [due to ice melt].

It is on this sledge journey to James Ross Island and Vega Island that Nordenskjöld discovers Prince Gustaf Channel and determines that Mount Haddington sits on a volcanic island that is separated from the mainland by this channel. Equipped with provisions, and killing seals along the way, he and Jonassen and the sledge dogs follow this new-found channel (Nordenskjöld et al. 1905).

Meanwhile, as Sobral notes, both the men and the dogs at the winter station are once again saved from starvation through the finding and killing of a seal. It is Sobral's character to recognize the fortune of both the men and the dogs, being as concerned for the canines as he is for himself and his mates (Tahan 2016a, b).

5.14.30   October 5, [1903]

$$
7\,\text{am}
\begin{bmatrix}
-0.3 \\
-1.2 \\
+8.9 \\
737.2 \\
79\% \\
\text{Max} = +0.7 \\
\text{NE}\,9\,\text{m} \\
\text{Neb} = 10
\end{bmatrix}
\quad
2\,\text{pm}
\begin{bmatrix}
-12.85 \\
-12.85 \\
+19.8 \\
739.8 \\
\min = -14.7 \\
\text{SW}\,7.5\,\text{m} \\
\text{Neb} = 10
\end{bmatrix}
\quad
9\,\text{pm}
\begin{bmatrix}
-17.65 \\
-17.5 \\
+14.0 \\
739.65 \\
\text{SSW}\,12\,\text{m} \\
\text{Neb} = 10 \\
\text{snöfyk}
\end{bmatrix}
$$

Average wind speed = 10 m.
Max. wind speed = 13.8.
Min. wind speed = 3.0 m.

The NNE blew quite strongly until 7:30, when the SSW started, and it continued for the rest of the day. Snow fog all day.

It seems that those on the sledge trip have no luck with the weather.

5.14.31   *October 6, [1903]*

$$
7\,\text{am}
\begin{bmatrix}
-13.8 \\
-13.6 \\
+6.7 \\
736.5 \\
82.7\% \\
\text{Max} = -0.3 \\
\text{NE}\,1.5\,\text{m} \\
\text{Neb} = 10
\end{bmatrix}
\quad
2\,\text{pm}
\begin{bmatrix}
-9.25 \\
-9.4 \\
+17.0 \\
737.1 \\
81.8\% \\
-18.4 \\
\text{NE}\,1.5\,\text{m} \\
\text{Neb} = 10
\end{bmatrix}
\quad
9\,\text{pm}
\begin{bmatrix}
-4.2 \\
-4.7 \\
+14.1 \\
735.3 \\
75\% \\
\text{ENE}\,3.5\,\text{m} \\
\text{Neb} = 10
\end{bmatrix}
$$

Average wind speed = 3 m.
Maximum wind speed = 12 m at 12 pm.
There was 2 h of calm.

From 1 to 2:30 pm, thick fog and sometimes snowfall. From 2:30 to 8, almost clear sky with a splendid sun; then, a Cu-Ni started to form to the W, and little by little, it grew bigger until 12 pm, when nebulosity was 7. At 3:35, the SSW, which had been blowing since yesterday, calmed, and the NE started shortly after and continued for the rest of the day. Temperature was rising, and it reached 0° at 10 am, while a strong NE was blowing. The barometer is dropping. Since this night had promised to be clear and calm, I had planned to make some latitude observations, but the wind prevented me from doing so.

5.14.32   *October 7, [1903]*

$$
7\,\text{am}
\begin{bmatrix}
-8.0 \\
-8.0 \\
+8.3 \\
735.75 \\
\text{Max} = -0.4 \\
\text{NE}\,1\,\text{m} \\
\text{Neb} = 10
\end{bmatrix}
\quad
2\,\text{pm}
\begin{bmatrix}
-13.05 \\
-13.25 \\
+16.0 \\
737.4 \\
\text{min} = -13.2 \\
\text{WSW}\,5\,\text{m} \\
\text{Neb} = 10 \\
\text{snöfyk}
\end{bmatrix}
\quad
9\,\text{pm}
\begin{bmatrix}
-19.5 \\
-19.3 \\
+13.8 \\
739.15 \\
\text{SSW}\,11\,\text{m} \\
\text{Neb} = 10 \\
\text{snöfyk}
\end{bmatrix}
$$

Cloudy sky all day, with fog and snow. Variable wind from 1 am to 11 am. From 12 to 2 pm, W and WSW, and SSW for the rest of the day.
Average wind speed = 5.5.
Max. wind speed = 11 ms at 11 pm.
Min. wind speed = 1 m at 7 am.

5.15   What a Monotonous Life!

5.15.1   October 8, [1903]

$$
7\,\text{am}
\begin{bmatrix}
-19.6 \\
-19.45 \\
+6.4 \\
740.15 \\
\text{Max} = -5.2 \\
75\% \\
\text{SSW } 11.5\,\text{m} \\
\text{Neb} = 10 \\
\text{Snowing}
\end{bmatrix}
\quad
2\,\text{pm}
\begin{bmatrix}
-17.7 \\
-17.6 \\
+14.4 \\
740.1 \\
72.5\% \\
\text{min} = -20.5 \\
\text{SW } 11.5\,\text{m} \\
\text{Neb} = 10 \\
\text{Snow fog}
\end{bmatrix}
\quad
9\,\text{pm}
\begin{bmatrix}
-18.0 \\
-17.7 \\
+17.3 \\
740.3 \\
76\% \\
\text{SSW } 9\,\text{m} \\
\text{neb} = 10
\end{bmatrix}
$$

Average wind speed = 11.0.
Max. wind speed = 13 m.
Min. wind speed = 8.2.
All day, SSW and SW blowing, with a little of snow and fog. Neb = 10.
Today, observations of magnetic variations were made.
What a monotonous life!
The barometer has been dropping all day, although very slowly.

The young Sobral craves the excitement of fossil-hunting, an activity in which his mentor is engaging on this very day. The monotony of weather-recording at the station, while Jonassen takes his place on the sledge excursion, must have weighed heavily on Sobral's mind.

5.15.2   October 9, [1903]

$$
7\,\text{am}
\begin{bmatrix}
-14.7 \\
-14.5 \\
+6.8 \\
737.3 \\
78.8 \\
\text{Max} = -14.5 \\
\text{WSW } 4\,\text{m} \\
\text{Neb} = 10
\end{bmatrix}
\quad
2\,\text{pm}
\begin{bmatrix}
-12.45 \\
-12.55 \\
74\% \\
+15.9 \\
737.5 \\
\text{min} = -13.2 \\
\text{NW } 0.5\,\text{m} \\
\text{Neb} = 10
\end{bmatrix}
\quad
9\,\text{pm}
\begin{bmatrix}
-10.7 \\
-10.5 \\
+14.7 \\
735.8 \\
87\% \\
\text{NNE } 1\,\text{m} \\
\text{Neb} = 10
\end{bmatrix}
$$

Average wind speed = 3 m.
Maximum wind speed = 7 m.
There was 3 h of calm.

All day, sky overcast with Ni and S-Cu, and snowing (little amount). Snow consisting of precious crystals. Some say that snow falls shaped as perfect crystals only when the temperature is below [-] 20 degrees. But today, for example, with temperatures around [-] 10°, I saw snow fall in the shape of the most perfect crystals.

At 10, the SW calmed, and from 3 pm to 12 at night, the NNE and the NE blew.

Today, Bodman and I went to look for the skin of the seal that he had killed the other day. It's about two hours from the station in the eastern part of the strait that separates Seymour Island from Snow Hill. The sledge was pulled by Fia's four puppies, who are the dogs that are left at the station. Despite being only six months old, they easily pulled the skin of the seal as well as one of us seated on the sledge. We found a dead ossifraga [giant petrel] close to where the skin was. Because it had its wings cracked and its body hurt, we supposed that the dogs were the ones who killed it. Without doubt, it had eaten so much fat that when these [dogs] went to catch it, it couldn't fly. According to Bodman, who walked around the hills, you can see there is an area of water free of ice that goes from Cape Gage to Cockburn Island.

5.15.3   *October 10, [1903]*

$$
7\,\text{am}
\begin{bmatrix}
-16.25 \\
-16.1 \\
+7.8 \\
735.55 \\
81\% \\
\text{Max} = -7 \\
\text{SSW } 6.5\,\text{m} \\
\text{Neb} = 10
\end{bmatrix}
\quad
2\,\text{pm}
\begin{bmatrix}
-16.2 \\
-16.1 \\
+16.8 \\
738.2 \\
76.9\% \\
\min = -17.1 \\
\text{SSW } 10\,\text{m} \\
\text{Neb} = 10
\end{bmatrix}
\quad
9\,\text{pm}
\begin{bmatrix}
-15.45 \\
-15.45 \\
+11.5 \\
739.3 \\
76.9\% \\
\text{SSW} \\
\text{Neb} = 10
\end{bmatrix}
$$

Average wind speed = 6.5 m.
Max. wind speed = 10.5.
Min. wind speed ____ [blank]. There was an hour of calm.

Snow fog all day, sometimes snowing. Until 3 am, it kept blowing from the NNE. At 4 am, it was completely calm. The SSW started between 4:30 and 5 am. At 9 pm, the WSW started to move and to grow calm.

5.15.4   *October 11, [1903]*

$$
7\,\text{am}
\begin{bmatrix}
-14.6 \\
-14.5 \\
+4.7 \\
740.85 \\
81.8\% \\
\text{Calm} \\
\text{Neb} = 10
\end{bmatrix}
\quad
2\,\text{pm}
\begin{bmatrix}
-11.5 \\
-11.8 \\
+17.0 \\
744.35 \\
\min = -16.9 \\
81.6\% \\
\text{calm} \\
\text{Neb} = 7
\end{bmatrix}
\quad
9\,\text{pm}
\begin{bmatrix}
-19.4 \\
-19.3 \\
+14.6 \\
745.7 \\
\text{WSW}\,2\,\text{m} \\
\text{Neb} = 1
\end{bmatrix}
$$

Average wind speed = 0 m.
Max. wind speed = 2 m.
There was 16 h of calm.

Sky overcast with fog all morning. It started to clear up in the afternoon, and at 9 pm, only a bank of fog could be seen to the NE.

The temperature dropped in the afternoon; at 12 at night, it reached around $-22°$; and then, it started to rise again.

5.15.5   *October 12, [1903]*

$$
7\,\text{am}
\begin{bmatrix}
-15.2 \\
-14.95 \\
+6.8 \\
743.75 \\
\text{Max} = -0.5 \\
81.2\% \\
\text{Calm} \\
\text{Neb} = 10 \\
\text{S}
\end{bmatrix}
\quad
2\,\text{pm}
\begin{bmatrix}
-8.1 \\
-8.5 \\
+18.6 \\
743.5 \\
88\% \\
\text{Min} = -21.6 \\
\text{Calm} \\
\text{Neb} = \_[\text{blank}]
\end{bmatrix}
\quad
9\,\text{pm}
\begin{bmatrix}
-10.0 \\
-10.0 \\
+14.2 \\
742.75 \\
86.8\% \\
\text{Calm} \\
\text{Neb} = 4
\end{bmatrix}
$$

Average wind speed = 0.2 m.
Max. wind speed = 3 m.
There was 16 h of calm.

From 7 am until midday, it blew from the NE and NNE; the rest of the day was calm, and there were some weak gusts from the N and NE. All day until 8 pm, overcast sky: fog in the morning and stratus in the afternoon. At 8 pm, it started to clear up, and at 10 pm, nebulosity was just 1.

At 4 pm, I went out to make the meteorological observations, and, as usual, I took a look at the ice of the strait, when, suddenly, my attention was drawn to an object around a mile away in distance, whose color contrasted with the white surface. At first, I thought it was a stick; I thought that Bodman had put it there. But it seemed to be moving and growing thicker, until, finally, because of its movements, it occurred to me that it was a penguin. I ran inside, and after getting glasses [binoculars], with whose help I confirmed my belief that I had encountered one of these good fellows, [and] followed by Ekelöf and Åkerlundh, I walked toward it with the intention of hunting it, and after a little chase, it was our prisoner. At first, when it saw us, it walked toward us calmly, without speeding up or slowing down its march, and standing erect, showing off its beautiful plastron [white chest or breastplate], walking straight.

Maybe he was going to meet some of his fellows, but when he was at around 50 m from us, the barking of one of the puppies took him out of that thought. The dog was walking in circles, barking with his tail between his legs and his hair standing up on end, without daring to get close. It was the first time that he had seen a penguin. The penguin stopped, but as soon as he saw that we were surrounding him, and that we were too big to be [part] of the same family as him, he started to show signs of vivid nervousness, and, as I tried to catch him, he lay on his belly, and the race began. It's incredible, the speed that these animals, who seem to be so heavy, are able to reach. It was an "Aptenodytes forsteri" [Emperor penguin] and a pretty small one, because it wasn't more than 80 cm high (I'm going to measure it accurately). It weighed 60 lb. After a lot of hard work, we finally transported it to the station, where, after taking photographs of it in every position possible, it was placed in a barrel. Tomorrow, we are going to kill it. With the goal of seeing from which direction it had been coming, I went to follow the trail. I walked following it for about an hour. The tracks came from the NE, between Seymour Island and Cockburn; the entire way, with the exception of some 200 m from the place where we had caught it, it had traveled on its belly. Maybe to travel great distances, they use only this method. When they travel like this, they move with the help of their flippers and their legs, so the trail has the width of the body in the middle, where the snow is completely marked. On this furrowed surface, you can see the marks of the legs and, to the sides, the marks of the flippers like this [draws the tracks], more or less.

On this day, while Sobral and company have their close encounter with an Emperor penguin, Nordenskjöld has a close encounter of his own, with the three survivors from Hope Bay, whom, quite interestingly, he also at first believes to be giant penguins. Spotting Andersson and Duse in the distance, covered in black soot from head to toe, and wearing homemade wooden glasses to protect their eyes, Nordenskjöld finally realizes that these "penguins" are actually humans, although "to what race of men these creatures belong" is a mystery to him. He even originally thinks of them as "savages, reminding one of Australian aborigines, or some other low race of human beings" (Nordenskjöld et al. 1905: 307–310). (Reading this specific passage from his writing, one would almost surmise that Nordenskjöld viewed the human race as a classification of different orders of higher and lower beings.) Speaking to Nordenskjöld in English, the two from Hope Bay finally convince him of their identity and take him and Jonassen to the third survivor,

Grunden, who is cooking in a tent nearby. The five men decide to christen (so to speak) the place of their meeting, near Cape Gordon, "Cape Well Met."

5.15.6   October 13, [1903]

	7 am		2 pm		9 pm
	−14.8		−7.1		−10.5
	−14.3		−7.95		−10.45
	+ 10.6		+ 18.0		+ 15.8
	746.1		749.5		752.8
	Max = −6.8		68.5%		74%
	82%		Min = −17.5		SW 10 m
	Calm		Calm		Neb = 0
	Neb = 1		Neb = 8		

Average wind speed = 3.2.
Maximum wind speed = 10.2.
There was 12 h of calm.

All morning, complete calm and clear sky. Despite the fact that the temperature was −10°, water ran down the hillsides. How nice it is to get warm in the sun, with a blue sky, hearing the water running merrily, surrounded by this magnificent nature!

I made a meridian observation of the sun. Close to 12, cirrus and Ci-Cu started passing by from the SW, and weak gusts began to blow from the NW. At 2, neb. = 8, and at 2:20 pm, the SW started to blow. At 7, it began to clear up, and at 9 pm, neb. = 0. According to Bodman, some spaces of water are seen to the E and between Cockburn and Cape Gordon. The poor penguin has been in the barrel all day. Its death has been postponed for 24 h so that if the photographs, which are being developed tonight, come out wrong, we will be able to take new ones tomorrow.

5.15.7   October 14, [1903]

	7 am		2 pm		9 pm
	−9.7		−1.1		−1.3
	−9.8		−2.8		−2.5
	+ 8.1		+ 16.3		+ 17.0
	751.85		748.3		742.15
	77%		61%		70.5%
	Max = −5.4		Min = −15.0		

[The remainder of this page is left blank.]

5.15.8   October 15, 1903

[This diary entry is missing.]

5.16   But Where Did These Three Come From?

5.16.1   October 16, [1903]

[In the top right corner of the page are the following notes, hastily scribbled]: 12th. Jonassen of _______ [unreadable]. Rescue. Sledge. 1 month. 9 men.

Today has been a day of surprises, and nice surprises for me. After having frukost, I was looking through the glass of the window, when I saw that the people from the sledge trip were coming. But something else drew my attention. It wasn't only the ones who had set off from here, but several more. So, to the call that I made: Hey, here comes the sledge! I added: But they are several! More than the ones who left!

As soon as everyone heard me, they ran to the beach. I was the last one to come out, and I couldn't find my hat. And since I was almost certain that the ones coming with Nordenskjöld were from the "Antarctic," and that they were coming to get us, I didn't want to leave without my hat. But when, running, I got to the beach, I found that the three strangers who had come with Nordenskjöld didn't look like "rescuers," but like "rescued." The clothing, the face, the hair, anyway, everything they had was covered with a thick layer of soot and grease. They had a figure worse than that of a coal miner who has been working in the coal for several days without washing himself or that of those who clean chimneys. They depicted the picture that Nansen draws of himself and Johansen after the winter in Franz Josef Land. In one of them [these strangers], I immediately recognized Duse. In another one, after some hard work, Grunden (one of the sailors from the "Antarctic"), and with the third one, we introduced ourselves to each other, he was Dr. G. Andersson [Johan Gunnar Andersson]. Duse, who was the first one I said hello to, answered me with some words that greatly relieved me. In Argentina, everything is fine, there has been no war, and negotiations with Chile are on a peaceful path.

But where did these three come from? Why do they have this look of coal miners with their hair all disheveled and kind of smeared with a mix of grease and soot? Questions were posed in an astonishingly quick succession, and the poor guys managed to answer them as quickly as possible. On December 29 of 1902, the "Antarctic" anchored in the NE part of Louis Filippe Land, leaving a disembarking party composed of Dr. G. Andersson, Lieutenant Duse, and the sailor Grunden, with enough provisions for a 3-week sledge trip and 1 month of provisions for 9 men. The aim of the shore party was to go from the place where they had disembarked, traveling across the ice, to "Snow Hill" station, to look for us, and then to go back to wait for the "Antarctic" in the place where they had disembarked.

At the same time that this attempt was being made by sledge over the ice, the "Antarctic" would sail along the E coast of Snow Hill with the same objective. The ice was in a very bad condition that year; a compact mass stood in the way, hindering all possible communication between the ship and us. When the shore party was placed on land, _________ [unreadable], they say, Larsen was sure that he would be able to pick them up again, because he believed that the condition of the ice would improve day by day. [The following two sentences are crossed out]: But this did not happen. The ice piled up more and the shore party _________ [unreadable] that to Louis Filippe Land, to Joinville Land, the Antarctic didn't appear.

The three who had disembarked [from the ship] set off on an excursion toward Snow Hill, up to Sydney Herbert Bay, and there they were stopped by water[1]. They went back to the departure point, and, when days and months passed without the "Antarctic" appearing, they saw that their destiny was to spend the winter there[2]. The place was quite appropriate for that purpose. In that location, penguins form one of the most populous rookeries known on that coast. There were two types: Pygoscelis adeliae [Adélie penguin] and Pygoscelis papua [Gentoo penguin]. The hut was built as follows: First, they built some type of room, a square room, made of stones. Immediately, they placed the tent inside, then they constructed a roof, and they sealed the roof and the walls with a wax that they had taken to cover the provisions. They built the kitchen at the entrance, in the space between the tent and the stone wall.

(1) They weren't stopped by clearings in the ice but by the water that was on it, caused by the melting of snow. And, according to what they say, there was so much water that there wasn't any place to set up the tent. Only in March did they think that the ship would not come, and so, ever since then, they began to feed themselves with products from the land (penguins and seals), economizing the provisions as much as possible.

(2) The trip lasted 15 days. In the middle of February, they were back in the same place where they had disembarked.

I believe that they could have easily been able to make it to the "Snow Hill" station in April, but surely they didn't do it for some reason; probably, they were afraid that they could worsen our bad situation, because they thought that we had very little provisions.

As for the conditions of the ice in the surroundings of the strait of Joinville, they were excellent, not only during the summer, but also during all of the winter. Therefore, the non-return of the "Antarctic" can be explained in several ways: First, that the conditions of the ice at a distance from where they could see were bad, not allowing the "Antarctic" to return; second, that it became trapped in the ice and that it still is after having spent the winter there, or that it got freed too late to return. All of the provisions that the "Antarctic" had were not enough for one year, with the exception of some things, like bread. But with seals and penguins, which they could certainly hunt on the ice, they would be able to perfectly complete their provisions. Nordenskjöld, who had actually found a strait that goes around Mount Haddington, was in the mouth of it, near the place called Cape Curri [Corry] on the English navigation chart, when he encountered the three of them. They (Andersson, Duse, and Grunden) had gotten underway on September 29 toward our station, and because they thought that along the coast of Cape Gordon there would be ice-free water that would not let them continue directly, they penetrated the mouth of the strait where Nordenskjöld was and went up to Cape Gordon to see what route they could take. While Andersson was heading toward Cape Gordon to explore the ice, Duse went to the south of the same [Cape Gordon], finding a strait that links Sydney Herbert Bay with the strait where Nordenskjöld was. Duse believed that the route was possible through there, and so he went back to communicate this to his colleagues. When he was about to get underway, he spotted N and J [Nordenskjöld and Jonassen] going toward the E with the purpose of reaching Paulet [Island]. Duse fired off two shots with the Mauser pistol in order to draw their attention and at the same time shouted at them as loudly as he could; when Jonassen saw them [Duse and Andersson] coming, he thought they were natives, and he told Nordenskjöld to grab the Mauser gun so he could defend himself. Not only did Jonassen think that they were natives, but also the dogs wanted to run away. And they did, being stopped only with much difficulty. Nordenskjöld, who first spoke to Duse, didn't recognize him. Duse said: How are you? Good day. Don't you recognize me? Nordenskjöld: Good day. No. Nordenskjöld couldn't recognize Andersson, either. Once the necessary explanations from both parties were given, they [Duse et al.] put their baggage on the dog sledge, abandoning the other sledge, and they continued [together] through the strait to Sydney Herbert Bay, and from there to the station. The little strait is about 2 kilometers wide, and the big one is 24 km. The day of the encounter was October 12. Taking them (the three) thirteen days to complete less than half of the journey with their heavy sledge. During one of those days, when it had rained and their feet and legs had become completely soaked, Grunden's toes on his left foot became frozen, and Duse's little toe on his right foot froze. After a lot of rubbing, they managed to reestablish the circulation, but large blisters formed, causing them to suffer a lot during the rest of the journey. According to what they say, if they hadn't found Nordenskjöld, they would have needed another 10 days to complete the remainder of the trip, which, with the dogs, took only 3 and a half days. Before reaching the station, Peridota started giving birth en route during the journey, and Amager, who was behind her, was eating the puppies as they were falling out. How sorry I was that I didn't have plates to put in my camera to take photographs of the newly arrived. Bodman did it, developing some plates that came out perfectly well. Then, they [the three newcomers] cut their scruffy hair and took a good bath, transforming themselves into civilized people in less than two hours. The questions didn't cease, and we posed them to the "new" ones eagerly. They say that a Norwegian professor has invented a canon that shoots with electricity, etc., etc.

The impression that they bring about the conditions of the ice is that, according to what they can see, the ship that comes will have free passage.

To feed themselves, they had killed 500 penguins and 21 seals. They bring the nice weather; the water runs torrentially everywhere. Temperature above 0°. Everything seems to foretell that our liberation will come soon.

$$
7\,\text{am}
\begin{bmatrix}
-3.65 \\
-4.25 \\
+10.8 \\
746.75 \\
89.5\% \\
\text{Max} = +5.4 \\
\text{Calm} \\
\text{Neb} = 10
\end{bmatrix}
\quad
2\,\text{pm}
\begin{bmatrix}
- = +3.7 \\
- = 0.0 \\
\text{Hy} = 52.2 \\
\text{min} = -6.5 \\
\text{E}\,15\,\text{m} \\
\text{neb} = 9
\end{bmatrix}
\quad
9\,\text{pm}
\begin{bmatrix}
-3.5 \\
-4.8 \\
+14.2 \\
743.0 \\
77.2\% \\
\text{calm} \\
\text{Neb} = 10
\end{bmatrix}
$$

Average wind speed = 1.5.
Max. wind speed = 4 m.
Min. wind speed = ____ [blank]. There was an hour of calm.

It is remarkable to read the retelling of this story written in real time, as it was told to, and documented by, Sobral immediately after this extraordinary encounter took place. His notes and three-page documentation in his diary faithfully recount the encounter of the Hope Bay survivors with Nordenskjöld and the reunion of the returning parties with the Snow Hill expeditioners. His rendition of the meeting matches that of Nordenskjöld's as recounted in his book (Nordenskjöld et al. 1905), with the addition of Sobral's own humorous observations.

The nearly Chaplinesque scene of Sobral spotting the coming party, and then running back for his hat as the others race ahead of him, is a memorable one and again gives insight to Sobral's fastidious character—he did not want to get rescued without his hat. This detail is repeated by Nordenskjöld, who later writes that, after the dogs began barking and running toward the newcomers, Sobral was the first to "catch sight" of the party, but Bodman was the first to run down to greet them (Nordenskjöld et al. 1905).

Sobral's anxiety about his country being at war is finally calmed by Duse, who, in response to his query, delivers the news that all is well. Duse was the expedition member who had chided Ekelöf in Port Stanley, when the doctor had rudely told Sobral that he preferred the Chilean people to the Argentines. It is poetic closure that he now delivers the news of Argentina's state of peace to the anxious young under-lieutenant, whose first thought is with his country. In Nordenskjöld's book, however, Nordenskjöld states that it was he who put Sobral at ease, painting a tableau of the excited reunited men explaining their respective situations to each other, "whilst I hasten to inform Sobral of the important news that peace is established between the Argentine Republic and Chili" (Nordenskjöld et al. 1905: 314). Sobral's sincere concern was known by his comrades, hence the urgent attempt to allay his fears. Also, Nordenskjöld's gratitude toward Argentina is evident in his mentioning this point in his book.

It must be noted that this entry is the only one recently in which Sobral does not give the weather data first. The historic meeting and reunion with the Hope Bay expedition members takes priority.

Sobral compares this meeting to another historic meeting—that of the Norwegian explorers Fridtjof Nansen and Hjalmar Johansen with British explorer Frederick George Jackson in June of 1896 at Franz Josef Land, which Sobral refers to in his diary as "Tierra de Francisco José." Nansen had successfully drifted across the Arctic Ocean in the Fram *in 1893 and crossed the Arctic ice on foot in 1895, reaching the farthest north of that time—86° 3' N (*Markham *1896). After several weeks in pursuit of the North Pole, he and Johansen double backed, spending another 14 months crossing ice and water, and living on seals, walruses, polar bears, and sled dogs. By the time Nansen and Johansen encountered Jackson, they looked as wild and unkempt as the three expeditioners from Hope Bay. Johansen would later accompany Roald Amundsen to the Antarctic on the South Pole expedition. And, interestingly, although he would never go to the Antarctic, Nansen would later conjure up the miraculous events surrounding the Swedish-Argentine Expedition's meeting during his own lectures about Antarctic exploration.*

*Sobral reports that the Hope Bay party killed 500 penguins in order to survive; this number does not exactly match that given by J.G. Andersson, who would later write that 700 penguins were killed (*Nordenskjöld et al. *1905).*

*The final irony here is the unfortunate fate of the Emperor penguin at Snow Hill. According to Nordenskjöld, that "good fellow," as Sobral had called him, was roasted and served for dinner on this very night (*Nordenskjöld et al. *1905). Originally intended as the main course in a feast marking the two-year anniversary of the expedition's leaving Sweden, the Emperor was now eaten in celebration of the reunification of the two parties from Hope Bay and Snow Hill. The Aptenodytes Forsteri that Sobral had so carefully studied and measured and photographed, that he had chased and caught on his belly, was now residing in the bellies of the men who had studied him.*

5.16.2   *October 17, [1903]*

$$
7\,\text{am}
\begin{bmatrix}
+0.6 \\
-1.45 \\
+19.5 \\
740.0 \\
\text{Max} = +5.3 \\
61\% \\
\text{NNE}\,0.5\,\text{m} \\
\text{Neb} = 0
\end{bmatrix}
\quad
2\,\text{pm}
\begin{bmatrix}
+1.7 \\
+0.6 \\
+13.4 \\
734.05 \\
83\% \\
\text{Min} = -5.1 \\
\text{NE}\,15\,\text{m} \\
\text{Neb} = 10
\end{bmatrix}
\quad
9\,\text{pm}
\begin{bmatrix}
+2.0 \\
+0.6 \\
+16.9 \\
730.0 \\
80.5\% \\
\text{NNE}\,14\,\text{m} \\
\text{neb} = 9
\end{bmatrix}
$$

Average wind speed = 7.5 m.
Max. wind speed = 15.5 m
Min. wind speed = ____ [blank]. There was 2 h of calm.

Strong wind from the NNE blowing all the afternoon.

5.16.3   October 18, [1903]

$$
7\,\text{am}
\begin{bmatrix}
+1.75 \\
-1.25 \\
+14.9 \\
734.3 \\
53\,1 \\
\text{Max} = +5.8 \\
\text{Var} \\
\text{Neb} = 1
\end{bmatrix}
\quad
2\,\text{pm}
\begin{bmatrix}
+6.25 \\
+2.0 \\
+17.0 \\
736.5 \\
40.5\% \\
\text{Min} = +0.3 \\
\text{NNW} \\
\text{Neb} = 3
\end{bmatrix}
\quad
9\,\text{pm}
\begin{bmatrix}
+3.4 \\
+0.45 \\
+15.8 \\
736.1 \\
45\% \\
\text{var} \\
\text{Neb} = 6
\end{bmatrix}
$$

Average wind speed = 4.0.
Max. wind speed = 8.5 m.
Min. wind speed = 0.5 m.
Variable winds almost all day.

5.16.4   October 19, [1903]

$$
7\,\text{am}
\begin{bmatrix}
+1.0 \\
-0.2 \\
+14.1 \\
733.5 \\
84\% \\
\text{Max} = +7.1 \\
\text{NNE}\,6\,\text{m} \\
\text{Neb} = 9
\end{bmatrix}
\quad
2\,\text{pm}
\begin{bmatrix}
-0.2 \\
-2.75 \\
+15.2 \\
737.3 \\
59\% \\
\text{Min} = -1.4 \\
\text{WSW}\,6\,\text{m} \\
\text{Neb} = 9
\end{bmatrix}
\quad
9\,\text{pm}
\begin{bmatrix}
-4.6 \\
-6.05 \\
+18.0 \\
741.2 \\
62\% \\
\text{SSE[crossed out]calm} \\
\text{Neb} = 3
\end{bmatrix}
$$

Average wind speed = 5.0.
Max. wind speed = 9 m.
Until 8 am, winds from the N, NNE, and NE; since that time, SSW, SW, and WSW.

5.16.5   October 20, [1903]

$$
7\,\text{am}
\begin{bmatrix}
-5.75 \\
-7.15 \\
+13.8 \\
747.1 \\
67.8\% \\
\text{Max} = +2.7 \\
\text{S}\,6.5\,\text{m} \\
\text{Neb} = 0
\end{bmatrix}
\qquad
2\,\text{pm}
\begin{bmatrix}
+0.1 \\
-2.3 \\
+15.6 \\
750.45 \\
60\% \\
\text{calm} \\
\text{Neb} = 0
\end{bmatrix}
\qquad
9\,\text{pm}
\begin{bmatrix}
-2.7 \\
-3.9 \\
+18.2 \\
752.2 \\
\text{calm} \\
\text{Neb} = 4
\end{bmatrix}
$$

Average wind speed = 2.5.
Max. wind speed = 7 m.
There was 8 h of calm (from 2 pm to 12 at night).
In the morning, SSW and SW blowing. Clear sky all day.

5.16.6   October 21, [1903]

$$
7\,\text{am}
\begin{bmatrix}
-3.8 \\
-5.2 \\
+13.3 \\
750.3 \\
\text{Max} = +0.7 \\
76\% \\
\text{C} \\
\text{Neb} = 10
\end{bmatrix}
\qquad
2\,\text{pm}
\begin{bmatrix}
+5.5 \\
+1.4 \\
+16.7 \\
745.75 \\
\text{Min} = -6.8 \\
46.5\% \\
\text{C} \\
\text{Neb} = 10
\end{bmatrix}
\qquad
9\,\text{pm}
\begin{bmatrix}
+0.5 \\
-1.55 \\
+18.0 \\
741.35 \\
63.8\% \\
\text{ENE}\,2\,\text{m} \\
\text{Neb} = 5
\end{bmatrix}
$$

Average wind speed = 3.0 m.
Max. wind speed = 9 m.
There were some moments of calm.
Overcast sky almost all day; predominating winds from the ENE, NE, and NNE.

With the arrival of the newcomers from Hope Bay, the daily routine at Snow Hill is rejuvenated, and Sobral is preoccupied with new activities. Never does he abandon, however, his scientific measurements and his dutiful recording of them.

5.16.7 October 22, [1903]

$$
7\,\text{am}
\begin{bmatrix}
0.0 \\
-2.0 \\
+18.5(?) \\
739.1 \\
62.6\% \\
\text{Max} = +5.5 \\
\text{Calm} \\
\text{Neb} = 7
\end{bmatrix}
\quad
2\,\text{pm}
\begin{bmatrix}
+1.95 \\
+0.5 \\
+21.5 \\
740.6 \\
64.4\% \\
\text{Min} = -6.3 \\
\text{SW}\,6\,\text{m} \\
\text{Neb} = 10
\end{bmatrix}
\quad
9\,\text{pm}
\begin{bmatrix}
-1.0 \\
-2.9 \\
+16.4 \\
740.7 \\
61.8\% \\
\text{C} \\
\text{Neb} = 1
\end{bmatrix}
$$

Average wind speed = 3 m.
Max. wind speed = 9 m.
There was 2 h of calm.
Sky almost overcast until 3 pm.

From 5 to 10, clear sky, but it became overcast again at 11 pm. From 10 am to 6 pm, a hot SW blew.

5.16.8 October 23, [1903]

$$
7\,\text{am}
\begin{bmatrix}
-0.7 \\
-1.5 \\
+10.2 \\
726.2 \\
\text{Max} = +3.6 \\
88\% \\
\text{N}\,9\,\text{m} \\
\text{Neb} = 10
\end{bmatrix}
\quad
2\,\text{pm}
\begin{bmatrix}
+4.55 \\
- \\
+16.0 \\
712.50 \\
61.2\% \\
\text{Min} = -3.6 \\
\text{Var}\,2\,\text{m} \\
\text{Neb} = 9
\end{bmatrix}
\quad
9\,\text{pm}
\begin{bmatrix}
-2.45 \\
-2.8 \\
+21.0 \\
716.9 \\
97.5 \\
\text{WSW}\,15\,\text{m}
\end{bmatrix}
$$

Average wind speed = 7.5.
Max. wind speed = 15 m.
Min. wind speed = 1.5 m.

The NE, which had begun to blow yesterday night at 11 pm, started to increase in strength, little by little, reaching its maximum of 10.5 m at 10 am and decreasing right away. The barometer is dropping swiftly: From 7 am to 2 pm, it decreased around 12 mm, sometimes 2 mm per hour. It began to rise at midday. At 6 pm, the wind from the SW started, and then, it kept blowing from the WSW, SW, and SSW until midnight. The temperature, which was around +5° at 6 pm, is dropping. The sun shone between 2 and 4 pm.

5.16.9   October 24, [1903]

$$
7\,\text{am}
\begin{bmatrix}
-6.5 \\
-7.2 \\
+12.8 \\
714.5 \\
\text{Max} = +6.1 \\
77\%
\end{bmatrix}
\qquad
2\,\text{pm}
\begin{bmatrix}
-7.5 \\
-8.3 \\
+19.3 \\
719.65 \\
\text{Min} = -11.25 \\
75.5\%
\end{bmatrix}
\qquad
9\,\text{pm}
\begin{bmatrix}
-9.1 \\
-9.8 \\
+21.3 \\
721.85 \\
70\%
\end{bmatrix}
$$

Average wind speed = [blank].
Max. wind speed = [blank].
Min. wind speed = [blank].
All day, wind blew from the SW.
[Crossed out]: Two days ago

On the 21st of this month, Dr. Andersson, Bodman, and Jonassen went on an excursion to Cockburn Island, returning yesterday. The excursion was pretty happy, because Andersson found some very interesting fossils that show that Cockburn is of a more modern formation than Snow Hill. Bodman went up to the top of Cockburn, making hypsometric observations which indicated that the island has a height of 450 meters, approximately. The fossils that Andersson found are marine [species]. On the island, there were approximately 700 penguins and more than 1,000 cormorants. They killed a lot of them, bringing them to the station. According to what he [Bodman] saw from the top of Cockburn, the conditions of the ice can't be better. The Gulf of Erebus and Terror is free of ice, almost completely; toward the E, you can't see any ice.

With these hopeful words of not seeing any ice, and having the possibility of an ice-free Gulf of Erebus and Terror that could open the way to a rescue, so ends the chronological entries portion of José María Sobral's diary.

Figures 5.1, 5.2, 5.3, 5.4, 5.5, 5.6, 5.7, 5.8, 5.9, 5.10, 5.11, 5.12, 5.13, 5.14, 5.15, and 5.16.

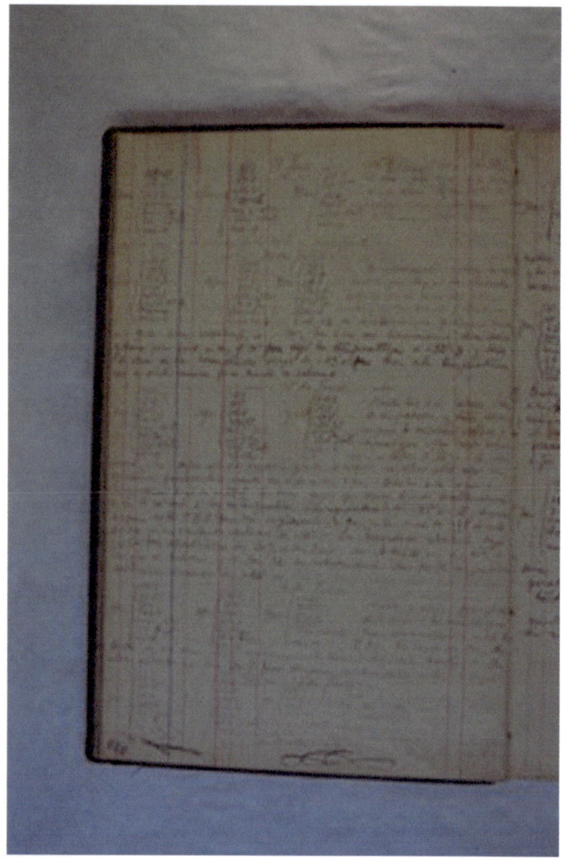

Fig. 5.1 A page from José María Sobral's diary during the expedition's second winter, 1903, chronicling Sobral's Antarctic adventures and scientific work. Sobral turned his voyage diary upside down to document the second overwintering. (Departamento de Estudios Históricos Navales, Archivo Histórico, El Diario de Alférez de Navío José María Sobral.)

Fig. 5.2 José María Sobral at work "shooting the sun" during the freezing temperatures and wild weather of the Antarctic winter. Photographed by his meteorological mate Gösta Bodman. (Photo from *Antarctica: or Two Years Amongst the Ice of the South Pole*, by Otto Nordenskjöld et al., 1905, page 293.)

Fig. 5.3 Gösta Bodman and one of the sledge dogs standing outside of the house on Snow Hill. Photographic image by José María Sobral. (Departamento de Estudios Históricos Navales, Archivo Histórico, Archivo Fotográfico, Ex-0039 a, Expedición Sueca—Dibujos de Sobral.)

Fig. 5.4 Sledge dog with Emperor Penguin. Photographic image by José María Sobral. (Departamento de Estudios Históricos Navales, Archivo Histórico, Archivo Fotográfico, Ex-0040 r, Expedición Sueca—Dibujos de Sobral.)

Fig. 5.5 Skis and Canadian snow shoes, used and recommended by Sobral. Photographic image by José María Sobral. (Departamento de Estudios Históricos Navales, Archivo Histórico, Archivo Fotográfico, Ex-0039 q, Expedición Sueca—Dibujos de Sobral.)

Fig. 5.6 An Antarctic Penguin (Pygoscelis), one of the thousands encountered by the expedition, many of whom gave their lives for the humans. Photographic image by José María Sobral. (Departamento de Estudios Históricos Navales, Archivo Histórico, Archivo Fotográfico, Ex-0039 d, Expedición Sueca—Dibujos de Sobral.)

Fig. 5.7 Harnessing the sled dogs for a sledging journey. Photographic image by José María Sobral. (Departamento de Estudios Históricos Navales, Archivo Histórico, Archivo Fotográfico, Ex-0040 n, Expedición Sueca—Dibujos de Sobral.)

Fig. 5.8   A Weddell Seal. Seals provided sustenance for the expedition members during the two winters. Photographic image by José María Sobral. (Departamento de Estudios Históricos Navales, Archivo Histórico, Archivo Fotográfico, Ex-0040 f, Expedición Sueca—Dibujos de Sobral.)

Fig. 5.9 Snow Hill Glacier on the left, Mount Haddington on the right, and Lockyer Island in the front, with sled dogs lying in repose in the forefront. Photographic image by José María Sobral. (Departamento de Estudios Históricos Navales, Archivo Histórico, Archivo Fotográfico, Ex-0040 ñ, Expedición Sueca—Dibujos de Sobral.)

Fig. 5.10 Penguins protecting their eggs and defending their nests. (Photo from *Dos Años Entre los Hielos, 1901–1903*, by José María Sobral,1904, page 218.)

Fig. 5.11 The three newcomers from Hope Bay (Bahía Esperanza) at Snow Hill, October 1903, still covered in the grease and soot accumulated from living in a hut for nine months. José María Sobral wrote in his diary that he wished he had had plates in his camera so he could also capture their appearance photographically. *Left* to *right* Toralf Grunden, Johan Gunnar Andersson, and Samuel A. Duse. Photographed by Gösta Bodman. (Photo from *Antarctica: or Two Years Amongst the Ice of the South Pole*, by Otto Nordenskjöld et al., 1905, page 315.)

Fig. 5.12 Hope Bay (Bahía Esperanza), where Johan Gunnar Andersson, Samuel A. Duse, and Toralf Grunden overwintered in 1902–1903—also called Louis Philippe Land by Sobral, as seen in the caption he wrote on this original hand-tinted photo print. Photographic image by José María Sobral. (Departamento de Estudios Históricos Navales, Archivo Histórico, Archivo Fotográfico, Ex-0039 p, Expedición Sueca—Dibujos de Sobral.)

Fig. 5.13 The way to Hope Bay (Bahía Esperanza). This is the view that José María Sobral and the expedition most likely saw on their voyage along the Antarctic Peninsula. (Photograph by Mary R. Tahan.)

Fig. 5.14 The nunatak and penguin colony at Hope Bay (Bahía Esperanza), where Johan Gunnar Andersson, Samuel A. Duse, and Toralf Grunden overwintered and sustained themselves on penguins and seals. (Photograph by Mary R. Tahan.)

Fig. 5.15 A re-constructed representation of the hut at Hope Bay (Bahía Esperanza) in which Johan Gunnar Andersson, Samuel A. Duse, and Toralf Grunden resided over the winter. This structure is preserved by the personnel of Argentina's Base Esperanza (seen in the background). (Photograph by Mary R. Tahan.)

Fig. 5.16 The hut that Carl Anton Larsen, Carl Skottsberg, and company built on Paulet Island, in which the crew and scientists took refuge from February to November of 1903. (Photo from *Dos Años Entre los Hielos, 1901–1903*, by José María Sobral,1904, page 319.)

References

Markham AH (1896) Arctic exploration. In: Report of the sixth international geographical congress, Held in London, 1895, eds. Secretaries of the Royal Geographical Society,. London, John Murray, 177–201

Nordenskjöld N Otto G, Andersson JG, Skottsberg C, Larsen CA (1905) Antarctica: or two years amongst the ice of the South Pole. The Macmillan Co, New York

Skottsberg C. (1961) Personal letter from Carl Skottsberg to Gustavo Giró, dated 8 August 1961. Private collection of María Edelia Giró and Edelia "Puchi" Giró, viewed by the author in February 2017, in Ushuaia, Tierra del Fuego, Argentina

Sobral JM (1901–1903) Expedition diary of José María Sobral, Swedish Antarctic Expedition of 1901–1903. Departamento de Estudios Históricos Navales, Archivo Histórico. El Diario de Alférez de Navío José María Sobral. Buenos Aires: Archivo D.E.H.N.—A.R.A

Sobral JM (1903a) Conferencia del alférez de navío José M. Sobral [Conference by under-lieutenant José M. Sobral, sponsored by the Naval Center, and read in the Politeama Argentino on December 19, 1903]. In: Boletin del Centro Naval, Tomo XXI, ed. Capitán de fragata Juan I. Peffabet, 485–512. Buenos Aires: Taller Tipográfico de la Escuela Naval Militar

Sobral JM (1903b) Handwritten notes for speeches and for official report to the Minister of the Navy. Departamento de Estudios Históricos Navales, Archivo Histórico, Colección Sobral. Box 2, Envelope 4, Manuscrito referido al viaje a la Antartida y Conferencia en el Centro Naval 1903. Buenos Aires: Archivo D.E.H.N.—A.R.A

Sobral JM (1904) Dos años entre los hielos, 1901–1903 (Two years amidst the ice, 1901–1903). Buenos Aires: Imprenta de J. Tragant y Cia., Bolivar 319

Tahan MR (2016a) Paw prints in human Antarctic history. Scientific Committee on Antarctic Research (SCAR). The 34th SCAR Biennial Meetings and 2016 Open Science Conference in Kuala Lumpur, Malaysia. Abstract and poster. Abstract Book, 20–30 August 2016, page 1017. ISBN 978-0-948277-32-0. http://www.scar.org/scar_media/documents/outreach/communications/SCAR_OSC_2016_Abstracts.pdf

Tahan MR (2016b) Paw prints in human Antarctic history: the use of sled dogs in Antarctic exploration. Latin American Antarctic Historians. The XVI Encuentro de Historiadores Antarticos Latinoamericanos in Buenos Aires, Argentina. 28—29 October 2016. Paper and presentation. http://dna.gob.ar/xvi-encuentro-de-historiadores-antarticos-latinoamericanos

Chapter 6
The Amazing Rescue

Abstract The rescue of all three parties of the Swedish-Argentine Antarctic Expedition by the Argentine Corvette *Uruguay* occurred in November 1903, and the return of the expedition to Buenos Aires took place in December 1903. From October 25 through December 2, 1903, the summary of events includes Sobral's trip to Seymour Island with Andersson, wherein Sobral adds his and Andersson's names to a sign left by Captain Larsen in 1892 during his first trip there on the *Jason*; the reading of this sign by Captain Julián Irízar and Lieutenant Jorge Yalour of the Corvette *Uruguay* on November 7, and the subsequent discovery of the two reunited parties at Snow Hill by Captain Julián Irízar on November 8; the arrival of Captain Carl Larsen and part of his third party from Paulet Island at Snow Hill on the evening of that same date; the rescue of the remaining crew at Paulet Island; the retrieval of collected fossils and specimens left by Nordenskjöld and Andersson at Seymour Island and Hope Bay; the voyage back to Argentina; and the public reception of the expedition and rescuers at the Buenos Aires port on December 2, 1903.

Some true facts are stranger than the wildest fiction.

After Sobral had written his final diary entry, a sequence of remarkable events took place—a chain of events that, if even one link had been missing, would not have led to the historic rescue at which we can now marvel.

The arrival of the newcomers from Hope Bay had breathed new hope—and new life—into the six expeditioners stranded at Snow Hill, including Sobral, who renewed his vigorous study of all things Antarctica. Although they feared they would never again see their ship the *Antarctic*, they did hold hope that another ship would come for them. The group had redoubled their efforts to collect fossils, conduct geographical and geological studies, and chart the territory which was opening up around them, and they had continued their sledging trips, now with the addition of Andersson, Duse, and Grunden.

"No ice." These basically had been Sobral's last words in his diary on October 24, 1903. It was the day that Sobral had reported on Andersson, Bodman, and Jonassen's sledge trip to Cockburn Island during October 21st to the 23rd, to collect fossils and to reconnoiter the terrain and the ice.

© Springer International Publishing AG 2018

M.R. Tahan, *The Life of José María Sobral*, Springer Biographies,

https://doi.org/10.1007/978-3-319-67268-7_6

Two days later, on October 26th, Sobral himself departed on a sledge trip, along with Nordenskjöld, Andersson, and six dogs, to Seymour Island. The journey was planned as a two-week excursion to continue his magnetic observations, carry out geological studies, and add to their now very rich collection of fossils. This excursion was one of several performed during this time, while the weather was changing for the better, and as the scientists continued their search for geological and paleontological samples in the midst of their uncertain situation. Their destinations for their continuous studies were Seymour, Lockyer, and Cockburn islands.

On this particular journey, while seeking fossils in the northeastern part of Seymour Island, where vertebrate fossils had been found previously by Nordenskjöld during his excursions there, Sobral and Andersson came upon a sign post that had been left by the *Antarctic*'s Captain Larsen when he had first landed at Seymour Island 11 years prior, in 1892, during his whaling expedition in the *Jason*. The coincidence of finding this sign, with Larsen's name and the date of his visit etched into it, is one worth mentioning. What happened next, however, was a keystone to the series of events that followed. As Sobral wrote in his book to describe the event (Sobral 1904: 299):

> … walking one day with Andersson, we found a stick that Larsen had planted 11 years before, that said "Jason 1892"; I wrote a little below, "Andersson, Sobral, October 1903", without imagining that that insignificant sign was going to give us so much to think about 3 or 4 days later.

His spontaneous decision to etch his name and Andersson's name, along with the date of their visit, into the sign, would result in the rescue for which the men were so eagerly yearning.

Sobral took full delight in this sledge journey. Being out in the field much more suited him than being back at the house on Snow Hill. He seemed to come even more alive during this expedition and participated in his colleagues' search for fossils and geological samples as well as in his own magnetic measurements. On October 31st, there was a storm, but as of November 1st, the weather cleared, and the sea looked to be completely free of ice in the direction of the northern quadrants. "What a joy to see the water again! Watch the waves!" wrote Sobral later in his book. He took pleasure in "the cries of the seals and penguins that pass in groups on the white icebergs, moved by the impulses of the current of tide!" (Sobral 1904: 301). When Nordenskjöld left camp on November 2nd to briefly return to the station at Snow Hill, Sobral and Andersson stayed in the field, continuing their excursions and living together comfortably in the tent. They worked and cooked and, most importantly, ate—a favorite pastime of Sobral, foodie that he was. But an accident on the 5th brought an end to this happy life. On that day, it was Sobral's turn to cook (Sobral 1904: 302):

> … we had killed a young seal and collected the blood, with which we thought to make one of the most exquisite *Antarctican* sweets. With the blood was beaten a little flour, and this would be fried in seal fat. I had set to melt the fat, and amused myself in undoing the flour pellets that had formed, when I do not remember what it was that Andersson wanted to do with the pan, which, placed in the fire, contained the boiling fat. Whatever it was, the fact is

that when I looked back, it was to see that the pan had spilled its contents on his hands. Our joy was interrupted just as the feast was in its beginnings. Immediately I wrapped his hand with cotton that I happened to have, and we went back to the house in search of remedies. Goodbye seal blood snacks! Goodbye charms of camp life! And precisely the day when the penguin eggs had begun to arrive!

Sobral and Andersson made a mad dash for the house at Snow Hill and encountered Nordenskjöld along the way, who had been on his way back to camp. Together, the three returned to the station to tend to the injury. Unfortunately, Dr. Ekelöf had just left on a sledging excursion with Duse and Jonassen to Lockyer Island, to do cartographical and bacteriological work. But fortunately, Andersson's burn, though covering a large area, was not serious, and Nordenskjöld was able to dress it. He and Bodman "heard such an enticing account of the good time Andersson and Sobral had been having for the last few days" that Bodman decided to go with Åkerlundh to the island the next day to gather the penguin eggs (Nordenskjöld et al. 1905: 500). And so the two went to Seymour on November 7th. To Sobral, the ultimate injury was that Bodman and Åkerlundh got to enjoy the sweet feast that he and Andersson had begun (Sobral 1904).

6.1   On Paulet Island

While the nine expeditioners at Snow Hill continued their scientific work and their culinary efforts, the now 19 members (after the death of Ole Christian Wennersgaard) on Paulet Island had devised a strategy to be rescued. On October 31st, the same date that Sobral, Andersson, and Nordenskjöld were on their field trip at Seymour Island, and around the same day that Sobral was etching his name under Larsen's name on the sign that Larsen had left 11 years ago, Larsen himself set off on a boat from Paulet toward Snow Hill. It was one of the boats they had salvaged from the *Antarctic* and had pulled over the ice to Paulet when taking refuge there.

Larsen took with him five companions: Reinholdz, K.A. Andersson, G. Karlsen, Olsen Ula, and A. Andersson. They crossed rough waters, and dodged ice floes and icebergs, in a dicey journey that would take them eight days to reach the wintering station. Stopping along the way at a penguin colony on one of the islands, on November 3rd, Larsen and company collected penguin eggs for food for themselves and for those at the wintering station and killed a seal for their own sustenance. Pursuing this basic survival did not, however, deter them from continuing to look for fossils, and to collect lichens and moss for Skottsberg. On November 4th, Larsen reached Hope Bay and the stone hut which had housed Andersson and his companions. Here Larsen borrowed the tarpaulin and a pole from the roof to use for his own boat. The party of six sat out a storm here in Andersson's old haunt and then set forth again on November 7th toward Sidney Herbert Bay (Nordenskjöld et al. 1905).

6.2   Meanwhile, Back in Buenos Aires

Dr. Francisco P. Moreno, the scientist who had been instrumental in organizing Sobral's participation in the expedition, had grown greatly concerned when Austral summer had come and gone with no sign of the *Antarctic* reappearing (Rabassa 2003). By Austral autumn, May 1903, with still no word from the Swedish expedition, he appealed to President Julio Roca to take action, to find the missing expedition that now so famously included a Navy lieutenant from Argentina. With the support of the Navy Minister Onofre Betbeder, a rescue plan was devised wherein, initially, the Argentines were to join forces with the Swedes to retrieve the stranded expedition. The Argentine corvette *Uruguay* was selected as the relief ship, and departed from Buenos Aires on October 8, 1903. But the whaler *Frithjof*, which was anticipated to rendezvous with the *Uruguay* in Ushuaia, did not arrive from Sweden in time. And, indeed, none of the other international relief expeditions that had been initiated were able to make the journey in a timely manner. And so, Captain Julián Irízar and the *Uruguay* set about rescuing the men on their own (Nordenskjöld et al. 1905). On November 1, 1903, the date scheduled for the rendezvous that never happened, the *Uruguay* set course for the Antarctic. The heavy metal canon ship had been retrofitted specifically to encounter the ice. It had been a relatively quick retrofitting, but it had been done with as much resource as possible, and it would have to do.

6.3   A Happy Confluence of Arrivals

Let us set the stage, then.

It is November 7, 1903.

Sobral, Nordenskjöld, Andersson, and Grunden are at the wintering station at Snow Hill, Sobral and Andersson having just returned from an enjoyable field trip to Seymour Island.

Duse, Jonassen, and Ekelöf are just arriving at the house on Snow Hill from their sledging expedition to Lockyer Island.

Bodman and Åkerlundh are on Seymour Island collecting Sobral's coveted penguin eggs.

Larsen and his companions from Paulet are in a row boat making their way to Snow Hill.

And Commander Julián Irízar and the *Uruguay* have just anchored off of Seymour Island.

Let the reunions begin.

Captain Irízar and Lieutenant Jorge Yalour disembark from the *Uruguay* and search the northeastern part of Seymour Island. They find a sign post with the names "Sobral" and "Andersson" etched into the sign, along with the date "October 1903." It is Sobral's sign that he had spontaneously carved, which leads them to determine that they have found the Swedish expedition (Sobral 1904; Rabassa 2003). Continuing, the two Argentine naval officers come upon the tent where Bodman and Åkerlundh are camping during their quest to collect the penguin eggs that Sobral had so enticingly described. The two expeditioners are in the tent. The Argentine men speak to them. The expeditioners cannot believe their eyes.

It is now November 8th. We are at Snow Hill. It is during the morning.

Four dark shapes walk across the ice toward the house. Are they penguins? As there are only two men currently not in the house, the appearance of four figures is mystifying. Are they crewmembers from the *Antarctic*? Most of the occupants run out of the house to look. Sobral, however, does not believe that they are men, so he remains in the house. This must be another refraction, another optical illusion. But as he sees more excitement, and an increased pace among his colleagues, he realizes that these indeed must be men arriving. And he shoots out of the house, trying unsuccessfully to put on his skis in mid stride, then tossing them aside as he runs on the ice in his boots (Sobral 1904).

Åkerlundh's familiar features take shape. He is in the lead, and he is the first to speak to them, giving voice to the immortal words that answer their speechless questioning. Nordenskjöld quotes him as saying, "There's an Argentine ship out there, but they have heard nothing of the *Antarctic*!" (Nordenskjöld et al. 1905: 503).

Sobral, still behind everyone else, only hears the first part of the news, and he is overjoyed to learn of the Argentine ship's arrival, but tempers his response, cognizant of both the question of the *Antarctic*'s fate and the feelings of his non-Argentine companions. He would later write (Sobral 1904: 306–307):

> What I can say is that, at that time, I felt proud of my country, I felt proud to be a compatriot of those who had come here with the *Uruguay,* and if from my lips there did not emerge the most thunderous "hurray" ever heard by the ice or by the men, it was because I understood that, what for me was a reason for immense selfishness, for others naturally implied much less...

Although he was not certain that his colleagues felt this, Sobral was sensitive to the fact that this was an Argentine rescue of a Swedish expedition, and, putting himself in their place, he stated that, theoretically, "I would logically suffer a disappointment in receiving help from foreigners, expecting it from those of my country" (Sobral 1904: 307).

When he heard that there was no news of the *Antarctic*, Sobral doubled-down on his calm composure (Sobral 1904: 308):

> I at that time felt as happy as a man can feel, all my aspirations were filled, because the success of the *Uruguay* was a triumph for my country. In life, the moments of complete satisfaction are short, too short; the news that, since November 1902, nothing had been heard of the *Antarctic*, cooled our joys. Those who were on that ship were our friends, and precisely because we were saved, thinking about their probable loss bittered our soul.

Faced with the fact that the *Antarctic* and its people could be lost forever, and the decision of whether or not to attempt to wait for them, Nordenskjöld somewhat froze into momentary inaction, but later reported that it was Sobral who made a stand (Nordenskjöld et al. 1905: 504):

Sobral – the one who now had most reason to feel himself at home – was the first to say that all our efforts must at once be directed to the discovery of the fate of the *Antarctic*. The decision, however, had to be left to the commander of the relieving expedition.

That commander, Irízar, was known by Sobral from his days on the *Sarmiento*, when the two had traveled together during the training ship's first voyage, as well as from Sobral's year of service on the cruiser *Patria*, during which time he served under Irízar's orders (Sobral 1904). Lieutenant Yalour, also, was a friend of Sobral through their time together at the Naval School. He felt doubly grateful that he had been rescued by "compatriots and friends" (Sobral 1904: 308). And he appealed to the group of rescuers and rescued to immediately search for survivors of the *Antarctic*.

Irízar's decision was obvious. He made it very clear that he wished to leave right away, and urged Nordenskjöld to prepare to depart on the *Uruguay*. The expedition members, however, were torn between leaving immediately and waiting to search for their companions from the *Antarctic*. They commenced to pack, however, gathering, first and foremost, their precious collections of scientific samples, and then their instruments and personal belongings (Nordenskjöld et al. 1905).

It was now night time. While most of the men at Snow Hill packed, Duse and Grunden accompanied the two Argentine officers back to their ship, taking them via a sledge to which Jonassen had harnessed the dogs. It was Duse's intention to squeeze in one more cartographical study before leaving (Nordenskjöld et al. 1905).

For his part, Sobral was both proud of his Argentine compatriots for their tremendous achievement in reaching the stranded expedition, and reluctant to leave without his *Antarctic* compatriots. He was immensely touched by his country's efforts, and he was also very loyal to his expedition mates. And he was overjoyed to receive a letter from his parents, brought to him by the crew of the *Uruguay*. This little note from home brought him a feeling of warmth and a strong desire to return (Sobral 1904). But his *Antarctic* friends were out there somewhere. The decision to leave was not an easy one.

Fortunately for Sobral, and for all, Captain Larsen made the decision moot. His surprise appearance on that Antarctic summer night, along with five of his Paulet Island companions, lent a sense of surrealism to the setting. Speechless, the men watched as the six *Antarctic* crewmembers approached them from the shore. They had rowed and sailed their way to Snow Hill, and had landed at the ice seven hours previously, marching 15 miles over the surface toward the station, and arriving just in time to meet their expedition mates before they boarded a rescue ship. Larsen was thrilled to hear of the Argentine ship. And as Duse and Grunden returned from having escorted the Argentine officers to their ship, the reunion was completed. The men at Snow Hill truly could not believe that all that they had wished for had come true (Nordenskjöld et al. 1905).

Sobral took this welcomed "new surprise" in calm and philosophical stride (Sobral 1904: 309):

> If, ten minutes later, they would have told me that I had to stay for another few years in Snow Hill, not one of my nerves would have contracted, and I am sure that my heart would not have accelerated its number of heartbeats.

> All organized beings adapt to the environment in which they have to live, disappear or die;
> even the most delicate constitution would have to conform to strong emotions or cease to exist.

It was a bittersweet reunion, as the inhabitants of Snow Hill learned of the death of the sailor Wennersgaard, and the loss of the *Antarctic*. But it was overall a joyous one, in that all parties had been reunited in this astounding manner. Larsen brought with him letters from 1902, that the expeditioners at Snow Hill hungrily read (Sobral 1904). The next morning, Larsen visited the Argentine ship and was welcomed enthusiastically. He would later write (Nordenskjöld et al. 1905: 377):

> … I met with the heartiest reception imaginable on the part of Captain Irízar and all his
> officers. No introduction was needed, Captain Irízar at once guessing that I must be Larsen
> and embracing me. It was really affecting to be thus greeted by the representatives of a
> foreign nation.

It must be noted here that, according to Nordenskjöld, prior to this meeting, Irízar had sent word to Nordenskjöld via his second-in-command, Lieutenant Ricardo J. Hermelo, that he had decided that he and the *Uruguay* would indeed search for the *Antarctic* and its survivors before leaving the area (Nordenskjöld et al. 1905).

The next two days were a whirlwind of packing and preparations to leave the station that Sobral and his companions had called home for the previous two years. During this time, Sobral traveled 25 km from the station to the west coast of Seymour Island to retrieve a magnetic device that had been left there. On his way back, while passing the strait between Snow Hill and Seymour, he "could see, for the first time in two years, the blue and white flag, at the top of the *Uruguay* (Sobral 1904: 310). This would be a theme he would revisit later.

The nine Greenland dogs who had been a part of this Antarctic home were put into action, as they pulled sledge load after sledge load of priceless fossil specimens and important scientific equipment from the house at Snow Hill to the Argentine ship *Uruguay* anchored approximately 7 km away. Both Sobral and Larsen commented on their impressive performance. Larsen would later write that "It was a real pleasure to see that the animals could draw such heavy burdens" (Nordenskjöld et al. 1905: 577). Sobral especially emphasized their stamina, adaptability, and ability to pull sledges under very difficult circumstances on various terrain (Sobral 1904). It is another clue to Sobral's character that he made these observations about the dogs during the culmination of his Antarctic experience. He would go on to write about their importance in his notes and writings after his return from the expedition.

Sobral, his leader Nordenskjöld, and his expedition companions finally vacated the house on Snow Hill. They shut the door behind them, and, together with the dogs that pulled the last sledge load, marched across the ice to the ship that awaited them.

The last sledge load was brought on board the ship. But, unlike in some other expeditions, the dogs who had pulled the sledges were not left behind. Sobral had discussed the dogs quite frequently with Nordenskjöld during their time together at the station. They had devised a plan to donate the dogs to the Argentine

Government and identified Observatory Island at New Year Island as the best place to keep them. Nine dogs remained at Snow Hill, and it was Sobral's desire to take them with him, and, with Nordenskjöld's blessing, bring them to New Year Island. "These dogs occupy very little place in a ship," he rationalized. "They are all tied together in any place at the bow." The nine that would be transported on board the *Uruguay* "were perfectly well in three square meters, so the small inconveniences that can arise in having to provide the food and cleaning where they live, disappear before the huge benefit they bring to their owner" (Sobral 1904: 312). And so the faithful dogs were faithfully brought on board the ship that would take the expedition away from the home it had known for two winters.

Once on board, Sobral reveled in the royal treatment he and his companions received from the crew of the *Uruguay*. "Flexible and perfumed" new clothes awaited him (Sobral 1904: 317). The previously dapper under-lieutenant, who had had to cobble his boots from seal skin, and sew his rough clothing together over the course of the past two years, could now throw those rude old rags into the current. He felt he could present himself suitably when the remainder of the Swedish scientists boarded the ship—his personal pride and confidence must have been bolstered by this outwardly aesthetic and comfortable adornment. His Argentine fellows treated the expeditioners as kings. They gave up their cabins for them. And, standing in guard formation, and offering a welcoming cheer, raised the Swedish flag as the scientists boarded the Argentine ship.

As he left Snow Hill aboard the corvette *Uruguay* on that 10th of November 1903, Sobral was filled with a certain yet modest satisfaction that he and his companions had completed their mission despite the harsh circumstances in which they had found themselves. But above all, he felt an immense sense of pride and joy that fellow Argentines, in an Argentine naval ship, had rescued them all (Sobral 1904).

A year later he would write (Sobral 1904: 312):

> I did not feel any emotion when leaving the little house that had so faithfully sheltered us for two years… I felt that I was leaving it with all indifference, as one leaves a hotel where he has lived for two days. But now… I feel nostalgia for Snow Hill and I want to go back, I need to go back.

The nostalgic dream was one that would never be fulfilled for Sobral.

6.4 The Retrieval

After their departure from Snow Hill, the expedition members on the *Uruguay* went to Seymour Island to deposit provisions for the next expedition, leaving with that depot a letter explaining the course of events and actions that the *Uruguay* and *Antarctic* crews had taken. On Seymour, the expeditioners also retrieved the additional fossil samples that had been left there and brought those on board the ship for transport back to Sweden. Sobral lamented having to leave some of his own boxes of fossils, as well as the seal skulls he had so carefully preserved. He had intended to give these to a museum in Argentina, but was now forced to leave them behind back at the house on Snow Hill (Sobral 1904).

After collecting Nordenskjöld's fossils on Seymour Island, Irízar and the crew set a course for Paulet Island to rescue the remainder of the crew and scientists of the *Antarctic*, arriving there in the early hours of the morning on November 11th. The Paulet contingent was asleep in their long stone hut. They had spent the previous four days vigorously collecting 6000 penguin eggs to sustain themselves. The *Uruguay* blasted off a signal—a sound not heard by the expeditioners in a long time—and the sleepy party members turned out of their hut. The gratitude and delight of the 13 emaciated men—and at least one surviving cat—who eagerly embraced them there touched the hearts of all on board the Argentine corvette. The reunion was a tearful and poignant one. The complement of the late *Antarctic* boarded the *Uruguay* as a few of the crewmembers erected a crucifix at the grave of the fallen sailor, Wennersgaard. (Sobral 1904; Nordenskjöld et al. 1905).

After bringing the stranded expeditioners on board, Irízar and Nordenskjöld left additional provisions in a depot on the island, along with a notice written in English (Sobral 1904: 320):

> On board the Uruguay ship, belonging to the Navy of the Argentine Republic. The 11th of November of the year one thousand nine hundred and three.

> The undersigned captain of the *Uruguay* on his voyage to the Antarctic regions, to relieve the Swedish expedition directed by Dr. Nordenskjöld, wintering in Snow Hill, having picked up in this point Dr. Nordenskjöld, Lieutenant Sobral, and their companions, came to this Island for embarking the crew of the *Antarctic* wrecked in Erebus and Terror Gulf where they had passed the winter. After having taken on board this wrecked crew, we continue to Buenos Aires.

> We intent in leaving Paulet to sail through the strait between Joinville and Louis Filippe, picking up there the fossil collection left by Dr. Andersson in his wintering place, following from that station in demand of New Year Island passing E. of King George Island.

> In this place we have left a depot of provisions specified in the adjoining list for the use of any needed person.

> We have left also another on the depot on the eastern part of Seymour Island.

> – Julian Irízar, commander.

Sobral, the least recognized expedition member on the journey to Antarctica, was now receiving top billing. But Sobral himself remained modest. And he took great pleasure in the reunification that was taking place all around him (Sobral 1904: 321–322):

> On board were nothing but smiling faces; all the unpleasantness had passed, and hearts were throbbing to think that it would soon be possible to embrace loved ones back home, that the sun would warm up more intensely, and that the air would be anointed with the perfume of plants and flowers.

Sobral was most heartened to see his old companion Skottsberg as he was rescued from Paulet Island. He was also awed to see the fiercely beautiful icescape of Hope Bay where Andersson and his mates had survived. For, after Paulet, the ship stopped at Esperanza/Hope Bay, where Nordenskjöld and Andersson went ashore to collect Andersson's plant fossils and geological samples, eyeing the simple stone structure that, now devoid of its roof and inner tent, looked like a sad pile of stones, but which had harbored three stranded expeditioners and sheltered them over a harsh winter. Surrounding the hut were the ever-present penguins, surveying their land and home, and, for the moment, no longer prey for hungry men and dogs. (Nordenskjöld et al. 1905).

Sobral marveled at the tenacity and perseverance of the two parties who had been stranded at Paulet and at Hope Bay. He wrote (Sobral 1904: 321):

> In spite of the hard life they endured, scientific observations were not neglected during their wintering; Skottsberg had made a complete collection of the mosses and lichens of the island, and Andersson made interesting zoological observations.

> I was truly satisfied and happy to see that a ship of our own land, had brought from a precarious situation those working men, worthy of better luck, deserving of having returned in the same *Antarctic* loaded with the fruits of so much privation, so much constancy and goodwill. At last, everything was ready, and the entire Swedish expedition was on board the *Uruguay*, under the protection of the flag of my country, which, for the first time, fluttered at the impulse of the frozen Polar breezes.

6.5   The Return North

Over the next several days, the *Uruguay* traveled through Bransfield Strait and then crossed the Drake Passage, where the valiant ship fought hurricane-style winds and tumultuous waves. The turbulent seas and high winds proved so treacherous that the ship was partially dismantled, its top mast breaking off, causing, in Sobral's words, an "honorable scar" received from "enraged nature" in the midst of the ship's "victory" (Sobral 1904: 325).

On November 18, the *Uruguay* stopped at Staaten Island so that Irízar, Nordenskjöld, and the others could send telegrams to Argentina and Sweden, notifying the leaders there of their return. While most of the scientists visited the main island, Sobral and Bodman took a small boat from the ship to New Year Island, in order to compare their magnetic instruments and check on the progress of the Observatory they had visited two years prior while on their way to Antarctica. Sobral and Bodman took with them the nine Greenlander dogs who had survived with him at Snow Hill. Five females and four males were presented to the Argentine personnel who worked on New Year Island. Sobral tested the dogs on the new terrain himself, making sure that they could negotiate the grassy areas. He reported that the dogs could walk "perfectly well" in the parts where the grass was not long, such as along the path from the house to the beach, and that they would be "of great utility" in carrying food and equipment for the men from the landing to the Observatory (Sobral 1904: 327). It was Sobral's way of finding a home for the trusty sledge dogs who had helped him at Snow Hill, and of contributing to the ongoing magnetic work of the Argentine scientists at the Observatory.

It was here that he also marveled at the completed Lighthouse that had been under construction during his trip south. Now fully functioning, it shone a beacon of passage to the ships traveling toward the Antarctic and was called by Sobral "A work that honors not only the national Navy but the entire country" (Sobral 1904: 326).

After two days of bad weather on the island, which allowed Sobral to eat his fill of sheep stew, Sobral and Bodman rowed back out to the *Uruguay* that had now arrived to collect them (Sobral 1904). The ship continued north on November 20. By the 22nd, it had reached Santa Cruz, and the inhabitants of Snow Hill had their first peek—in two years—into a town with homes and a working population (Nordenskjöld et al. 1905). Sobral was grateful for the welcome extended by the town, and for the efforts of the telegraph operator, who worked nonstop to spread the news to the world regarding the rescue and return of the Swedish-Argentine Antarctic Expedition (Sobral 1904).

6.6 The Jubilant Entrance into the Buenos Aires Harbor

On November 30, 1903, the stalwart *Uruguay* entered the brown-colored, mighty river that is known as Rio de la Plata. Here, in a little bay, the ship stayed to make repairs and spruce itself up before sailing up to the entrance of the city of Buenos Aires.

But Buenos Aires came to the ship.

Boats of all shapes and sizes, Naval personnel of all ranks, cruised up to the corvette to pay their respects. They brought with them news from home for the weary expeditioners (Nordenskjöld et al. 1905).

The ship remained anchored there until December 2nd, when it made its way up the river to the docks, entering the city of Buenos Aires like a victor coming home. The slow steam turned into a stately procession, as other ships joined the *Uruguay* in its grand entrance, accompanying the expeditioners in their victorious return, and saluting them with flags and cheers and whistles. Along the docks, and indeed in every conceivable building along the river's edge, the people of Buenos Aires turned out to welcome the members of the Swedish-Argentine expedition and the crew of the Argentine corvette who had helped bring them to safety.

Among those individuals there to greet them were Argentina's President Julio Roca, and Navy Minister Onofre Betbeder. Sobral's former commander on the training ship *Sarmiento*, his advocate for participation in the Antarctic expedition, and one of the architects of his rescue, now stood on the quay to greet the young Under-Lieutenant, who had returned from the ice of the Antarctic to his beloved home land of Argentina.

Humble even at this momentous occasion, Sobral was thankful for "the brave crew of *Uruguay*, who have made the name of our country known, respected, and admired throughout the world," and expressed his "gratitude to all the people for the undeserved honors that touched my soul" (Sobral 1904: 328).

He was away from the ice, and in the embrace of his home.

Figures 6.1, 6.2, 6.3, 6.4, 6.5, 6.6, 6.7, and 6.8.

Fig. 6.1 A depiction of the Argentine corvette *Uruguay*, sailing off the coast of Antarctica, arriving to rescue the Swedish Antarctic Expedition, in November 1903. (Departamento de Estudios Históricos Navales, Archivo Histórico, Archivo Fotográfico, U-0212 i, Corbeta *Uruguay*. En la Antártida. Oleo Demartino.)

Fig. 6.2 The officers and crew of the rescue ship *Uruguay*, sent to search for the expedition members of the *Antarctic*. (Departamento de Estudios Históricos Navales, Archivo Histórico, Archivo Fotográfico, U-0109 a, Corbeta *Uruguay*—Rescate Tripulación del *Antarctic*. Jefes y Oficiales.)

Fig. 6.3 The corvette *Uruguay* and its rescue crew departing from Buenos Aires to Antarctica, given a farewell by Argentine President Julio Roca and Minister of the Navy Onofre Betbeder. (Departamento de Estudios Históricos Navales, Archivo Histórico, Archivo Fotográfico, U-0109 b, Corbeta *Uruguay*—Rescate Tripulación del *Antarctic*. Partida—Pte Roca y Mtro Marina Onofre Betbeder.)

Fig. 6.4 A rendering of the meeting between the officers of the corvette *Uruguay* and the Swedish Expedition members at Snow Hill, depicting Captain Julián Irízar shaking hands with Otto Nordenskjöld as José María Sobral (*center*) and Lieutenant Jorge Yalour look on. (Departamento de Estudios Históricos Navales, Archivo Histórico, Archivo Fotográfico, U-0109 d, Corbeta *Uruguay*—Rescate Tripulación del *Antarctic*. Tripulación de la Corbeta *Uruguay*—tripulación del *Antarctic*.)

Fig. 6.5 Captain Julián Irízar (*center*), of the rescue corvette *Uruguay*, visits the Swedish Expedition house at Snow Hill, as the expeditioners prepare to pack. Standing to the left is Otto Nordenskjöld, and to the right is José María Sobral, with the sled dogs in the foreground. The photographer was Gösta Bodman. [Departamento de Estudios Históricos Navales, Archivo Histórico, Archivo Fotográfico, L-0058 e, Casilla de madera Isla Snow Hill. Refugio Expedición sueca. 1901–1903. Fotografo Bodman (Nordenskjöld, Irízar, Sobral).]

Fig. 6.6 The first portrait taken of José María Sobral upon the expedition's return from Antarctica. Sobral is pictured here on the corvette *Uruguay* during its stop in Banco Chico for repairs before entering the Buenos Aires harbor. Despite the metaphorical "battle scars" that can be seen on Sobral's face, he still maintains his calm appearance and characteristic poise. (Departamento de Estudios Históricos Navales, Archivo Histórico, Archivo Fotográfico, P-0140 g, Sobral, José María. Alférez. Primer retrato tomado en Banco Chico al volver de la Expedición a la Antártida.)

Fig. 6.7 Under-Lieutenant José María Sobral and crew members of the rescue ship *Uruguay* upon their return to Buenos Aires from Antarctica. (Departamento de Estudios Históricos Navales, Archivo Histórico, Colección Sobral, Box 2, Envelope 3, newspaper cuttings album Raúl Pinero.)

Fig. 6.8 José María Sobral, reunited with his brother in Buenos Aires, upon his return from the Antarctic in December 1903. He was finally home with his family, but missed his adopted home of Antarctica. (Departamento de Estudios Históricos Navales, Archivo Histórico, Colección Sobral, Box 2, Envelope 3, newspaper cuttings album Raúl Pinero.)

References

Nordenskjöld NOG, Andersson JG, Skottsberg C, Larsen CA 1905 Antarctica: or two years amongst the ice of the South Pole. The Macmillan Co, New York

Rabassa J 2003 Estudio preliminare. In Dos años entre los hielos, 1901–1903, José María Sobral (reprint of original book published in 1904), 1–22. Buenos Aires: Eudeba, Colección Reservada del Museo del Fin del Mundo

Sobral JM 1901–1903 Expedition diary of José María Sobral, Swedish Antarctic Expedition of 1901–1903. Departamento de Estudios Históricos Navales, Archivo Histórico. El Diario de Alférez de Navío José María Sobral. Buenos Aires: Archivo D.E.H.N. - A.R.A

Sobral JM 1904 Dos años entre los hielos, 1901–1903 (Two years amidst the ice, 1901–1903). Buenos Aires: Imprenta de J. Tragant y Cia., Bolivar 319

Chapter 7
Recommendations from a Seasoned Antarctic Explorer

Abstract Personal recommendations, historical overviews, and statistical facts regarding Antarctic exploration were written by José María Sobral toward the end of his time in Antarctica, during the Swedish-Argentine Antarctic Expedition of 1901–1903. These writings from his diary include a passage dated around October 8, 1903. This treatise on Antarctic exploration contains Sobral's views and summaries regarding his experience in the Antarctic, an overview of historical expeditions conducted by other countries, and suggestions and recommendations for future Argentine expeditions to Antarctica.

Polar exploration became a science and a passion for José María Sobral. He had been bitten by the Polar bug. He would not let go. And he began his lifelong romance with Antarctica by spelling out, in his faithful diary, a good amount of what he had learned on this expedition, and from his readings of other explorers' writings.

And so, in the middle portion of his diary—between the account of the sea voyage, and the documentation of the second winter—are seven pages filled to the brim with personal recommendations, historical overviews, and statistical facts (Sobral 1901–1903, diary pages 48 through 54).

In these writings, Sobral foreshadows the locational strategies selected by Robert Falcon Scott and Ernest Shackleton in 1901–1904. He foresees the success of using dog-pulled sledges, which Roald Amundsen would go on to do in order to secure his South Pole victory in 1910–1912. And he voices his belief in Argentina's responsibility to further explore Antarctica, an effort that the country would embrace in his lifetime and beyond.

Part of this diary passage is dated October 8, 1903, but more likely the entire segment was written toward the end of his time in Antarctica, probably just before or just after being rescued—November 8 would be a closer date.

In this diary passage, Sobral writes a treatise on Antarctic exploration, including his personal views and firsthand summaries regarding his experience in the Antarctic, an overview of historical expeditions conducted by other countries, and his own suggestions and recommendations for future Argentine expeditions to

© Springer International Publishing AG 2018

M.R. Tahan, *The Life of José María Sobral*, Springer Biographies,
https://doi.org/10.1007/978-3-319-67268-7_7

Antarctica. Some of this is a precursor to his notable speech that Sobral would later deliver for the Naval Center in Buenos Aires in December 1903.

7.1   Sledge Trips and the Superiority of the Sledge Pulled by Dogs

7.1.1   Pages 48, 49, and 50

Without a doubt, sledge trips should be a part of the program of every polar expedition, whether they are from a station or from a ship. When the ship is stopped along its way by the ice, that same ice offers a great path for the sledge. It's clear that if the ship is frozen in the middle of the Polar sea, as the "Belgica" was, for example, then an expedition leaving by sledge must not count on [having] the ship in the future. That expedition must make land or perish; that is, it has to make a trip in the same manner as Nansen's expedition toward the [North] Pole, with his retreat to Franz Josef Land. In the Polar sea of the south, it seems that the ice does not have great movements [drift], as it does in the north. But still, these movements are enough for a sledge party, who is on the retreat, to lose its ship and find itself condemned to perish from hunger and cold. The spaces of water from where the ship could go to the [South] Pole are three: 1st, between Victoria Land and Graham Land, separated by more than 200° of longitude; 2nd, between Graham Land and Enderby Land, separated by approximately 105° of longitude, and third, the smallest of the three between Knox Land and Kemp Land, that is, some 43° of longitude. The three routes through water offer great risks, much greater than in any part of the North Pole, precisely due to this "indifference" to movement that the ice of the south seems to have, and especially because of the coldness of the southern summer. It will be said that if the ice has little movement, in summer, after wintering, a ship will probably find free water and will be able to make its way back. But that is not the case. It is very possible that, if in the summer during which the ship entered the ice, there was free water around, then the next year the ice will form a compact field perhaps up to 10° further to the north. We have tangible examples to illustrate what I say; on Weddell's trip in 1825, he reached 74° 15′ S and 34° 17′ W, the sea toward the south being completely free of ice, with only some ice mountains in sight. Sixteen years later, Dumont d'Urville attempted to penetrate into the Weddell Sea, reaching only 64° of latitude, and in 1841, at the meridian where [James] Weddell had been, [James Clark] Ross couldn't make it any further than D'Urville. But more to the east, through 15° W, he reached 71° 30′ S. Ross, in 1841, discovered Snow Hill Island, Seymour Island, and the southern part of Louis Filippe Land, disembarking at Cockburn on January 1st. Then, he tried to head toward the south, not making it any further than to the south of Snow Hill Island, his progress being impeded by compact clumps of [ice] floes; he found himself compelled to retreat toward the north, not having reached further than

64° 30′ south. On November 18, 1892, Captain [Carl Anton] Larsen could disembark at Seymour Island, and the following year, in the month of December, he
was able to reach 68° 10′ S, to the E of Graham Land, discovering King Oscar
Land, and going back again toward the north, where he disembarked at Cape
Seymour.

The Swedish Expedition, in January of 1902, could only make it to the east side
of Graham Land at the height [latitude] of the Polar Circle. In the middle of
January, they were able to disembark at Seymour Island.

In 1903, on January 9, this region of Cape Seymour was surrounded by a field of
ice toward Louis Filippe Land and by a compact group of glaciers to the east. These
differences are also found in the northern hemisphere, but on a much smaller scale.

If an expedition tried to get into the region where the "Belgica" wintered, they
probably wouldn't reach a higher latitude than the one reached by the Belgians.
And if a sledge party sets off from the ship (the way it must be done if one wants to
accomplish an important mission), it necessarily has to be supported by a series of
depots placed in the western part of Graham Land: from Alexander Land, or more
to the south if possible, to the Belgica Strait. These depots, however small, have to
be relatively numerous, because with them, the work of departure and retreat
becomes easier, giving it [the journey] much more speed. If the wintering takes
place closer to Victoria Land, then the retreat will probably be on the "ice barrier"
toward Victoria Land. As for on-land stations, naturally, the closer to the pole that
the observations are made, the more interesting they will be.

The British have certainly chosen the best route to obtain a good result. But I
think they could take better advantage of the effort of an expedition by dividing it
into two parties. One of them, "the land party," would spend the winter at the foot
of Mount Erebus; in the spring, another small party would go toward the south onto
the "*ice barrier*." The party from the ship that would winter in Wood Bay, like the
British are going to do, would go onto the "*inland ice*" toward the magnetic pole.
I think it is very unlikely that any sea to the west of Victoria Land would be found
free of ice, as some fear; this [happens] in the summer, and I think that it is
impossible in the spring. It is said that the "*ice barrier*" that begins at Cape North,
in the meridian part of Victoria Land, doesn't have a great height, making it more
accessible than Mount Erebus, which is higher. A station set up at that point could,
with the same probabilities of success as that in Wood Bay, allow an attempt at the
discovery of the magnetic pole. This is located, according to the British Admiralty,
at 73° of latitude south and 147° E of Greenwich, that is, 300 miles away from
Wood Bay, and about 500 miles away from Cape North and Pointe Géologie in
Adélie Land—from Wood Bay, which is almost to the west of Pointe Géologie, to
the SSE, and from Cape North to the southwest.

The most convenient method of exploration is the sledge pulled by dogs,
because, traveling with them, the maximum amount of distance can be reached in
the minimum amount of time, also exposing the minimum number of men.

The superiority of the sledge pulled by dogs over that [pulled by] men has been
proven by their magnificent explorations: The Duke of the Abruzzi, [Fridtjof]
Nansen, [Robert] Peary, and other travelers of the Arctic.

7.1.2   Page 51–Supplement I

From "*Observations Météorologique de l'Expedition Arctique Suedoise 1872–1873.*" Taken from page 110 of this book.

[Sobral describes the table of average temperatures listed in the book, in French. He also compiles a chart comparing the monthly average temperatures recorded at the respective Arctic locations of various expeditions from 1821 through 1859, including those of William Parry, John Ross, and Francis Leopold McClintock, and those taking place in Point Barrow, Port Leopold, and Beechey Island.]

7.1.3   Page 52

From "An Expedition to Greenland" by Prof. A.E. Nordenskjold [Adolf Erik Nordenskjöld, Otto Nordenskjöld's uncle].

[Sobral describes the average temperatures recorded, the times and latitude/longitude, and the conversion to centigrade (accomplished by Bodman) for the 13 years listed in the chart that Sobral composes on this page.]

7.1.4   Page 53

[On this page, Sobral creates an intricate and extensive chart listing the monthly average temperatures of 25 expeditions in the Arctic, Greenland, Canada, Russia, and Scotland, including their respective wintering stations, latitude/longitude, altitude, years recorded, and hours of observation.]

7.2   To Finally Penetrate into the Heart of the Frozen Zone

7.2.1   Page 54–October 8, 1903

Even supposing that the British and German expeditions obtain the utmost success, that, for example, their physical, biological, and geological work is the most complete possible, that their geographical discoveries are crowned with the best of events; that is, assuming that these two expeditions have fulfilled their program missions to the letter, the virgin field to be explored in the southern regions will still be enormous. It's true that those places are visited in part by whaling ships and sealers, and they will continue to be so, as long as they [the whale and seal hunters] obtain some profit from the hunting that those seas and frozen lands offer. But we do not have to rely on that to satisfy the demands of science; it is only the ships that are equipped with the most modern observational instruments, and managed by specialists, that will finally fill, little by little, the blank pages of the Antarctica encyclopedia. It is evident that an expedition directed to any point of the Antarctic will bring a very rich and interesting collection of data, many of which, after being worked on, will open new paths and will provide new methods to finally penetrate into the heart of the frozen zone.

It seems that Sobral gave great thought to these recommendations while still in Antarctica; he would later use these in his speeches and writings.

First and foremost, this treatise is a call for science, not commercial enterprise. It is Sobral's expression of his belief that the scientific knowledge gleaned from the conscientious exploration of Antarctica will far outweigh the commercial profit gleaned from the harvesting of its natural resources.

Reference

Sobral JM (1901–1903) Expedition diary of José María Sobral, Swedish Antarctic Expedition of 1901–1903. Departamento de Estudios Históricos Navales, Archivo Histórico. El Diario de Alférez de Navío José María Sobral. Buenos Aires: Archivo D.E.H.N, A.R.A

Chapter 8
The Speech for the Naval Center in Buenos Aires: A Call to Antarctic Action

Abstract The following includes a transcript of the speech and lecture that was presented by José María Sobral at the Teatro Politeama Argentino in Buenos Aires on December 19, 1903. The event was sponsored by the Naval Center (Centro Naval) and benefited the Argentine Naval League (Liga Naval Argentina). In the speech, Sobral recounts the challenges and discoveries of the Swedish-Argentine Antarctic Expedition, describes life at Snow Hill and along the Antarctic Peninsula, and sets forth goals for future Argentine expeditions and accomplishments in Antarctica.

When the Swedish-Argentine Expedition made its triumphant return to Buenos Aires on December 2, 1903 aboard the corvette *Uruguay,* a mass of people enthusiastically turned out to greet the arriving heroes. Men and women, young and old, filled the streets and accompanied the expeditioners as they were driven in horse-drawn carriages from the old dock at La Boca to the stately and ornamental *Centro Naval* Naval Center building at the corner of fashionable Calle Florida and thriving Avenida Córdoba (Nordenskjöld et al. 1905).

Exactly one week later, Nordenskjöld, Lieutenant Yalour, and Skottsberg were invited to speak at the Argentine Geographical Society to regale their audience with tales of the expedition and stunning descriptions of their discoveries (Nordenskjöld et al. 1905).

Modest Sobral prepared for his own moment in the limelight, which was scheduled for December 19, 1903. He was invited by the Centro Naval to present a lecture at the Teatro Politeama Argentino, a respectable and decorous theatrical space on the cultural and lively Avenida Corrientes.

Equally decorous as the venue in which he would make his presentation, the under-lieutenant painstakingly prepared copious notes and outlines for his speech (Sobral 1903b). Drawing from his diary entries, he wrote an extensive account of daring accomplishments and historical discoveries that would draw the listener into his Polar world.

His speech gives details of a typical day in Antarctica; it reflects on the dangers that awaited the men, dogs, and wildlife on the continent; and it provides a summary of the wealth of scientific knowledge gained from the expedition. It is an exemplary achievement of distilling the many scientific discoveries of geology, oceanography, meteorology, and geology made by the Swedish expedition, and placing them in context with earth history and with future advancements.

Shining a spotlight on his character, Sobral's speech emphasizes the courage of the scientific men in exploring the frozen Polar regions, and the importance of sustaining these efforts for his country. Surprisingly, it also offers a peak into what could be interpreted as Sobral's own possibly prejudiced attitude toward those he terms as the "savage," as opposed to the "civilized" individual. From his words, he seems to support the conquest of the native peoples' lands in Patagonia. (A military campaign that resulted in the killing, capturing, and displacement of the indigenous people during the late 1870s was reportedly spearheaded by none other than the then-Army General Julio Argentino Roca, who, now as Argentina's president, personally welcomed the Swedish-Argentine expedition upon its return from Antarctica). Sobral's words echo those of Nordenskjöld, who, in his book, refers to Australia's Aboriginals as a "low race" (Nordenskjöld et al. 1905). Not to mention echoing Ekelöf, who had insulted the Argentine people and their mixed "infinity of races" straight to Sobral's face during their short stay in Port Stanley (Sobral 1901–1903). Here, in his speech, Sobral seems to champion the virtual disappearance of "the savage"—"*el salvaje*," as he seemingly refers to the indigenous people—and to justify the stripping away of the native dwellers' home and territory for the sake of "civilization." Coming from an individual who believed he had been treated less than fairly during the expedition, indeed at times as racially inferior, due to his ethnic background, these words, if interpreted thusly, are rather shocking. Perhaps he was influenced by the politics of the time; in analyzing his words, one must also take into consideration the general attitudes and actions of the era.

Sobral now also seems to appeal to the commercial interests of his audience, citing reasons of commerce, of profitability, and of enticing riches to be found in the natural resources and wildlife of Antarctica, as good reasons to carry out scientific studies and explorations of the Polar region. Perhaps this is expected, given the times and the motivation.

In the last paragraph of the speech, Sobral opens the theatrical curtain ever so slightly on a portion of his life and his innermost feelings, giving the audience a sense of his profound depression and loneliness in Antarctica—part of which, again, stemmed from the treatment he received at the hands of some of his Scandinavian companions. It is with these words that he ends his speech, begging pardon for the harsh conditions of the Antarctic that he has poetically described, and explaining his patriotic fervor as a tonic to quell "the frightening isolation of those regions."

The speech was published in 1903 in the *Boletin del Centro Naval* (Sobral 1903a). The following, translated into English from the original Spanish, is the full text of Sobral's speech.

8.1 Conference [Lecture] by Under-Lieutenant José M. Sobral

8.1.1 Sponsored by the Centro Naval [Naval Center] and Read in the Politeama Argentino [Theater] on December 19, 1903

8.1.1.1 For the Benefit of the Liga Naval Argentina [Argentine Naval League]

Sirs:

It is an honor, undoubtedly undeserved, this one that the President of the Naval Center has deigned to grant me, inviting me to give a lecture on the Polar campaign, in which, representing my country, I have had the satisfaction of taking part.

My attitude would be bold if, in presenting myself to the enlightened audience who listens to me, I did not count in advance on their indulgence and the kindness of which so many evidences I have received since my return to Buenos Aires.

For I do not bring to you, gentlemen, anything but my good will, stimulated by the desire to repay, even in part, the generosity of this noble people, who has awakened in me feelings that in vain the word of a man would attempt to interpret, when there is no eloquence, which is the voice of the heart.

8.1.2 Scientific Interest

Ladies:

Sirs:

It is often said: What is the interest of a Polar expedition? What benefit can it bring to humanity?

To some, a Polar expedition may mean money; to others, only progress in science. For the profane in these things, to know what kind of formation these lands belong to, the knowledge of their topography, their fauna and flora, and, in short, the complete geography of them, does not mean anything; it is only a waste of time and a lot of useless sacrifices; and in the face of these observations, which are not based on arguments of value, perplexity imposes, because it is impossible to refute the tenacity of the unbeliever.

Not for a man of science, but for anyone who has an idea of the importance of these matters that I have enumerated, the simple fact of exploring the unknown, justifies the expenses of a trip, and even the sacrifice of lives; but when it is considered that these observations, these studies, can yield gold, when it is known that some of them can greatly facilitate the expeditions of exclusive commercial character, then those who previously shrugged, will also have to agree with the benefit of them.

And of this I will offer a palpable example. Tell me: If the meteorological regimes of these regions were well known, if the strength and direction of the currents of these seas were known, would not that facilitate the voyages of the sealers and whalers, and of all those who with any objective go there?

Obviously yes, it would.

And if these regions are not exploited scientifically, how can we know that there is not something there, maybe a lot, that can be used for profitable purposes?

8.1.3 Commercial Interest

The reasons of purely commercial interest are immense, and their importance exceeds any calculation. In letters that I have published at another time, I have drawn the attention of my country to the enormous riches that we can obtain from the Polar regions; and today, I insist, in the presence of this distinguished audience, wishing that our public powers draw their attention to those inexhaustible mines that are waiting for nothing but a little resolve, to be converted into great riches; to our youth, to my contemporaries, to direct their energies toward those regions, and thus we will extract from their frozen insides scientific truths that will enrich the minds of the scholars, and positive truths that will compensate the efforts and the fatigues of those who seek them. Fishing and hunting in the South Seas constitute this source of wealth, and no country is in a better position than Argentina to successfully undertake this enterprise.

Let us take possession of those lands and dominate those seas, now uninhabited, but that are at the same latitude as others in the north, where populous cities flourish, and from which civilization has sent us samples like Nordenskjöld, Larsen, and Nansen. Years will pass, new generations will succeed us on the stage of life, and when the population of my country is counted by hundreds of millions, those happy Argentines will see their flag flutter in Polar towns; and there, when the May sun stands face to face with the Aurora Australis, those who contemplate that superb spectacle will acclaim the names of the present generation, and will tear down the huge snowdrifts to raise in their place the marble blocks that will serve as a pedestal to the glory of Argentina. Outlined as they are, the motives that these Polar expeditions may have, let us concentrate on the one that motivates this conference, and take as the beginning of the narration the day of the greatest sorrows and emotions that I have felt during the two years of Polar wintering.

8.1.4 In Snow Hill

That day was February 21, 1902, the day when we waved goodbye for a year to our comrades who returned on the *Antarctic*, leaving us on Snow Hill. When crafting these lines, when preparing this conference, my whole being shudders remembering the emotions of that instant, comparable only to that of December 21 of the previous year, when I left Buenos Aires in the direction of the Pole; and, for its more faithful narration, I am copying what my diary literally says:

"Goodbye, *Antarctic*, I am bidding you farewell with sadness, because you are the bond that joins us with civilization and that breaks with your departure; I hold you in affection, a lot of affection, because you are the bearer of my thoughts for my homeland and for the loved ones who await me in it! From today on, we are cut off from the world, the solidified water and the long nights with their horrible torments will form an impassable barrier for the human, and here, in this dreadful solitude, only six of us remain; without any change, looking at the same faces and places, we will fight boredom and nostalgia, if they come, as they come, devoting ourselves to work, to complying with the program of the expedition, and we will repress our heart, to silence, upon birth, every feeling other than that of the duty entrusted to us, and of our own preservation, preparing ourselves in that titanic struggle to overcome the rude and savage attacks that nature, in its horrible nakedness, launches against us with all the power of its disordered elements."

"We are left alone, and each and every one of us must provide for our needs. The night comes, and it is not possible to fall asleep, because of the insomnia that takes over us, thinking of that comrade who went away, taking with him our caresses and hopes and the last tears that, through time and distance, are shed for affections."

"Snow falls in abundance; the earth and objects, covered with a thick cloak of that white and beautiful dress, form a beautiful contrast with the dark green of seawater."

"We are working."

8.1.5   First Trips

To regulate our way of life, it was necessary to know the surroundings of our camp, and where our radius of action would extend to, and, to that effect, several explorative and study expeditions were organized and carried out.

Because winter was coming, and ice formed in many places, before the explorations by sea were terminated, it was decided to make a boat trip to Lockyer Island, with the purpose of establishing there a store of provisions that could serve as a point of support during spring excursions, and we initiated it on Saturday, March 11, 1902; Nordenskjöld, Jonassen, and I, provided with food for six days, left the station carrying the supplies we wanted to deposit at Lockyer.

I will mention some details of this exploration by boat, so that we can see the difficulties that are always present and that must be overcome quickly, to avoid the penalty of being defeated. Our harnessed team and equipment consisted of five dogs, a kitchen, a reindeer-skin sleeping bag for three people to sleep in, and a small tent.

The work that navigation demanded was divided in turns among the three of us, in this way: an hour of rowing, and half an hour of rudder for each person. Like all things, at first, everything went very well, but after a while, the new ice began to set an obstacle before us; the advance through that new ice was like this: A man standing at the prow would hit the surface with a stick and would break the ice, whose thickness was three centimeters, and that way, we would open our furrow, just as men do with the ax in the jungle, cutting down the tree that stands in the way. That ice was sometimes thicker, and therefore more resistant; then it could no longer be broken with the stick, and this relatively comfortable system would then be replaced with another one, which, as well as being more effective, was also more dangerous: We would hold on to the gunwale of the boat, and we would jump onto the frozen sea, which would break when receiving the weight of our body; this exposed us to suffering sea bathing, not pleasant indeed in that circumstance, because the temperature was low, and we did not have more clothes to wear other than what we had on us.

Difficulties would increase in direct proportion to our progress; the tide was favorable, and because of it, our march was fast. A single ice bank joined Snow Hill, Lockyer Island, and Haddington Island, and the ice floes that were floating at sea were swept by the tide at a speed of 3 miles, clashing one against the other, and making a deafening noise; and continually we were forced to maneuver and to row with all energy to save our boat. Pressures were very strong, each clash between two ice floes produces a noise that causes alarm, and to give an idea of the force that develops in these collisions, it suffices to know that the edges would raise and enormous mounds would form on each floe. If our boat or any other was taken by two of those colossi, its destruction would be inevitable and immediate.

We worked until 6 p.m. without being able to reach the island, for the heap of ice floes kept us two miles away, and we decided to camp and plant our tent on that ice field attached to Haddington Island.

The night was quiet, and on the next day, early, with a northern breeze curling the surface of the water, and with a bright sun in all its splendor, we had lunch merrily; and, promising ourselves a happy journey, we set off at full sail, trying with new energies to do what on the previous day had been impossible. The temperature was relatively high, and the new ice that had accumulated along the bank disappeared due to the heat and the wind. We thought that on the Snow Hill side the crossing would be possible, but it was not. Nordenskjöld and sailor Jonassen climbed a very tall iceberg, to reconnoiter the surroundings from there; because they took the two boat hooks, I was not able to keep the boat moored, for the pieces of ice to which it was moored fell, and the boat, drawn by a strong current, began to drift, and I called with a loud voice to my comrades, who could not hear me, and although I maneuvered with my oars the best I could, and did not lose the calm demeanor so necessary during these cases, the boat was carried away: On one side was the great *iceberg*, which my companions had climbed, and on the other side the small ice floes and new ice. One of the dogs, frightened by the crashes of the boat against each of these icy masses, threw himself into the sea; after swimming a little, he returned to the boat with pitiful howls, and, despite the critical situation, I left the oars and pulled on board the poor animal, who was at that moment my partner in danger. At last, I managed to get closer to the iceberg, just as Jonassen approached;

jumping on a piece of ice, he reached the boat and came to my aid, and now being two, it was easy for us to master that situation, which with every minute that passed had become more difficult. Seeing the impossibility of approaching the island, it was decided to deposit the depot at Haddington, and at two o'clock we unloaded our supplies onto the ice, and prepared the sledges to carry the provisions to the land, which was about five kilometers away from us. In the vicinity, we saw some penguins, which are the graceful inhabitants of those places, and some seals belonging to the class called Leopard seal (Ogmarhionus leptonix).

We deposited the depot in a cape, which, since then, we called Cape Depósito; we secured the boxes with large stones, and precautions were taken so that the provisions would remain in good condition until spring. It was very hot, the temperature had risen to 7° above zero; water would flow in torrents down the mountainsides, and, to quench our thirst, we used the primitive procedure of lying down on the ice field and drinking that rich and life-giving liquid.

Upon return, we moved boat, supplies, etc., into the ice field, and about 50 m from the water, we set up the tent. The barometer was dropping rapidly, and the thermometer remained high. In order to secure our tent against the wind, the boat was moored nearby on the southwest side; on the sides, the boxes of provisions were placed, also tied up; and we positioned the door to the north. After dinner, we got into our bags and slept peacefully, without dreaming about the unpleasant awakening that we would have, because, during our sleep, a horrible death was stalking us. It was 7 a.m., the wind was blowing with great force, carrying with it enormous masses of snow; we were still sleeping, when sailor Jonassen, who was to my right, exclaimed: Water in the tent! Despite our astonishment, and not understanding what was happening, we got out of the bag; the three of us, having the same difficulties, and trying to put on our boots, crawled out of the tent. My boots were as hard as stone, and I could not put them on; after making a great effort, I could finally do it, and we immediately began to salvage our equipment. If there had been a delay of a few moments, and without the alarm given by the sailor, all three of us would have been buried in the depths of that sea.

The cause of the water entering the tent was the strong wind that, little by little, had broken the edge of the ice field until it reached our tent.

The temperature was 16° below zero, and hurricane wind was blowing, violently dragging a lot of snow with it, so it was not the most pleasant thing to remain outdoors in that weather. We placed our equipment 300 m from the water, and we spent all day without the tent, walking on the ice to get warm. At midday, we were able to heat up some water, and, although it was not very hot, we drank a little *kernnicas* with delight, and that was our first and only meal throughout the day. At night, we set up the tent, and we had to get into our bag fully dressed and even wearing our boots, for that bag had a lot of snow inside of it. A finger of my hand became frozen, and the sufferings I went through that night were undoubtedly the most effective remedy applied to my finger. The next day, the morning was clear, and the wind was still blowing at a speed of 15 m per second. After breakfast, we loaded our boat, beginning our return. The first hour of the journey was the most labor-intensive one; to the north of our starting point, the tongue of a glacier was emerging from the ground, and the wind was pushing us toward it, for which reason we could not set sail: Jonassen and I rowed, while Nordenskjöld took on the rudder, and by means of rowing, we overcame that urgent difficulty. We were free of the glacier, from which we were glad to separate, when we set sail with all the curls, and with a tailwind, we made our way to our station. The swell was very little due to the great density of the water of the sea surface, which remained in liquid state only due to the lively agitation of its molecules.

In our absence, Bodman had worked on the installation of magnetic instruments, and Ekelöf on that of his bacteriological devices; our little house was transforming into a nice nest for the winter.

Ekelöf was prepared to amputate many fingers, but luckily he rarely had the opportunity to render his services to us. In the last storm, Ekelöf's canoe was carried by the wind, and only a few fragments of it were found scattered on the beach. During the rest of the month of March, a kennel was built: It was a box four meters long, divided into two compartments, one for the dogs from the Malvinas [Falklands], and the other for the Greenlanders. The new ice was strong enough to bear the weight of a man, so much so that on March 26 we went for a walk on it: It produced the same effect as walking on soft leather.

Young sea ice is completely different from freshwater ice; while the latter has the fragility of glass, the former is very elastic; the cause of this elasticity is the concentrated brine that adheres to the crystals when it solidifies with water, and from which it is impossible to separate. An example of the elasticity of the young ice is when it is formed between two icebergs and there is a small movement of approach between the two; instead of breaking, thanks to its elasticity, it takes on a wavy form, and sometimes it seems as if the waves had been surprised by the cold, acquiring a solid state.

8.1.6 *Distribution of Time*

On April 1, 1902, according to the plan formed, magnetic observations began, and from that day on, our life was always the same. Work absorbed most of our time, and undoubtedly, this is a good disposition, because less time was spent on setting free our imagination, which can usually do so much damage when launching itself on wings of fantasy. This lessened the dangers we did not yet know, caressing only all that was optimism, and increased [our thoughts of the experiences] we had been through, which, after contemplating how they were defeated, we could then measure and appreciate their gravity.

Work was distributed as follows: Nordenskjöld was in charge of everything pertaining to geology; Ekelöf [was in charge] of our health and bacteriological observations; Jonassen was in charge of dogs and sledges. In addition, Jonnasen rendered the most diverse services, thanks to his ability as blacksmith, carpenter, and cobbler. Bodman and I had weather observations. Äkerlundh was our cook.

Life in our station was most uniform. It will suffice for me to describe a [typical] day so that you have an idea of what happened during the two years.

8.1.7   Domestic Life

At 9 a.m., Äkerlundh would pass by our room yelling, the meal is ready! And at the same time, he would fill our basins with water. After personal grooming, at 9:30 a. m., we would sit at the table to have *fruckost* made up of a dish that was very variable the first year, and that was composed of herring or oats with coffee; but the second year our *fruckost* was, in general, seal meat. After *fruckost*, at 10 a.m., the smokers would light their pipes, and we would all go on our own to work. At 2:30 p.m., we would meet to have *middag*; this consisted of two dishes. At 5 p.m., we would have a cup of coffee, and at 9 p.m., a meal or *kvall*, consisting of a dish, and tea or cocoa. Those who did not have the first guard shift would go to bed at around 11 p.m., and would read until 12 p.m. or 1 a.m. Night guard shifts were covered by Nordenskjöld, Bodman, Ekelöf, and myself. The first guard shift was from 10 p.m. to 1 p.m. [am], the second at 3 a.m., and the third one from 5 to 9 a.m. From 9 a.m. to 9 p.m., Bodman and I were in charge, taking turns. The objective of these guard shifts was to make hourly observations of the phenomena for which we had no data recorders, such as nebulosity, direction of clouds, and direction of the wind.

Every two or three days, they would fill a few barrels with ice brought from the snowdrift, and that ice would melt in the kitchen for food and washing, and in a tank that was in the dining room for drinking. For the stove and the kitchen, the first year we used coal, and the second [year we used] seal grease, for we had run out of coal.

In general, there was a fairly pleasant temperature inside the house. For fuel economy and for fear of fire at night, the stove was turned off and the temperature would drop a few degrees below zero, but once the cooker and the stove were turned on, the temperature would rise rapidly. We would bathe every 15 days in a tub. To this aim, a fire was made in the kitchen, heating the water in some pots, and we would parade in there in turns. Soap lasted until the last moment; only when one is about to run out of that article and has to economize it, does one understand the enormous importance of it. Humidity inside the house was very significant; a lot of ice would form on the walls, roof, and floor, and whenever the temperature went up a little, it [the ice] would fall in the form of rain. The layout of our beds was the same as an onboard cabin; those of us who had the upper bunks, we were the ones who suffered the consequences of that thawing. We had a gramophone, and sometimes, on Sundays, some philharmonic [group] would make us hear its repertoire.

The sun was rising more and more along the horizon, and the long winter nights were approaching: The storms were getting stronger, the cold becoming more intense. Birds were already very rare, and when they passed by, they did so in a northerly direction, toward warmer latitudes, and when accompanying them in their swift flight with our saddened looks, we confided all our hopes to them. In the months of July and August, we made several sledge trips around the area, and, although the temperature of 40° below zero was not very pleasant, we began each with pleasure, because they broke the monotony of our life, and meant a change in our daily habits. And we saw with satisfaction that, after each of these excursions, joy was reborn, and conversations were more cheerful and numerous. For us, the passing of a bird or the finding of a seal, during winter, was a great event; and, since our conversations had extinguished a long time ago, we would hold on to any of these events, commenting on them in every way and for as long as possible.

In general, what would entertain us the most, and serve as the topic for some discussion, was atmospheric phenomena: The first thing that one asked upon getting up was, what is the weather like today?

In the middle of August, we killed the first seal, and, with pleasure, we ate the fresh meat—after several months of feeding ourselves only canned food. In my conceptual opinion, seal meat is very good food.

Our observations were carefully written down, and their accuracy was guaranteed by the greatest care and attention we devoted to them.

8.1.8 To the South

Spring was approaching, and with it, new breaths and new hopes animated us; the days were getting longer, and the propitious time for the sledge excursions, that we so very much wanted, was getting closer. Beginning in mid-September, preparations began for a trip to the south, which would be aimed at exploring King Oscar Land.

What worried us most, therefore, were provisions; these were taken out of their tin boxes in order to lighten their weight, wrapped in paper, and then bagged in canvas sacks to preserve them from snow and moisture.

On sledge trips, the clothes one wears must be as light and as warm as possible. In the provisions, the maximum nutrition should be sought with the minimum in weight and volume, and the more the variety, the better.

However insignificant these details may seem, their importance is capital, and they can only be appreciated when one has the need to make use of them.

The reason why sledge travel is limited to spring is because you cannot go out on these excursions, with good profit, without enough light; and because, in general, during the summer, sea ice—especially if you cross a region somewhat distant from the coast—dislocates, and great cracks open, making it very difficult to advance, even with the help of canoes. Of course, when the trip is made on land ice, on the so-called "*inland ice*," the case is not the same.

On our journey to the south, we took two sledges: One, the strongest, intended for carrying the largest load, was pulled by five dogs; and the other one [was pulled] by Nordenskjöld and me.

We had a very lightweight tent, and two sleeping bags: One for two men, which was used by Nordenskjöld and Jonassen, and which was made of reindeer skin; and the other one for one man, which was used by me, and which consisted of two thick blankets, in the form of a bag, put into a canvas lining. One of the most important elements of the equipment was, without a doubt, our stove: It consisted of the Swedish heater *"primus,"* in which, by means of the heat produced by an alcohol flame, oil is transformed into gas before it burns. It was placed in a brass wrapping that preserved it from the wind and made the most of the heat. On September 30, after the load had been placed on the sledges, we set off; the other three comrades— Bodman, Ekelöf, and Åkerlundh—accompanied us for a short stretch of the way, and there we parted, making our way to Cape Depósito, where we were to complete our provisions. This was another separation that halved each group.

From the station to the cape, the ice had such an adequate surface for traveling that, after eight hours of marching, we had traveled the twelve miles that separate them [the station from the cape].

The march and life during sledge trips are also quite monotonous, especially if you go far from land. The person in charge of making lunch would get up at 7 or so, and set about cooking, while the other two would make observations. The cook would bring snow, and melt it in two saucepans; pemmican was made in one of them, and coffee in the other. To make pemmican soup, the dough is cut into very small pieces and heated in water, not needing to boil.

Coffee was made in a special way; to get the maximum benefit from it, it would boil for ten minutes. Lunch was completed with biscuit and butter.

The *"primus"* or heater, with the help of which we cooked, must be handled with great care when lighting it, because sometimes we had been about to burn down our tent as a result of carelessness. When the pemmican soup was ready, we would all three gather around the pot while coffee was being prepared, and we would have lunch, discussing what had happened the previous day, and we would talk about the probable distance to traverse at the present time. If the state of the weather had such a great influence in our lives at the station, it was much more so here, for upon it depended the possible distance of travel, and the comfort of the travel. After lunch, we would fold our sacks, take them out, and sweep and shake our tent, stripping off as much as possible the snow and other things that would adhere to it and that would always represent an excess of weight. Then the whole load was fastened onto the sledges, the dogs were harnessed to them, and, taking the lead, Nordenskjöld and I, we would set off.

The ice did not have mounds in the part we traveled through, but, despite that, the advance was generally quite difficult, due to the furrows that the strong winds make in the snow, and which sometimes would reach a height between thirty and forty centimeters. Not only stormy days prevented us from going, but also even completely calm days that had fog. It was on these days that one truly suffered, not only physically, but also in morale.

These mists, as I have observed, apparently consist of little balls of snow, which fall in relative and reduced quantity, but which in fact are composed of an enormous amount of microscopic snow crystals, which not only impede the visibility of the objects at some distance, but, because everything there is white, and the only thing that forms a contrast is ourselves, there are no shadows or reliefs. Hence, one has the most erroneous idea when wanting to discern the size of an object placed at a small distance away; a crate, for example, located at about ten or twenty meters away, sometimes seems to be the size of a house, and other times the size of a matchbox. The direction, in those cases, would have to be indicated with a compass. So, our muscles would suffer greatly due to snow furrows, because on many occasions, where one believed to see a height and lifted the foot to step onto it, there was a well in which one inserted the leg, and vice versa, we would see a well, and it was actually a height that we would stumble into.

In addition to the refraction effects already indicated, there were other beautiful ones in which an object appeared double, that is to say, it multiplied. One day, on the way back, we had one of those snow fogs; there was a fresh breeze from the south; with the practice of many days of traveling with fog, we no longer needed the compass, and we directed our course guiding ourselves by the collapse of snow furrows. Apparently, about ten feet away, and as if on a small hill, we saw a blackish object that we first took for a bird, then for a feather, and, finally, we were convinced that it was a seal. Then, for the first time, we saw one of the most enchanting phenomena observed in these latitudes: The seal was creeping toward us like a reptile, with the help of her flippers, and above her there was the exact same image appearing, and it was making the same movements. When it came to our side, we realized that it was a specimen of "Lobodon Carcinophagus," which probably was heading toward the neighboring *iceberg*, to have a good lunch of crabs. The sledge of the dogs was carrying a weight of around two hundred kilos, and ours about eighty, but on the third day of travel, twenty kilos were transferred from our sledge to the dogs'. For their feeding, a special pemmican was carried; but we soon ran out of it, and, as no seals were found in the latter part of our journey, we had to give them ours.

From the second day, we began to see signs of land to the south, but they would disappear again. Although the temperature was a few degrees below zero, during the travel, thirst was unbearable; it would sometimes become a horrible suffering. As we walked, we would take snow, but it seemed that, instead of alleviating our thirst, it increased it. All these sufferings were in silence. It is clear that it would have been enough to melt some snow so that the malaise would have disappeared, but by doing so, oil would have been spent. There were delays and, therefore, our journey lengthened. When we suffered from thirst and fatigue, our imagination, instead of weakening, worked actively; I would imagine that there, in the distance, I could see Avenida de Mayo with its cafés and innumerable tables full of people, who drank large glasses of ice water; and at other times, like forming a *pendant* to the picture of reality, I could see a little house surrounded by large trees and a stream meandering in the vicinity, and then, with a nervous movement, I would give a strong pull on the sledge so as to arrive sooner at our goal, and the only consolation I had was the thought that this continuous towing could not be eternal.

Almost all Polar explorers have suffered from thirst, and that is why it is recommended to use ebonite jars, filled with water, and to put them on one's chest, protected by clothing, so as to avoid its freezing; undoubtedly, if we had had them, they would have been very useful.

Another aspect of sledging [one must keep under consideration] is when there is a storm—something that in those regions is most common. Then, you lock yourself in the tent and spend the day in your sleeping bag, talking, sleeping, or mending your clothes and shoes. The ration is reduced to a minimum, because when one does not work, one has no right to eat. The poor dogs, curled up, endure all the rigors of the weather, and only when the snow covers them, and by making an effort, do they rise and shake themselves for the moment, to return to the same position. Sometimes, after having lain for a long time, they melt the snow with the heat of their body. There comes a time when, either because the temperature drops, or because of inaction, they are not so hot; then their wet coat freezes, forming a single mass with the ice, and the poor dog is in that dangerous situation until someone, digging him out of the snow, cuts his hair with a knife, releasing him. If it goes unnoticed, he can freeze to death. On October 7, in the afternoon, we reached the so-called Christensen Island, and set up our tent on the ice, a few meters away from it. At night, the dogs went hunting; they found a seal with a calf, and killed the baby seal. The next day, while Nordenskjöld and Jonassen were on an excursion to the nunatak, I stayed near the tent to observe a noonday sun for the calculation of latitude.

Observations of corresponding altitudes were also made. On this day, we are not moving: The whole equipment will be dried and observations will be made. After concluding the astronomical observations, we went out hunting; our objective was to kill a baby seal, because their flesh is delicate. About two hundred meters from our tent, there were four or five seals with calves, and one of them was our prey.

It seems that birds are starting to come, because, as soon as we walked away with the flesh of the baby seal, we saw an Antarctic Megalestris that pounced on the remains; this is the most carnivorous bird in the Antarctic. After lunch, I went for a walk around the old crater, and it seems that Christensen and Sindemberg [Lindenberg], as well as all the other so-called Seal Islands, are nothing more than nunataks [*]. From the top of Christensen, you could see that an extensive snowdrift joined all the so-called Seal islands, and my opinion is that they are nothing more than volcanic nunataks that rise from a great plain that has very little height above sea level, and which is covered by that extensive snowdrift.

[*] Nunatak is a peak devoid of snow that rises in a glacier.

The front of the glacier is of the so-called Chinese walls, that is, a vertical front, and it stretches making some curves toward the direction of Sindemberg [Lindenberg], to then join the inland ice of Graham Land. As for the number and distribution of nunataks, it is very different from what was thought; navigation charts also include Sindemberg [Lindenberg] and Christensen, as active volcanoes.

Another thing that is striking is that, according to the most modern charts, Larsen did not see Graham Land from Christensen, which would only have happened if there were fog. From where I stood, one could perfectly see Mount Haddington, Graham Land, and there, toward the SW, Cape Framnäs, with Mount Jason.

As you ascend to the highest part of Christensen, you first find a two-hundred-square-meter platform or plateau, devoid of snow, and completely covered with fragmented basalt stones.

This plateau looks toward the NE, and to continue the ascent there, you must do so through the snow accumulated at the sides, because, to the SW, it is delimited by a perfectly vertical wall. Vegetation here is as rich or richer than in Snow Hill, apparently consisting entirely of lichens and mosses; this vegetation is very stunted, because the tallest plant is no more than two centimeters high. Probably, the cause of the vegetation being richer here than in Snow Hill is that the plateau, of which I have spoken, is quite protected from the prevailing winds and is continually subjected to the sun's rays. The highest part of Christensen is about 300 m high, and from there one can have a very good view of the surroundings. At the end of my excursion, I returned to the tent, the dogs had run with the speed typical of Greenlanders, and were lying beside the tent, ignoring the pieces of meat scattered around them.

On October 9, in the morning, a letter was placed on a pile of stones, and we started our march at 10 a.m., taking a NW direction, to look for a point from where it would be possible to ascend to the snowdrift. The march was difficult, especially for those who had to pull the sledge, for we were going through a maze of ice mounds filled with crevasses; to the left we had the snowdrift, and to the right a miniature ice ridge that follows a direction parallel to the front of the snowdrift and that surely was formed by pressures.

The day was unpleasant to travel because of the heat; the temperature at noon was 4° above zero, and not only we but also the poor dogs suffered horribly from thirst and fatigue, having to make repeated stops.

Thus, we marched toward Lindenberg, until 1:30, when we found a place where a pile of snow allowed the ascent, and there we were able to make our way to the SW.

Although we were climbing, the march was fast, because the snow was quite hard; we walked without sinking, and the sledges slid better.

At 7:30 p.m., we set up our tent; the temperature remained below zero, and the barometer was falling rapidly, all of which led us to forecast a southwester.

The next day, wind and fog from the southwest; the temperature dropped to 15° below zero, and the low temperatures and storms continued thereafter; many days we would spend inside the tent in cruel inaction: When a good day came, we would make the most of it.

On the 19th, the temperature dropped to 20°. With these colds, the march is very pleasant, and no one either wants or thinks about making stops, because, with a soft breeze, if you are not wearing a woolen suit, the cold and very penetrating air mortifies you.

On the afternoon of the 13th, the Seal nunataks disappeared, and toward the south, there was a hill covered with an ice shell; we only see in it some perfectly vertical rock edges that do not allow the accumulation of snow.

On October 14, my diary says: The cold weather continues, we march and work with pleasure; the footwear I wear are common boots, which cause me great mortification because of the extraordinary hardness they acquire due to the cold, and of course, in that instant, I think how inadequate my footwear is for those latitudes, and I recognize the superiority of that of reindeer skin, which, in addition to being strong and warm, is soft and light. This footwear allows the use of more grass in which to wrap the lower limbs, so with it feet are kept perfectly dry, which is not the case with the others.

Although the temperature is only between 20° and 25° below zero, I have felt some cold during the night, and this is because when I lock myself in my bag, the water vapor that is exhaled when breathing condenses, and then freezes on blankets and canvas wraps, gives them the hardness of a stone. So after several days, so much ice has accumulated that it is impossible to close the bag, and consequently, very cold air penetrates into it. You can have an idea of the temperature inside my bag by knowing that the footwear I had placed inside it, in order to soften it up with the heat of my body, was as hard the next day as the day before, and that the snow that had accumulated in its joints had not melted in the least.

It is clear that being away from land and with these temperatures, personal cleanliness is as neglected as circumstances demand; there we could not follow the advice: "To the land wherever you go, do what you see," and we therefore would do what we could.

One must not think about washing oneself; and because of this necessary abandonment, which I believe to be hygienic in low temperatures, little by little, a layer forms on the face and hands, with the deposits of grease, soot, etc., that accumulates on them, in a way that it seems impossible to see them return to their normal state. Dining and kitchen utensils are not washed either, because water must be heated for this purpose, and on these trips, everything is economized, especially oil, which is the only fuel we carry. To handle the instruments of study and to write in our diaries, it is desirable that hands be the least dirty as possible, since we cannot demand for them to be completely clean, and so we make small snowballs, which melt with the heat of the hands when rubbing them, and with this liquid, the thicker layer that covers the hands disappears.

When we cook, which is always in the tent, an amount of steam condenses in the inner part of it, and with any movement that is made in it, a very unpleasant rain of ice crystals falls. From October 15th in the afternoon, until the 18th in the morning, we stayed in the same place because of a great storm. In view of this, we decided to leave our sledge; taking provisions for eight days, the sleeping bags, and the tent, we headed toward the SW, leaving the other sledge with provisions deposited in depot for the return. Here, we were imprudent, which could have imposed a very high cost: We left that depot without taking points of reference to some noticeable features of the coast so as to facilitate finding it. This depot was the only food we had, and it had to last until our return to the station, and if we had lost it and then could not find any seals, we would have either died from starvation or become anthropophagous.

After around an hour of marching, we were ascending in a remarkable way, the crevasses in the ice multiplied, and we had to pass over snow bridges, below which we could see the unfathomable abyss, in all its horrible beauty.

The white of the surface turned pale blue, which grew darker, passing successively through all gradations, until there, toward the bottom as far as the eye can see, it was the most pronounced blue.

At that time, one of my reindeer inner-boots broke, and as the temperature in the air was 20° below zero, my foot began to freeze; by receiving timely care, the damage that could have been produced was contained, but the left heel was completely frozen, and only after repeated and vigorous rubbing with snow did the circulation return.

At 6 p.m., we camped in a nunatak, 200 m higher than the position occupied in the morning.

During the march, I observed a beautiful mountain toward the WNW that was completely covered with snow, which I suppose is Mount Jason.

At four o'clock in the morning, gusts from the SW were so strong that, fearing that our tent would break, we broke camp and went to shelter ourselves in a well that forms leeward of the nunataks; but things did not improve with the change. We had, indeed, moments of calm, but the wind would suddenly blow from all sides, swirling with extraordinary force.

There, our tent shuddered, and we were astonished at how it resisted the furious gusts. At 1 p.m., it somewhat eased. Then a hole opened near the side of my head, and the wind came in, bringing with its icy gusts a breath of death. It was the challenge of nature enraged, perhaps for being surprised by the human eye, and the white goddess, protector of those regions, resisted the audacity of men, throwing at their chests—with the glove of challenge—the terrible punishment.

The temperature was 22° below zero, and the wind was so strong, that there were times when it was very difficult to remain standing. We tied up our equipment on the sledge, and we left that place, apparently the cradle of storms, looking for a better sheltered location, but it was not possible to find it. In those frozen fields, death ruled, and before it, we squared off to fight that battle. We had to fix our tent, because without it, we could neither cook nor sleep. As is known, one cannot sew with thick gloves, or without fingers being somewhat sheltered by other better temperatures and in calm conditions; that protection is non-existent with 22° below zero temperature, and while being caressed by that furious wind.

Jonassen undertook the sewing of the tent, while I created leeward protection with my body and the rest of the tent the best I could, and that poor man concluded his work with several frozen fingers, which were restored only after a month. The tent was placed on the flank of the nunatak, and, as the slope was very pronounced, we had to use the hoe that we had to make a small platform so as to be able to set up the tent.

This place was the best that could be found in those distressing moments, and, although the wind continued with the same violence, the danger of losing the tent disappeared, by means of the defenses that we had made within the ice, using that same ice.

We were running out of food for the dogs, and being hungry, they would eat what they could find, which, by the way, must have been very little. I caught them with a bag that had contained pemmican, and, guided by the smell, they had broken it from its bottom, licking the fat that remained in it, and the hunger that they suffered was such that, a moment later, I found them eating the harnesses.

On the 20th, it seems that the wind wanted to subside, and I undertook the ascent of the nunatak to have a view of the surroundings. There was a bit of fog; however, you could see quite well. The land seems to make a curved entrance to the W, and then returns to the SW and S, very much like Larsen's sketch. The land here has the same morphological aspect as that of Louis Filippe and that of the Belgica Strait: deep gullies full of snowdrifts, and contrasting with them, black nunataks with their basalt blocks.

Not a single fossil has been found here; the vegetation is apparently the same as at our station; no plant is seen to flourish, and a more desolate picture than the one present in these regions, where ice is the exclusive and absolute king, is impossible to conceive. Given the amount of provisions we had, and the probable number of storms we would have to endure, we had to go back.

On the 21st, even though the wind was blowing from SSW, we began our return. Slight snow was flying low from the ice, down the slope of the snowdrift and with a tailwind. Our march was very fast; Nordenskjöld went ahead, marking the road with a stick and inspecting the ice, in order to avoid any accident. But nevertheless, in spite of that, from time to time, we disappeared into one of those great crevasses at the sides of the sledge, and, thanks to the fact that we would hold on to the sledge with one hand, we could remain on the surface without rolling down those deep abysses.

At last, we came out from among them, and we lost the old sledge tracks that we had been following so far; and after some time of good marching, we thought we were in the surroundings of the depot that we had left on the other sledge, but we could not see it. If some marking had been made, as I have already pointed out, the finding of provisions would have been a very simple matter, but, under the circumstances we were in, it was quite difficult. For us, finding that depot was our salvation; it was a matter of life and death. So we were all eyes, until, as luck—which never left us—would have it, after a while looking for it, we were able to spot it a few meters to our right. A great deal of snow had accumulated on our depot, and the parts of the sledge that were not sunk into it were so white in color that it was difficult to distinguish it at any distance.

That day we did not continue the trip, and we spent it arranging our equipment well, to undertake the march the following day.

Longitude and latitude observations were made. The following days were of true suffering; either storms would keep us imprisoned in our tent, or we would have to march with continuous fog. Our first intention was to follow a more westerly direction than that which we had undertaken upon our departure, but bad weather forced us to take the shortest route, so we headed toward Christensen. We were all beginning to suffer in the eyes; during the march, I never wore glasses, because, the ones I had being of very poor quality, there was no purpose in wearing them. Nordenskjöld and Jonassen always wore theirs, but nevertheless, they were the ones who suffered the most. The days when you had to strain your eyes the most, and when you felt more pain in the eyes, were not the clear and sunny ones, but the ones filled with those fogs that so much delayed our march.

We still had pemmican to give to the dogs, but the ration was greatly reduced; they ate up the cardboard boxes that had contained meat extract, and, the day prior, they ate a whip.

Finally, on October 31st, we arrived in Christensen, and immediately after camping, we went hunting for seals. We killed a large one and the calf, which we distributed among men and dogs.

In general, to kill seals, we would not use guns; with a hard blow from a stick, we would split the skull, leaving them instantly dead.

On November 1st, we headed toward Lockyer Island. To the NE, two black spots were visible on the cloudy horizon: one was Cape Foster, and the other one was Lockyer Island. The fog in the afternoon grew thicker and thicker, but that did not diminish the speed of our march. At times, however, we were deceived by refraction phenomena, which made us lose some time. Going forward, we thought we saw an *iceberg*, and when we were only a few meters away, and we started to go around it to avoid it, we were convinced that it was a small ice uplift, which was half a meter high, at most.

The next day, the march was also made with a lot of fog, but, like the previous day, the march lasted eleven hours. Our fatigue was increasing, but our valor did not diminish.

On November 3rd, the SW continued to blow with snow, but luckily, we did not have to use the compass anymore, and we could perfectly see the land to the NE.

The temperature was around fourteen degrees below zero; the sky was clear around the zenith; but at low altitude, over the horizon, there in the SW, we could see the cumulus–nimbus that announce the proximity of the storm.

We are heading for the strait that separates Lockyer Island from Mount Haddington; our march is fast; we pull the sledge with all our strength, and so do the dogs that come behind us, driven by the whip of Jonassen, and perhaps because they already smell Snow Hill.

When one of the dogs, exhausted, reduces his pulling by just a little, leaving more of the burden to his companions, the whip works, and the poor brute, with a howl of pain, lurches with the sledge, and this supreme effort is rewarded only with a look of indifference from he who manages the dogs.

Large numbers of snow petrels (Pagodroma nívea) fluttered around us, the same as gulls (Larus dominicanus); but we could not see a single seal. The wind increased in intensity, raising with its force clouds of powdery snow; the temperature went down, and because we had a tailwind, the march was pleasant. At noon, we could see perfectly Cape Depósito, and we decided to make an attempt to not stop until we reached the station.

At 7 o'clock, we saw the island clearly, and we could perfectly see the basalt and the nunatak of the station. And we made our way to them. The wind was blowing with a force of seventeen meters per second, and a lot of snow was falling; we had a terrible hunger, and a vehement desire to drink something hot, as we had not had a snack since nine in the morning. On the way, we ate some chocolate. At 1:30 a.m., after sixteen hours of forced marching, we arrived at the station. That is to say, it had taken us three days to walk from Christensen to the station, whereas on the journey out to there, we had traveled the same distance of eighty miles in seven days of marching. The total distance traveled on that trip was three hundred and forty miles. Upon returning, we found everything perfectly well at the station. The observations had been made with the customary regularity, despite the shortage of personnel caused by our departure.

When we returned from this bumpy sledge trip, with a voracious appetite, we ate a lot, and drank huge amounts of coffee, tea, and water. What a pleasure to satisfy hunger and thirst!! On the subject of hunger and thirst, let's look at the weight differences that occur with forced exercise.

Before leaving on September 21 on this sledge excursion, we all weighed ourselves, and we did so again when we returned.

I had these differences: the day of departure, I weighed 152 lb, and on November 4th when we returned, I weighed only 135 lb; I had lost 17 lb in 44 days.

Fifteen hours later, I weighed 143.5; that is, in that short period of time, I had regained 8 lb of the lost weight.

The salient details of this small campaign, so full of accidents and hardships of all kinds, will give you an idea of what a sledge trip is like in those regions that form the empire of Polar challenges. Once again, the will and decision of men triumphed over the obstacles which, when advancing, they always find in their way; and once again, the energies of men, sustained by their hopes, during their vital fight against nature, dominated nature's challenges, bringing to the civilized world the shreds of its inclemency.

Ice and hunger were defeated on that day by constancy and resistance, and it is a great march, for not many have walked sixteen continuous hours without rest on the crystallized surface, skirting its mounds, traversing its currents, and jumping over its abysses.

The station was located at latitude 64° 22′, and 57° longitude west of Greenwich. Latitude calculation was made by meridian altitudes of stars observed with a prismatic circle over the artificial horizon. These observations were at my sole charge.

The calculation of longitude was made by lunar culminations, and it was under the charge of Mr. Bodman, with several determinations also made by myself.

Chronometers were set by meridian passages, or by altitudes observed with the prismatic circle over artificial horizon.

Astronomical observations were very difficult, especially in winter, with the rigors of cold. When the temperature is below 20°, and if the weather is not dry, the use of the theodolite instead of the sextant is recommended, because mirrors and glasses are continuously covered with a layer of ice that is very difficult to remove. In addition, there is always a lot of wind blowing there, in some cases making it impossible to observe the sextant, and the use of the theodolite is more advantageous.

8.1.9 Tides

Tidal observations were made throughout the month of June, one of the coldest and most stormy months. A tide scale was placed near the shore into a hole made in the ice; the installation was made on a piece of stranded ice, so there was never any scale displacement. Observations were hourly, and when the weather was good, they were made every five minutes, in the proximity of high and low tide. To perform the readings, of course, a lantern was taken, and in times of storms, we sometimes had to make two or three trips, for the lantern would go out. As the strongest storms were from the south, and the location where the tide gauge was placed was to the north, about 300 m from the station, in times of storm, while on the way back, we would have to go against the wind, and the lantern would go out. And because of the snow, which hurt our eyes when it was blowing at a speed of 30 m per second, we could not see the house, and therefore we would get lost; and sometimes we would spend a lot of time wandering around, without being able to find our home.

Mathematicians, until now, have only obtained an acceptable solution relative to the dynamic theory of tides, considering that the sea completely covers the spheroid, or that lands are distributed according to the parallels. And, as we know, the conditions of distribution of land and water on our planet are far from satisfying these conditions, but the place on earth that is closer to the conditions of the theory is undoubtedly the Antarctic, as it is an area of relatively small lands, surrounded by an immense sea.

So, this is why the observations of these phenomena are especially interesting from a purely theoretical point of view. In addition, there have not been any observations of this kind, taken of the south Polar sea, until our expedition.

Observations of terrestrial magnetism were made from April 1, 1902 to November 1, 1903.

Nothing has yet been deduced from these observations, and they have been made in accordance with the international program.

The strangest theories about the Antarctic climate had been made, but the one that had been accepted, until the *Belgica* expedition—which was the first to present a full year of observations, and which gave it a death blow—was that the Antarctic had an essentially marine climate, and that slight temperature variations and low pressures were its characteristics.

As for the distribution of land, the Antarctic is completely different from the Arctic.

The latter is a small sea surrounded by vast continents, while the former is a small continent, or an archipelago, surrounded by a great ocean.

And the theory about its climate was based on that. They would say: The Antarctic, with its small continental extension and its extensive seas, has a marine climate with relatively cool summers and mild winters.

At our station, we had the coldest summer ever observed in the world; the temperature during the hottest month was below zero; the winter of 1902 was also very cold, dropping the thermometer, in the month of August, down to 41.5° below zero. At 39°, the mercury would freeze, and we had to use alcohol thermometers.

As for the difference between the extreme temperatures that we have had, it is very large, and both have been in the same month: The coldest was 41.5° below zero, on August 6, 1902, and the hottest was 10° above zero in August of 1903; that is to say, a difference of 51.5°.

As for the atmospheric pressure, it is very low, but it is also subject to great variations. The maximum was 762 mm in April of 1902, and the minimum was 709.0 in the month of June of 1903; that is, 53 mm of difference.

The most common winds are those of S, SSW, SW, WSW, NNE, and NE, and the rarest are NW. There it always blows and usually in hurricane, but it does not mean that there are not calm and sunny days, during which one feels true pleasure to be in those places.

The air is almost always close to the point of saturation, and precipitation is very abundant. We had a rain gauge, but due to the strong winds, snow would not fall on it; these same difficulties have already been encountered in the north.

The force of the wind is very great, especially that of SW winds, which exceed 30 meters per second; with these speeds, the axes of our regulating anemometers would break, and the observations would be made every hour with pocket anemometers.

Ice temperature observations of glacier ice, sea ice, and earth ice were made at different depths.

The most complete works of the expedition are undoubtedly the meteorological ones, which are of special interest because they show the coldness and variability of a climate that the theorists had considered to be the opposite.

Meteorological observations extended from March 1, 1902 to November 8, 1903.

A geological survey was made of the entire region visited by the *Antarctic* and by us.

The collections of fossils are very rich and interesting: It is the first time that fossils of vegetables and vertebrate animals have been taken from the Antarctic. On Seymour Island alone, remains of enormous vertebrates, of rich vegetation, and of marine animals were found; on Haddington Island, as well as in Andersson's wintering place, in the Strait of Joinville, plant fossils were collected, and those in the latter location were much better preserved than those on Seymour Island, as leaf impressions were on slate, whereas on Seymour they were on sandy stones. On Cockburn Island, marine fossils were found as well.

Seymour Island is cut by numerous glens, where some snow is piled up, but there are no large snowdrifts in it. Although the height of perpetual snow is at sea level, the general impression of the island, for those who see it from a distance, is that of its being devoid of snow. And it is difficult to imagine the reason why, for example, part of Snow Hill is covered by a huge snowdrift that sinks its tongues into the sea, while Seymour and the N part of the island have little snow. Maybe it depends on morphological conditions. Of all the areas of those lands that I know, Seymour is the only one that is not covered by a layer of ice. On both Seymour and

Snow Hill, where the station is located, one can see large basalt walls following a SW–NE direction; on them, lichens and mosses grow. One of the richest areas of vegetation in that region is undoubtedly Cockburn Island.

How great is the difference between these desolate Antarctic lands and the Arctic lands! While in the south the vegetation is the most rudimentary, and only one blooming plant has been discovered by Racovitza—a naturalist from the Belgian expedition, in the north, at 80° latitude, the ground is covered with "Cassiope tetragona" and "Ceratium" of beautiful white flowers, and with the "Saxifraga," with its red buds, filling those places with joy.

The glacial lands of the north have a beauty that its antipodes does not lack: the so-called red snow. It has been observed on the islands of King George and Shetland, and in the land of Louis Filippe [Louis Philippe Land]. The snow takes on the red color from an alga "Sphoerella nivalis," of more or less ovoid shape. Two tons of red snow has been observed, one blood red and the other a brick color; the name of the alga that gives color to the latter is "Chlamydomonas lateritia."

8.1.10   The Two Ices

The ice observed in the Polar regions can be divided into two classes: land ice and sea ice. To the first belong the ice shells that cover the Polar lands, which are called "inland ice" (*inlandsis*).

Icebergs, or ice mountains, belong to the first group, while the *pack* belongs to the second.

The snow, falling and piling on that layer of ice, exerts a pressure, which has the effect of making flow, in between glens, kinds of rivers of ice, so to speak, that serve as drainage for the "inland ice."

From those rivers, due to the aforementioned reason, huge blocks of ice detach, which in general, and which is typical of the Antarctic, are of a tabular form.

In the Arctic sea, icebergs are, in general, of irregular form, full of peaks; but, however, there are tabular ones, and they are called *floebergs*, due to the assumption made by some that they are formed in the same way as floes (like sea ice) by the freezing of successive layers of seawater in the lower part. This has been proven impossible, as the sea, with its known colds, only freezes up to a certain limit of not more than three meters. There are some who have claimed that the *icebergs* of the Antarctic were formed in this way, but it suffices to see what snowdrifts are like in the south, and to observe their aspect, to be convinced that this is a mistake. As for the height of the *icebergs*, I will say that I have not seen any that were more than 100 meters; Cook saw in the south *icebergs* that were between 15 and 90 m; [George] Nares, in the expedition of the *Challenger*, observed them to be of up to 76 m of height. Their extension is sometimes very large, and they resemble true islands of ice.

Observations were made on the Snow Hill snowdrift; for this purpose, a number of sticks were placed nailed to the ice and oriented in a special way between two nunataks, and after placing in a plane the position of the sticks at different times, the direction and speed of the movement were obtained.

Oceanographic observations are also very important; a great number of soundings were made, and at the same time, animals were collected from all depths, as well as temperatures of the water.

The study of the plankton was given great attention, because it is of great interest within oceanography, since the quality of the plankton depends on the salinity and temperature of the waters, providing with its constitution information about the kind of waters where it is found, and their provenance.

According to what I have been told, the *Antarctic* made numerous soundings last year, which, because of the places where they have been made, are of special interest. There is a theory that claims that the mountain range extends to South Georgia, passing through Staaten Island, Burdwood Bank, and Shag Rocks; that the mountains of Louis Filippe Land also continue to South Georgia, passing through the Orkneys; and that the two continents are joined there. But bathymetric conditions do not agree with this theory, and, according to what I have been told, at the N of Shag Rocks, the *Antarctic* found the enormous depth of 6000 m.

Soundings were also made in the Belgica Strait and in Bransfield Strait, from Cape Murray to Joinville Island.

Several islands, straits, and lands were discovered.

The Swedish expedition, on its outward journey in 1902, proved that the land of Danco and that of Louis Fillipe are the same, and that the land of Trinidad is a small island. And on the last voyage of the *Antarctic,* it was found that Middle Island does not exist; a sounding was made in the position in which it was marked, and several hundreds of meters [of water] was found.

On the journeys made from the station, a coast never seen was traversed, between Mount Haddington and King Oscar Land. Mount Haddington, as well as Snow Hill, were found to be islands. A strait investigated by Nordenskjöld emerged from the N of Cape Foster and ended, more or less, where Ross's chart marks Cape Corry. Another much smaller strait joins the aforementioned with Sydney Herbert Bay, so that Cape Gordon is another island. In Joinville Strait, several small islands were discovered, as well as a narrow strait that cuts Joinville Island in two.

Topographical, geological, meteorological, and astronomical observations were made during the sledge trip to the south. The topographical and geological ones were under the charge of Nordenskjöld, and I was in charge of the others.

Christensen, according to my observations, is about twenty-two miles north of the position given by Larsen, and so is the rest of the coast of King Oscar Land. All in all, as can be seen from my exposition, no opportunity has been missed to do everything possible to advance science, and this, the pride of the modern age, owes much to the Nordenskjöld expedition.

Broadly speaking, I have narrated the life spent, for two years, in the Polar regions; and, very lightly, I have described those desolate and inhospitable lands.

The mysteries which these ices contain must be discovered, and no people, whatever their race and history, can advantageously compete with the Argentine people in this superb enterprise, which will elevate the name of our country, until it is inscribed with indelible characters in the list of great powers.

The fantastic Antarctic has already been beaten on its flanks, and in that titanic struggle that men have sustained against nature, the latter has compressed its rages and its rigors, ceded the keys of its first bastions, and, overwhelmed by the weight of defeat, has given part of its frozen dominions to the warm atmosphere of humanity.

A few years ago, our Patagonia was a legend; our southern seas inspired fright, and the regular and pleasant journeys that are made today to the Strait were at that time daring adventures that made one doubt the criterion of those who undertook them. A little resolution inspired our conquest of the territory where ignorance had prevailed, and a well-combined plan destroyed the frontiers of barbarism and of the rebel, and from that happy moment the savage disappeared from the Argentine soil, whose domains were stripped away, in the name of civilization.

Exactly the same thing happens with the Antarctic Polar ice cap, where, instead of sending armies of soldiers, we must take legions of scholars, and thus, in consonance with the national character, in a single time, in a single blow, we must break those perpetual snows, which will then never amalgamate, because the ardent rays of civilized life that feed the sacred fire of progress will prevent them from doing so.

The lightning ray has been harnessed, it is today taken where one wants; the time and the distances, which today are shortened by the transmitting of words through wire, have been suppressed, and even the waves have now been put at the service of progress; and as the century of our activity cannot slumber with the advancements that the past has witnessed, I already foresee the amazing progress that it will boast about, and with the dominion of the air, we will also see the definitive conquest of the Pole.

Ladies:

Sirs:

If I have tired you, if I have abused your benevolence, excuse me that there are no flowers at the Pole, no heat but that which springs from the bosom of the soldier, who on more than one occasion had to seek the colors of his country in the immaculate white of the ice and in the blue of its twilights, in order to cheer his spirit that had become disheartened by fatigue and by the frightening isolation of those regions.

References

Nordenskjöld NO, Andersson JG, Skottsberg C, Larsen CA (1905) Antarctica: or two years amongst the ice of the South Pole. New York: The Macmillan Co

Sobral JM (1901–1903) Expedition diary of José María Sobral, Swedish Antarctic Expedition of 1901–1903. Departamento de Estudios Históricos Navales, Archivo Histórico. El Diario de Alférez de Navío José María Sobral. Buenos Aires: Archivo D.E.H.N.—A.R.A

Sobral JM (1903a) Conferencia del alférez de navío José M. Sobral [Conference by under-lieutenant José M. Sobral, sponsored by the Naval Center, and read in the Politeama Argentino on December 19, 1903]. In Boletin del Centro Naval, Tomo XXI, ed. Capitán de fragata Juan I. Peffabet. Taller Tipográfico de la Escuela Naval Militar, Buenos Aires, pp 485–512

Sobral JM (1903b) Handwritten notes for speeches and for official report to the Minister of the Navy. Departamento de Estudios Históricos Navales, Archivo Histórico, Colección Sobral. Box 2, Envelope 4, Manuscrito referido al viaje a la Antartida y Conferencia en el Centro Naval 1903. Archivo D.E.H.N.—A.R.A, Buenos Aires

Chapter 9
After the Expedition: A Life of Science

Abstract An account of José María Sobral's life after returning from Antarctica, includes his decision to leave the Navy, his move to Sweden to follow his mentor Otto Nordenskjöld, his studies at Uppsala University, his becoming the first Argentine to earn a formal degree in geology, his return to Argentina, and his ground-breaking position as Director of the Argentine Geological Survey.

The bloom was off the flower … the flower that Sobral had failed to find at the South Pole.

For several months after the return of the Swedish Antarctic Expedition, the handsome and unassuming young Under-Lieutenant was in high demand, speaking at every organization meeting and every public event possible. Naval groups, civic societies, children's charities, educational organizations, all wrote to him pleading with him to speak at their institutions, for the benefit of their members and their beneficiaries (Correspondents 1903–1904). He delivered lecture after lecture about his time in Antarctica, making intricate notes to follow as he regaled his audiences with tales of his adventures, copying his diary entries over and over again, and now interspersing them with newfound wisdom and insight achieved after having completed the remarkable expedition (Sobral 1903).

He shared his story with many people and spoke with many groups, each time begging their pardon to bear with him as he relayed his experience as thoroughly as possible. In his speeches, he underlined the risks and the hardships of the expedition, but underplayed his own courage and conviction. He gave credit to the bravery of his companions and assumed a modest attitude toward his own accomplishments. He offered the narrative from his own point of view, but he asked the listener to look at the big picture, to take a holistic approach, to recognize the importance of the Polar regions and their interconnectedness to the world.

Most of all, he shared the intimacy of his passion: a newly discovered love for discovery, a desire to learn and understand the nature of the earth. His was now a mission to unearth the history of the planet's beginnings and to pursue the revelations of earth sciences.

© Springer International Publishing AG 2018

M.R. Tahan, *The Life of José María Sobral*, Springer Biographies,
https://doi.org/10.1007/978-3-319-67268-7_9

His first order of business was to write the book. After the speeches had been delivered in that last month of 1903, he set about completing his official book about the expedition, titled *Dos Años Entre los Hielos, 1901–1903*. It was the first book on Antarctica written by an Argentine. He published the book himself, not finding a major publisher to publish his work the way his colleagues Otto Nordenskjöld, Johan Gunnar Andersson, Carl Skottsberg, Carl Anton Larsen, and others had done (Rabassa 2003). He printed the book as a limited edition. And it was an immediate hit. Request after request was received to donate the book to societies, schools, libraries. He spent the first half of 1904 distributing the book and speaking at engagements in and around Argentina (Correspondents 1903–1904).

Sobral dedicated his tome to his parents. With the same tenderness that he had exhibited in writing about his mother and his father in his Antarctic diary, Sobral now made a loving dedication to them in his book (Sobral 1904):

> To my parents
>
> Who remind me of the most beautiful feelings of the heart: love, gratitude, and respect, I dedicate this book

In that book, he expressed his fervent desire to go once more to the Antarctic regions, that place where he had faced death many times. Though he had just arrived home, and though he had written most emphatically about the hardships he had endured, he wished to return to the white continent to explore even further (Sobral 1904:168):

> And yet, I would like to go through those sensations again. I would like to return for several years, to go into the freezing regions. I would hear again in that deathly silence the noise of the glacier, the hissing of wind and snow by the flanks of the tent; I would like to have to scrub my hands with snow to return them to life, and to march alongside a sledge pulled by twenty dogs in order to plant the flag of the mother country beyond the 80th parallel!

Sadly, José María Sobral would spend the remainder of his life striving fervently to return to the ice that he had so vehemently wished to escape during his second overwintering in Antarctica.

In his book, Sobral pays homage to the late Axel Ohlin, the scientist who had died after returning home early to Sweden during the expedition, and whom Sobral viewed as a fellow colleague who had "sacrificed himself for the sake of science" (Sobral 1904: 42). The tribute is included in Sobral's very first chapter citing the history of Antarctic exploration. He also pays homage to the loss of the *Antarctic* and one of its crewmembers Ole Wennersgaard. This later tribute is in the form of the final chapter of his book, which Sobral devotes to a personal account of the event as written by Carl Skottsberg, whose by-line is given as "Carlos Skottsberg" (Sobral 1904: 355).

Controversy courted Sobral from the very beginning, and the next-to-the-last chapter of his book reflects this. That chapter, titled "A Rectification," is a repudiation of a news article in which Sobral was misquoted from one of his speeches given at the end of 1903. In the chapter, Sobral addresses the article written by "Señor Enrique A.S. Delachaux" and appearing in *La Prensa* on January 1, 1904, in which Sobral is alleged to have quoted Otto Nordenskjöld in an opinion regarding "the hypothetical descent of the mountain range by the banks and island groups that extend west of the South American continent," specifically that "There is a theory that the mountain range is extended to South Georgia by the Isla de los Estados, the Burdwood bank, and the Shag Rocks; that the mountains of Louis Felippe Land also continue to the South Georgia passing through the Orkneys, where the two continents joined; but the bathymetric conditions do not agree with that theory and, as I have been told, north of the Shag Rocks the *Antarctic* found the enormous depth of 6000 m" (Sobral 1904: 335–336). Sobral went on to refute the implication that he had quoted his mentor about this and even exhibits his own open-mindedness about the theory, emphasizing that he had read some of this information from Johan Gunnar Andersson. He goes on to devote nine pages to the details and analysis of this theory. The chapter is another example of the compulsive need for accuracy that Sobral had exhibited during his Naval days and days in the Antarctic, especially when targeted toward scientific fact.

Sobral was still a Navy man and resumed his normal duties once he returned to Buenos Aires. He was inducted as an honorary member of the Centro Naval Club in 1903 (Centro Naval 1903). He also began to write and edit the naval publication *Boletin del Centro Naval*, publishing a series of lectures (Piazzi 1905). But he was smitten with science and simultaneously began to take courses at the Facultad de Ciencias Exactas, Fisicas y Naturales (School of Exact, Physical, and Natural Sciences) at the University of Buenos Aires. He successfully completed three courses and passed three examinations in coursework that would take him toward a degree in natural sciences, with a focus on geology. A doctoral degree in his field was not available at an Argentine University at that time, but he nevertheless strove to earn whatever degree would be possible. And he hoped that an opportunity would arise again to work in the Antarctic.

Despite repeated proposals, no opportunity presented itself to return to Antarctica, and the Navy did not seem eager to send him back any time soon. Moreover, Sobral's studies did not meet the agenda of the Navy at that time. The Under-Lieutenant appealed to the Navy about his desire to return to the South Polar region as well as requested that they allow him to study his newfound vocation of science. When the answer was in the negative, Sobral found himself forced to submit a request for a discharge, which he did in 1904.

The bloom was off in Argentina.

That same year, his Antarctic companions called to him. They had returned to Sweden to carry out studies of their Antarctic findings. And Nordenskjöld invited him to come and study with him. The Navy granted him his discharge. It came through at the very end of the year, on December 30, 1904. As of that date, Sobral was no longer in the Navy. He had made a life-altering decision. He would move to Sweden where he could study geology with his Antarctic mentor Nordenskjöld. The irony here is that the birthplace of those individuals—those handful of members of the expedition who had ostracized him—was the place which now welcomed him to a new home, within the embrace of geological studies, at Uppsala University.

Meanwhile, Argentina was in a love affair with Antarctica. A national palpable heartbeat had begun to pulsate, its object the great frozen region of the south. Technically, the rescue of the Swedish Expedition by the *Uruguay* was the first official institutional Argentine act in Antarctica. Emotionally, the great achievements and bravery of Under-Lieutenant Sobral, and the crew of the *Uruguay*, lit a passionate fire within the hearts of the Argentine population and fired up their imaginations. In response, when Nordenskjöld's friend Dr. William Speirs Bruce, returning from his own voyage to Antarctica on the *Scotia*, offered the Argentine government the meteorological observatory that he had established on Laurie Island in the South Orkney Islands in 1903, the Argentines jumped at the chance. Another orphan in terms of the lack of his country's support, Bruce and his Scottish National Antarctic Expedition of 1902–1904 had basically been shunned by the leadership of Britain's Royal Geographical Society. Argentina accepted Bruce's proposal to transfer ownership of the weather station and purchased it in early 1904, acquiring the station and its instruments, and naming it Base Orcadas. Argentina has continued to operate the station to this day, making it the earliest continuously populated settlement in the Antarctic, and the oldest weather station in Antarctica, and enabling important access to continuous meteorological data that dates back to 1903 (Rabassa 2003).

And so Sobral made the major move, sailing to Sweden in August of 1905, on board the steamer *Drottning Sophia*, and enrolling two months later at Uppsala University, where he passed the admission examinations and began studying in earnest to become a geologist. He also met and fell in love with a young woman—a Swede—who captured his interest as much as the science he had come to Sweden to pursue. Her name was Elna Wilhelmina Klingström.

In the following year of 1906, on the date of September 6, Sobral married his Swedish sweetheart. Ever a multitasker, he juggled full-time study and family life. He had four children with Elna Klingström Sobral over the course of eight years.

Studying fervently, and excelling in his courses, Sobral passionately pursued his University studies at Uppsala. He completed his dissertation, which was approved, and entered the graduate program in January 1908, receiving his graduate degree in geology (Uppsala 1908) and earning his doctorate in geology in June of 1913 (Sobral and Sobral 2014). His thesis, on May 28, 1913, was praised as demonstrating "the richness of the valuable and carefully made observations ... especially with reference to the chemical parts and the meritorious map" by the University's Thorulf Fries, who also wrote "I consider it necessary to recommend 'Approved with honors' as a note for this work" (Uppsala 1913). Sobral now had his wish—a Ph.D. in geology, mineralogy, and petrography, for which he had worked very hard. It was a hard-won achievement.

He could have remained in Sweden. He could have accepted a lucrative position in Canada (Sobral and Sobral 2014). But he decided to return to his homeland and to contribute his newly acquired knowledge, skills, and academic degree to the land of his birth.

Sobral returned to Argentina the following year, in 1914, once more triumphant —this time as the first Argentine to have earned a Ph.D. in geology—indeed, as the first to have earned any degree in geology. He immediately began working at the Dirección Nacional de Minas e Hidrología (National Directorate of Mines and Hydrology), today known as the Servicio Geologico Minero Nacional/SEGEMAR (National Geological and Mining Service Institute, or Argentine Geological Survey). It began as a humble job that turned into a prestigious position and a platform for his scientific work, as well as for his patriotic pursuit to protect Argentina in its rights to its resources. He did fieldwork throughout Argentina and in its surrounding countries, including Uruguay, Brazil, and Chile, the latter of which he worked with to establish boundary treaties (Marenssi 2007). One recalls the angst he suffered in Antarctica regarding a possible war between Argentina and Chile; now, he worked to allay such tensions and keep the peace. He proceeded to undertake some of the very first rock petrographic studies in Argentina and succeeded in compiling a geological map of Argentina in 1923. By the end of 1922, several years after he had begun working with the Directorate, he had assumed its directorship. As the first non-foreign director of the Argentine entity, he had also embarked on creating the register of mines and mining code, for which he spent several years of hard work and stubborn persistence, protecting what he strongly viewed to be Argentine resources (Sobral and Sobral 2014; Rabassa 2003). It seemed that the hero of the Antarctic was now occupying the highest science position in Argentina.

He also did a brief stint as a professor of mineralogy and geology at the University of Córdoba, from 1918 to 1919 (Marenssi 2007).

Sobral also continued to grow his family. Once back in Argentina, Sobral and his wife had five more children. The final tally was nine—one more child than the total number of Sobral and his siblings. Sobral had two girls and seven boys.

International honors abounded and were abundant. Sweden awarded him with the "Order of the Sword" on May 25, 1904, and the "Order of the Polar Star" (Stockholm 1904). He was invited to become a member of the Geological Association of Stockholm in 1908. His Antarctic colleagues Johan Gunnar Andersson and Carl Skottsberg named a new mineral after him, in 1919, to commemorate his admirable contributions to science—that rose-red gemstone was aptly called "Sobralite" (also known as Pyroxmangite) (Sobral and Sobral 2014). Sobral was invited to become a full council member of the Royal Geological Society of Gothenburg in 1926. He received the David Livingstone Centenary Medal from the American Geographical Society in 1930. That same year he was appointed Argentina's Consul General in Norway and lived for a year in Oslo, where he was also given full membership in the Norwegian Geological Society in December 1931 (Sobral and Sobral 2014). And he was awarded membership in the Swedish Society for Anthropology and Geography in February 1937 (SSAG 1937). These were but some of the prestigious awards bestowed upon Sobral. The Argentine Navy also awarded him the "Naval Cross for Distinguished Services."

Part of the reason for his brief assignment in Norway was a situation that had arisen in 1930 in which Sobral was accused of management irregularities while at the National Directorate of Mines and Hydrology. The government at the time was under the administration of President Hipólito Yrigoyen, and the country was seeing many changes (Sobral and Sobral 2014). Sobral's Director position with the Directorate was officially ended on March 31, 1931. As of May 31, 1930, he had been exonerated and dispatched to Norway as Consul. He returned to Argentina in June of 1931 (Sobral 2017).

Upon his return from Oslo to Buenos Aires, Sobral was hired by the petroleum company Yacimientos Petrolíferos Fiscales (Fiscal Oilfields, or YPF) as a field geologist, a position he fulfilled with the same intricate detail and compulsive record-taking that he had exhibited in Antarctica. It was what he loved to do—explore his country, and record and analyze his findings. Sobral worked as chief of the geological department. A diary kept by him during his fieldwork in Ñirihuau (Río Negro Province), in the year 1932–1933, shows his careful observations and drawings of land features, rocks, sediment, limestone, and basalt. Measurements, charts, sketches, and preserved specimens pressed within the pages of the diary again communicate the depth of his passion for this type of work (Sobral 1932–1933).

In 1935, a few years after he began his work with YPF, Sobral retired, and spent the remainder of his time with his family and his land. It is stated that upon his retirement, the Navy did not recognize the two years that Sobral had spent in Antarctica, as he had not been sailing on board a ship during that time (Sobral 2016). Nearly three decades later, the Navy would try to recognize Sobral's efforts, inviting him to be a part of the final voyage of the *Uruguay* as the little corvette was transferred to Buenos Aires to take its permanent resting place as a museum ship (see Overview).

During his later years, Sobral never ceased speaking and writing about Antarctica and about scientific and geological concerns. He never abandoned his intense study of the frozen continent, continuing to read other explorers' accounts deeply and voraciously, and analyzing and critiquing each one he came across. The method of questioning and of deep analysis had not changed since his diary days.

Viewing his volumes and volumes of notes from those years, one surmises that Sobral was still intensely sharp, and very particular, analyzing everything he saw in front of him and correcting everyone else's work—especially writings by other scientists and explorers. His personal notes and manuscripts include volumes of transcribed accounts from other explorers' reports and papers, filled with his own comments, questions, and corrections on the articles. He painstakingly copied each article in the original language in which it had been published. Then, he annotated it, quite liberally and emphatically, with his own observations. Therefore, one finds manuscripts in his handwriting in several different languages. He would effortlessly change from English to French to Swedish, often interjecting his own personal notes in Castilian Spanish and peppering his papers with Norwegian, Basque, and other languages. It is an enviable and seamless transition from one language to another. Moreover, he would vigorously underline passages of particular interest to him, which often included references to geological phenomena, comparisons of geographic locations, and detailed descriptions of dog sledging. These topics were close to his heart.

One example of these academic journeys is Sobral's vigorous comments on Admiral Richard E. Byrd's report regarding his flight over Antarctica in 1930, correcting the American's assumptions on the ice shelf, correctly identifying the Dry Valleys from Byrd's descriptions of the land, and marking with several exclamation marks the descriptions of imprisoned gas (most likely subglacial lakes), which truly seemed to capture Sobral's imagination (Sobral 1930–1949). Quite emphatically, he also notes that the Larsen Ice Shelf was first seen by Nordenskjöld. Another example is Sobral's underlining of Roald Amundsen's account, commenting on his use of "les chiens" to achieve success in his 1910–1912 quest to reach the South Pole (Sobral 1930–1952). And another is his compassionate notes for Robert Falcon Scott regarding the anguish of manhauling.

All in all, Sobral was quite prolific, writing and editing a plethora of articles and books between 1910 and 1959 (he had already begun to publish his work while at Uppsala in 1910). These published works included *Contribution to the Geology of the Nordingra Region* in 1913, *Problemas Hidrográficos en los Andes Australes* in 1921, *Geología de la Comarca del Territorio de La Pampa Situada al Occidente del Chadi-Leuvú* in 1942, *La Exploración del Continente Antártico* in 1949, and *Reseña Histórica de Bahía Esperanza y Noticias Sobre una Expedición a la Tierra de la Reina Maud* in 1952 (see Bibliography). He also maintained his community and science outreach efforts, donating his books, and continuing to give speeches for which he was thanked and lauded, receiving letters of gratitude from every corner of Argentina (Correspondents 1949–1959).

In addition, Sobral continued to defend his mentor Otto Nordenskjöld and to correspond with his Antarctic friends, specifically Carl "Carlos" Skottsberg.

During the last two years of the 1950s, a young up-and-coming army lieutenant named Gustavo A. Giró Tapper, who was stationed in Antarctica, and who viewed Sobral as one of his heroes, began planning two ambitious crossings of the Antarctic continent. As part of his training and preparations, he traveled to Norway in February of 1961 and took it upon himself to visit Carl Skottsberg in Gothenburg, Sweden, bringing the veteran Swedish explorer tidings from his old Argentine friend, José María Sobral. Skottsberg was thrilled to hear from his Antarctic and Uppsala colleague. Memories of Snow Hill, Paulet Island, and the *Antarctic* overwhelmed him (Tahan 2017).

Two months later, everything changed. The ever-active Argentine explorer approached his final horizon. José María Sobral died on his birthday, on April 14, 1961. He had just turned 81. As a person who had avidly kept almanac-worthy records, comparing and contrasting them for similarities and differences, he would have been intrigued to know that the date of his birth was the date of his death. He greeted that death, which he had averted so many times on the frozen icescapes of Antarctica, in the warmth of his home in Buenos Aires. He was buried in La Chacarita cemetery. He had lived his life pursuing both knowledge and understanding—a balance of the mind's intellect and the heart's acceptance.

Carl Skottsberg perhaps gave the most eloquent remembrance of Sobral. The scientist who had so beautifully eulogized the *Antarctic* now gave a fitting farewell to Sobral, full of poignancy, practicality, and humor. He wrote to Gustavo Giró, in Spanish, on August 8, 1961 (Skottsberg 1961):

When you paid us a visit – too brief – to deliver the greetings from our friend José María, who would have had the faintest reason to fear that in a few months death would separate us forever? I knew it right away because his wife sent me a cable and, shortly after, a letter. Sobral was the last of Nordenskjöld's six comrades in Snow Hill. In April 1960, cook Åkerlundh died, in May that same year, Ole Jonassen, the Norwegian who was in charge of the dogs. As for the ones from Paulet Island, besides who is writing, the zoologist K.A. Andersson, 85, and the third pilot Reinholdz, 87, are still alive. The three from Hoppets vik (J.G. Andersson, Duse, and Grunden) are dead. Your trip to Snow Hill and the South Coast interests me very much. Several years ago, a member of an English expedition informed me that Nordenskjöld's hut was half full of ice.

Sunk and maybe preserved by the cold are the annals of a comic newspaper called "Strix", which was used by Sobral to study our language. Now they are very rare and expensive. The idea to set up the hut to make it useful as a refuge for future explorers is excellent.

He signed the letter "Carlos" Skottsberg—perhaps the name that Sobral had endearingly called him.

The loss of Sobral was felt by the Antarctic community. Tributes poured in from around the world. Notices were printed in Buenos Aires. A year later, in April 1962, Skottsberg himself began compiling a series of photographs to illustrate a new edition of Sobral's book. He delighted in the news that Sobral's birth home had been decorated with "a commemorative plaque," and he was especially happy to receive a special issue of a magazine that featured Sobral's biography. Sobral's wife Elna and his son Alvar remained in contact with the 81-year-old Skottsberg, giving to him what he described as "cards and envelopes with stamps canceled at Argentine Antarctic Stations, a beautiful memento for me" (Skottsberg 1962).

Gustavo Giró later made an historic overland trek across the entire Antarctic Peninsula, from Andersson's Hope Bay/Base Esperanza to Margarita Bay/San Martín Base, in January to October of 1962. He then established a new scientific research station named after José María Sobral—Base Sobral—in April of 1965, timing its opening to coincide with Sobral's birthday. Situated on the Filchner Ice Shelf, and aptly serving as the southernmost Argentine station, Base Sobral was used as the starting point for Giró's historic trek across the ice to the South Pole in 1965. His was the first Argentine expedition to reach the South Pole over land (Tahan 2017). Fittingly, the memory and spirit of Sobral had motivated and inspired yet another "first" in both Argentine and Antarctic history.

The importance and significance of José María Sobral to Antarctic exploration, to Argentina, to the Argentine Navy, and to worldwide knowledge cannot be overstated. The symbolism and real-life impact of this individual are profoundly felt, both in his country and throughout the globe, whose southernmost Polar region holds in its memory the soul of one earnest and energetic young Under-Lieutenant named José María Sobral.

Figures 9.1, 9.2, 9.3, 9.4, 9.5, 9.6, 9.7, 9.8, 9.9, 9.10, 9.11, and 9.12.

Fig. 9.1 Alférez José María Sobral re-entered society in Buenos Aires and took part in community celebrations following his return from the challenging Antarctic Expedition. (Departamento de Estudios Históricos Navales, Archivo Histórico, Archivo Fotográfico, P-0140 k, Sobral, José María. Alférez. A la salida del Te Deum luego regreso Antártida)

Fig. 9.2 A national phenomenon at the time, and a topic of great discussion, Under-Lieutenant José María Sobral became the subject for cartoons depicting his heroic presence in Antarctica. (Departamento de Estudios Históricos Navales, Archivo Histórico, Colección Sobral, Box 2, Envelope 3, newspaper cuttings album Raúl Pinero)

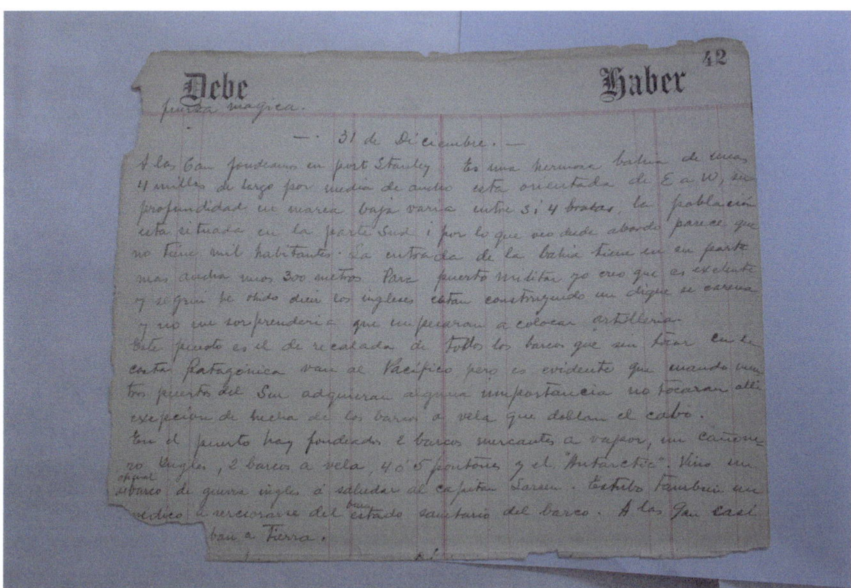

Fig. 9.3 Immediately upon his return from Antarctica in December 1903, José María Sobral began writing extensive notes and delivering speeches at various events and organizations. Pictured here is an excerpted page from his copious notes written in preparation for various speeches, invited lectures, and official report to the Minister of the Navy. (Departamento de Estudios Históricos Navales, Archivo Histórico, Colección Sobral, Box 2, Envelope 4, Manuscript referring to Antarctica trip and Conference at Naval Center)

Fig. 9.4 In 1904, José María Sobral was awarded with the "Order of the Polar Star" (pictured) and the "Order of the Sword," presented to the Argentine Antarctic explorer on behalf of the King of Sweden. (Departamento de Estudios Históricos Navales, Archivo Histórico, Archivo Fotográfico, N-0065 c, Sobral, José María. Alférez. Condecoraciones del gobierno sueco. Medalla)

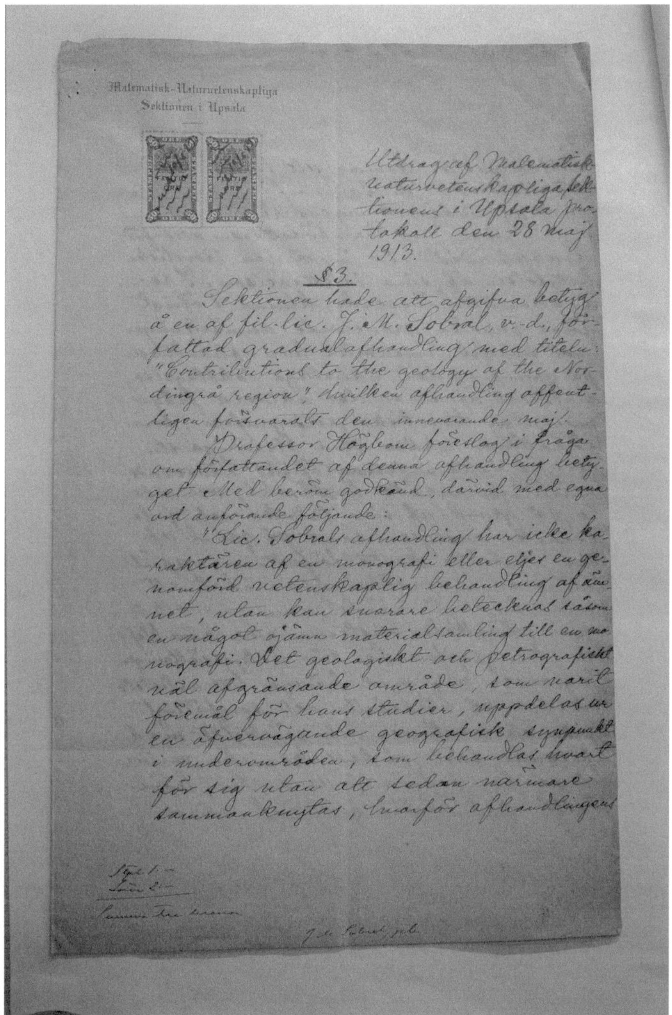

Fig. 9.5 José María Sobral's doctoral dissertation in Geology received critical analysis but also a positive review, approval, and honorary recommendation. Pictured here is the first page of the official thesis evaluation from Uppsala University in May 1913. Sobral received his Ph.D. in June of that year. (Departamento de Estudios Históricos Navales, Archivo Histórico, Colección Sobral, Box 2, Envelope 12, Graduation thesis evaluation, geology at Uppsala)

Fig. 9.6 Dr. José María Sobral—scholar, scientist, Antarctic explorer, and Naval officer. (Departamento de Estudios Históricos Navales, Archivo Histórico, Archivo Fotográfico, P-0140 f, Sobral, José María. Alférez)

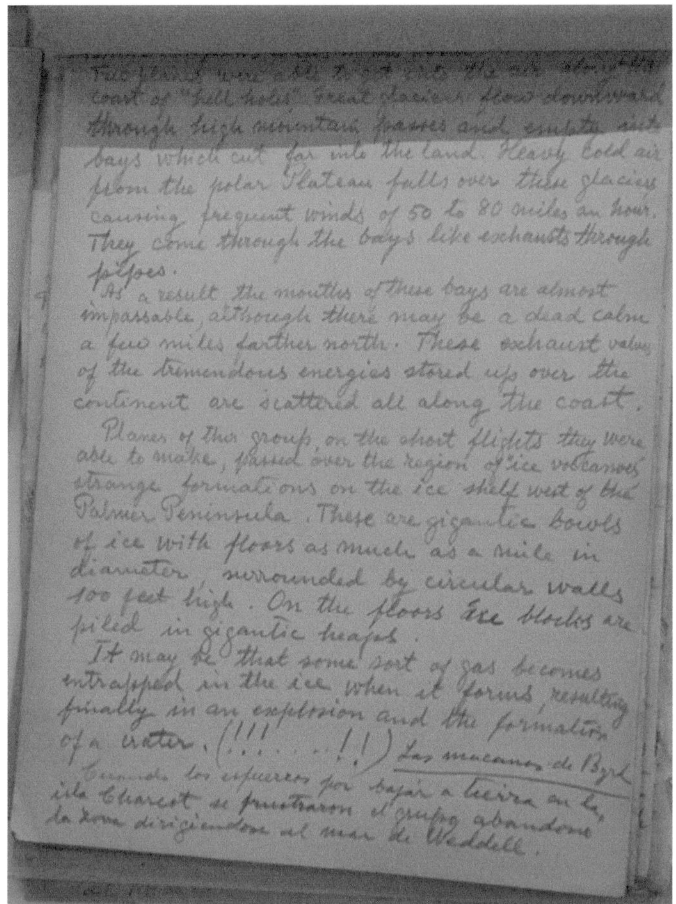

Fig. 9.7 An example of José María Sobral's notes about—and fascination with—other Antarctic expeditions and findings, in this case, his commentary on Richard E. Byrd's 1930 report regarding Byrd's air excursion over Antarctica. Note Sobral's enthusiastic exclamation marks, and his easy use of English and Spanish, switching effortlessly from one language to the other. (Departamento de Estudios Históricos Navales, Archivo Histórico, Colección Sobral, Box 1, Envelope 1, Manuscripts for conferences, publications, speeches)

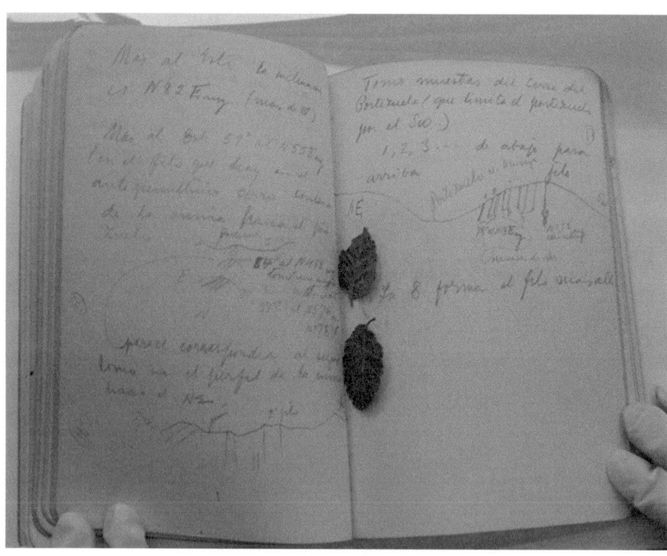

Fig. 9.8 The geology field expedition diary of José María Sobral, 1932–1933, written while he was under the employment of Yacimientos Petroliferos Fiscales (YPF). The diary begins with the journey to Ñirihuau, through Bahía Blanca and Patagones, in December 1932, where he studied the Río Negro territory and observed red sandstones, white limestones, basalts, and igneous rocks of andesitic type. The two pages pictured here portray his findings in January 1933. (Departamento de Estudios Históricos Navales, Archivo Histórico, Colección Sobral, Box 2, Envelope 9, Diary as leader of the geological division of YPF)

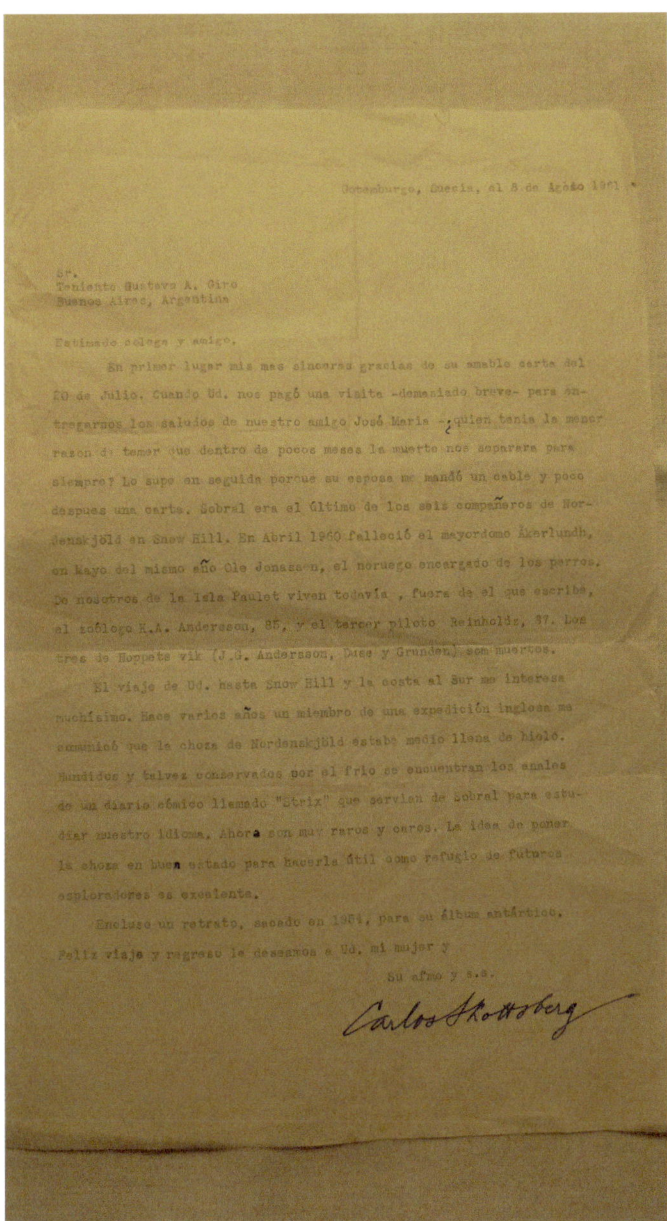

Fig. 9.9 The personal letter from Carl Skottsberg to Gustavo Giró, dated August 8, 1961, in which the veteran scientist and explorer laments the death of his Antarctic compatriot and fellow scholar José María Sobral. Lieutenant Giró was preparing to embark on his own historic Antarctic excursion, which would include the establishment of the scientific research station Base Sobral. Skottsberg's reminisces include how Sobral learned the Swedish language while in Antarctica using the comic book *Strix*. (Private collection of María Edelia Giró and Edelia "Puchi" Gamarino Giró, viewed by the author in February 2017, in Ushuaia, Tierra del Fuego, Argentina)

Fig. 9.10 The Naval Center—Centro Naval—sponsored the speech given by José María Sobral on December 19, 1903. Pictured here is the Centro Naval building today in Buenos Aires, Argentina. (Photograph by Mary R. Tahan)

Fig. 9.11 The corvette *Uruguay*, today permanently moored as an historical museum ship at Puerto Madero in Buenos Aires, Argentina. (Photograph by Mary R. Tahan)

Fig. 9.12 The long and productive life of Dr. José María Sobral, Doctor of Geology, and Navy Under-Lieutenant, has left the world with a legacy of Antarctic exploration, natural science, and human endeavor. (Close-up from a photograph by Gösta Bodman. Original photo—Departamento de Estudios Históricos Navales, Archivo Histórico, Archivo Fotográfico, Ex-0033 d, Expedición Nordenskjöld. 1902. En Snow Hill)

References

Centro Naval (1903) Diploma of honorary membership. Departamento de Estudios Históricos Navales, Archivo Histórico, Colección Sobral, Box 2, Envelope 13, Diploma of honorary membership of Centro Naval for José María Sobral, 13 December 1903. Buenos Aires: Archivo D.E.H.N. - A.R.A

Correspondents (1903–1904) Various personal correspondence received by José María Sobral. Departamento de Estudios Históricos Navales, Archivo Histórico, Colección Sobral, Box 3, Envelopes 1, 2, 20–24, 28–30, 35–44. Personal cards, letters, and mail, 1903 to 1904. Buenos Aires: Archivo D.E.H.N. - A.R.A

Correspondents (1949–1959) Various personal correspondence received by José María Sobral. Departamento de Estudios Históricos Navales, Archivo Histórico, Colección Sobral, Box 3, Envelopes 25–26, 31–32, 47–48. Personal cards, letters, and mail, 1949 to 1959. Buenos Aires: Archivo D.E.H.N. - A.R.A

Fundación Marambio. Alférez de Navío José María Sobral. http://www.marambio.aq/alferezsobral.html. Accessed 14 June 2017

Marenssi S (2007) Doctor José María Sobral (1880–1961): de los hielos antárticos al olvido argentino. Revista de la Asociación Geológica Argentina, versión On-line ISSN 1851-8249, Rev. Asoc. Geol. Argent. v.62 n.4 Buenos Aires oct./dic. 2007. http://www.scielo.org.ar/scielo.php?pid=S0004-48222007000400002&script=sci_arttext Accessed 6 Apr 2015

Piazzi, Atelio S (1905) Letter to José María Sobral. Departamento de Estudios Históricos Navales, Archivo Histórico, Colección Sobral, Box 3, Envelope 34. Letter forwarded by Piedrabuena Transport, Punta Arenas, June 1905. Buenos Aires: Archivo D.E.H.N. - A.R.A

Rabassa J (2003) Estudio preliminar. In *Dos años entre los hielos, 1901-1903*, José María Sobral (reprint of original book published in 1904), 11–46. Buenos Aires: Eudeba / Universidad de Buenos Aires, Colección Reservada del Museo del Fin del Mundo

Skottsberg C. (1961) Personal letter from Carl Skottsberg to Gustavo Giró, dated 8 August 1961. Private collection of María Edelia Giró and Edelia "Puchi" Gamarino Giró, viewed by the author in February 2017, in Ushuaia, Tierra del Fuego, Argentina

Skottsberg C (1962) Personal letter from Carl Skottsberg to Gustavo Giró, dated 19 May 1962. Private collection of María Edelia Giró and Edelia "Puchi" Gamarino Giró, viewed by the author in February 2017, in Ushuaia, Tierra del Fuego, Argentina

Sobral JM (1903). Handwritten notes for speeches and for official report to the Minister of the Navy. Departamento de Estudios Históricos Navales, Archivo Histórico, Colección Sobral. Box 2, Envelope 4, Manuscrito referido al viaje a la Antartida y Conferencia en el Centro Naval 1903. Buenos Aires: Archivo D.E.H.N. - A.R.A

Sobral JM (1904) *Dos años entre los hielos, 1901–1903 (Two years amidst the ice, 1901–1903)*. Buenos Aires: Imprenta de J. Tragant y Cía., Bolivar 319

Sobral JM (1930–1949) Original personal notes and manuscripts. Departamento de Estudios Históricos Navales, Archivo Histórico, Colección Sobral, Box 1, Envelope 1, Manuscripts for conferences, publications, speeches: notes on (Richard E. Byrd's) National Geographic Magazine report, August 1930, the conquest of Antarctica by air. Buenos Aires: Archivo D.E.H.N. - A.R.A

Sobral JM (1930–1952) Original personal notes and manuscripts. Departamento de Estudios Históricos Navales, Archivo Histórico, Colección Sobral, Box 2, Envelope 2, notes on Antarctica and Norway expedition: (Roald) Amundsen 1910-12 tomo II; and Iceberg—(Robert Falcon) Scott's last expedition vol. II. Buenos Aires: Archivo D.E.H.N. - A.R.A

Sobral JM (1932–1933) Geology Field Work Diary for YPF. Departamento de Estudios Históricos Navales, Archivo Histórico, Colección Sobral, Box 2, Envelope 9, Cuaderno—Expedition diary as the leader of the geological division of YPF. Buenos Aires: Archivo D.E.H.N. - A.R.A

Sobral Guillermo José (2016) Declaración dijo por Alvar Sobral. Statement said by Alvar Sobral, son of José María Sobral, and relayed by Guillermo José Sobral. Communication sent by Guillermo José Sobral to the author, received 2 December 2016

Sobral, Guillermo José (2017) Parte de la nota escrita por Alvar Sobral. Part of a note written by Alvar Sobral Klingström, 95 years old, the only living son of Doctor José María Sobral, and given to Guillermo José Sobral on 22 February 2017. Communication sent by Guillermo José Sobral to the author, received 26 Feb 2017

Sobral GJ, Sobral Jorge A (2014) Hablar en estos dias de Sobral. Articulo registrado—Hecho el depósito que marca la ley 11.723. Talk these days of Sobral, written 14 April 2014. Article sent to the author by Guillermo José Sobral, received 28 Oct 2016

SSAG (1937) Certificate of Membership. Departamento de Estudios Históricos Navales, Archivo Histórico, Colección Sobral, Box 2, Envelope 14, Diploma of membership in the Swedish Society for Anthropology and Geography. Buenos Aires: Archivo D.E.H.N. - A.R.A

Stockholm, Sweden (1904) Certificate of the Order of the Sword. Departamento de Estudios Históricos Navales, Archivo Histórico, Colección Sobral, Box 2, Envelope 11, Royal Knight of the Sword Order, King of Sweden. Buenos Aires: Archivo D.E.H.N. - A.R.A

Tahan Mary R (2017) Author's interview with the Gustavo Giró family – María Edelia Giró and Edelia "Puchi" Gamarino Giró, February 2017, in Ushuaia. Tierra del Fuego, Argentina

Uppsala University (1908) Sobral's acceptance into the graduate program. Departamento de Estudios Históricos Navales, Archivo Histórico, Colección Sobral, Box 2, Envelope 17, Faculty of Philosophy of Uppsala University, 31 January 1908. Buenos Aires: Archivo D.E.H.N. - A.R.A

Uppsala University (1913) Sobral's geology dissertation approval and evaluation. Departamento de Estudios Históricos Navales, Archivo Histórico, Colección Sobral, Box 2, Envelope 12, Certificate of evaluation of José María Sobral's graduation thesis in geology, with comments by Thorulf Fries, 28 May 1913. Buenos Aires: Archivo D.E.H.N. - A.R.A

Conclusion
The Strangeness Within, the Wildness Without

Today, in Antarctica, the winter station house at Snow Hill stands as a national museum under the management and preservation of the Argentine Antarctic Institute, providing a window into the world in which Sobral had lived and worked during those significant two years in the Antarctic. Remnants remain of the rough hut at Paulet Island, where the crew and scientists of the ship *Antarctic* had escaped death and had sought meager shelter. And a representation of the stone hut at Hope Bay is protected year-round by the personnel at Base Esperanza, offering a taste of the close quarters in which Andersson and his companions lived—the penguins still visit the hut every day, curious as to its purpose, and conscientiously tending to their chicks nearby.

Today, in Argentina, Sobral's house of birth in Gualeguaychú, Entre Rios, is a national historic building. The Centro Naval building maintains its elegant presence in downtown Buenos Aires. And the *Uruguay* remains moored as a museum ship at a dock in Puerto Madero, Buenos Aires, as does the *Sarmiento*. Boarding the *Uruguay* corvette, one marvels at how this little gunship had made it into the ice pack and back—a testament to the valiance of those who had navigated the craft. The ship had virtually visited another world and returned.

In Antarctica, Sobral, too, had felt like an alien on an alien planet. All was foreign to him—the formidable yet beautiful landscape, the freezing cold temperature, the different languages spoken all around him, the people who had accompanied him to this, the end of the earth, the science that attracted and intrigued him, and—not least of all—the ice: that never-ending sheet of pure solid liquid that moved yet was unmovable. He persevered to endure and quested to question and understand the strange, yet familiar nature that existed all around him. The young Argentine painstakingly learned the foreign words, the frozen feel, the ancient fossils, the living nature, the way of the animals, and the ways of the human beings.

In Sweden, Sobral found an adopted home and began a family. He dedicated himself to the study and mastery of science. And he armed himself with the tools to bring this knowledge back to his home country.

© Springer International Publishing AG 2018
M.R. Tahan, *The Life of José María Sobral*, Springer Biographies,
https://doi.org/10.1007/978-3-319-67268-7

In Argentina, Sobral pursued his passions, but the passion was not always reciprocated by his home country. And yet he continued to contribute to his home country's knowledge and practice of the science that was so important to him—the science he had explored and learned, the science that had brought him to the southern axis of the world.

Sobral's exploration and expedition exemplify our quest to find ourselves, to understand what is beyond our own being and our very own existence. It is the story of our search and our connection with nature, animals, the environment, the world, and ourselves.

A natural scientist, a poetic diarist, an innate explorer, Sobral was a true Antarctic pioneer. He traveled far to explore the unknown and to discover that which blooms from within.

Additional Reading

Bibliography of Books and Articles by José María Sobral[1]

Sobral, JM (1904) *Dos años entre los hielos 1901-1903*. Imprenta de J. Tragant y Cía, 364 p., Buenos Aires

Sobral, JM (1905) El futuro de nuestra armada. Boletín del Centro Naval 22(258):1057–1067

Sobral, JM (1905) Los viajes de la Presidente Sarmiento. Boletín del Centro Naval 22(255): 846–848

Sobral, JM (1910) On the contact features of the Nordingra Massive. *Bulletin of the Geological Institute* 9:118–128, Uppsala

Sobral, JM (1913) Contribution to the geology of the Nordingra region. *Almquist & Wiksells*, 178 p. y 12 laminas, Uppsala

Sobral, JM (1913) Contribution to the geology of the Nordingra region. Dissertation, Uppsala Universitat, 178 p., 12 pl., 1 karta 8, (inédito), Uppsala

Sobral, JM (1916) On a granite of Hemsön. *Sociedad Argentina de Ciencias Naturales, Physis* 2 (10):122–127, Buenos Aires

Sobral, JM (1916) Sobre cambios de nombres geográficos. *Sociedad Argentina de Ciencias Naturales, Physis* 2(11):290–291, Buenos Aires

Sobral, JM (1916) Homenaje de las asociaciones científicas argentinas a Sir Ernest Shackleton: realizado en el Prince George's Hall, 5 de octubre 1916. Sociedad Argentina de Ciencias Naturales, Buenos Aires

Sobral, JM (1917) La expedición sudpolar británica (1914-1916). *Sociedad Argentina de Ciencias Naturales, Physis* 3(13):75–81, Buenos Aires

Sobral, JM (1918) Sobre el piroxeno y la olivina de la diabasa de Ulfö. *Sociedad Argentina de Ciencias Naturales, Physis* 4(17):290–291, Buenos Aires

Sobral, JM (1918) Estudio petrográfico de algunas rocas argentinas. *Anales de la Sección Geología, Mineralogía y Minería del Ministerio de Agricultura* 2, 54 p., Buenos Aires

Sobral, JM (1921) Optical investigation of the new pyroxene Sobralite. *Bulletin of the Geological Institute* 18:47, Uppsala

Sobral, JM (1921) Problemas hidrográficos en los Andes Australes. Tixi & Schaffner, 41 p. y 2 cartas, Buenos Aires

Sobral, JM (1921) Some physiographic notes on the Sierra de Famatina. *Geografiska Annaler* 3:311–326, Stockholm

Sobral, JM (1928) La mina aurífera de San Ramón departamento de Tupungato [Provincia de Mendoza]. Segundo Congreso Nacional de Ingeniería. *Talleres Gráficos del Ministerio de Agricultura de la Nación*, 29 p., 22 figuras, Buenos Aires

---

[1]Partial list—portion per Guillermo José Sobral, sent to the author, received on 28 December 2016.

© Springer International Publishing AG 2018

M.R. Tahan, *The Life of José María Sobral*, Springer Biographies, https://doi.org/10.1007/978-3-319-67268-7

Sobral, JM (1941) Sobre el petróleo y el carbón. *Revista del Colegio Libre de Estudios Superiores* 9(10, 11 y 12-18-19):2264–2291, Buenos Aires

Sobral, JM (1942) Geología de la comarca del territorio de La Pampa situada al occidente del Chadi-Leuvú. *Boletín de Informaciones Petroleras* 212:33–81, Buenos Aires

Sobral, JM (1942) Geología de la región occidental del territorio de La Pampa. *1er Congreso Panamericano de Minas y Geología, Anales* 2:668–750, Santiago de Chile

Sobral, JM (1947) Cuando la corbeta Uruguay quedó desarbolada 15 de noviembre de 1903. *Yachting Argentino* 90: 3 p., Buenos Aires

Sobral, JM (1948) Epítome Geológico de la Antártida Occidental. *Revista Argentina Austral* 18 (208):25–33, Buenos Aires

Sobral, JM (1949) La exploración del continente antártico. *Revista Argentina Austral* 19(218): 5–11, Buenos Aires

Sobral, JM (1949) La Plataforma continental del Continente Antártico. *Revista Argentina Austral* 19(222):5–11, Buenos Aires

Sobral, JM (1950) El continente antártico y su exploración científica. *Anales del Instituto Popular de Conferencias* 5:68–104, Buenos Aires

Sobral, JM (1950) El hielo terrestre (Antártida Occidental). *Revista Argentina Austral* 20 (232 / 233):6–13, Buenos Aires

Sobral, JM (1952) Reseña histórica de Bahía Esperanza y noticias sobre una expedición a la Tierra de la Reina Maud. *Revista Argentina Austral* 22(250):6–9, Buenos Aires

Sobral, JM (1952) La exploración del continente Antártico. Síntesis y examen de las observaciones registradas hasta el día. *Revista Argentina Austral* 22(254):4–13

Sobral, JM (1952–1953) El hielo terrestre (Antártida Oriental). *Revista Argentina Austral* 22 (254):4–12, Buenos Aires

Sobral, JM (1953) La exploración del continente Antártico. Síntesis y examen de las observaciones registradas hasta el día. *Revista Argentina Austral*, 22(259):4–12

Reading for Further Information

Rabassa, J, Borla M.L (eds) (2007) *Antarctic Peninsula and Tierra del Fuego: 100 years of Swedish-Argentine scientific cooperation at the end of the world.* In: Proceedings of "Otto Nordenskjöld's Antarctic Expedition of 1901-1903 and Swedish Scientists in Patagonia: A Symposium", held in Buenos Aires, La Plata, and Ushuaia, Argentina; Taylor & Francis Group London, March 2–7, 2003

MIX
Papier aus verantwortungsvollen Quellen
Paper from responsible sources
FSC® C105338

If you have any concerns about our products,
you can contact us on
ProductSafety@springernature.com

In case Publisher is established outside the EU,
the EU authorized representative is:
Springer Nature Customer Service Center GmbH
Europaplatz 3, 69115 Heidelberg, Germany

Printed by Libri Plureos GmbH
in Hamburg, Germany